Production Technology of Underutilized Vegetable Crops

Savita • Monisha Rawat • Vrince Vimal
Editors

Production Technology of Underutilized Vegetable Crops

Springer

Editors
Savita
Graphic Era Hill University
Dehradun, Uttarakhand, India

Vrince Vimal
Graphic Era Hill University
Dehradun, Uttarakhand, India

Monisha Rawat
Department of Horticulture, School of Agriculture
Lovely Professional University
Phagwara, Punjab, India

ISBN 978-3-031-15384-6   ISBN 978-3-031-15385-3 (eBook)
https://doi.org/10.1007/978-3-031-15385-3

This Springer imprint is published by the registered company Springer Nature Switzerland AG
The registered company address is: Gewerbestrasse 11, 6330 Cham, Switzerland

Smart Farm

Preface

Vegetables are a vital food source containing carbohydrates, protein, fat, minerals, vitamins, and fibers, which play a critical role in the human diet. Vegetables are considered protective food because they protect our body from several diseases and make the immune system potent, detoxifying carcinogens and reducing muscular degeneration. Apart from the health benefits, vegetables also play an essential role in the economy because of their short growing span and ten times more yield than cereals. The demand for vegetables has risen with the high-income generation, while vegetable production has increased enormously. On the other hand, underutilized vegetables can be used to combat food insecurity because they are also high in nutrients. Their production is limited compared to major vegetables due to a lack of knowledge in farmers regarding the benefits of growing underutilized vegetables. Most vegetables are annual and take a small span to grow, so more than one crop can be produced within a year. Vegetable cultivation is a labor-intensive system which results in more employment. Lack of proper information on the nutritional and medicinal value of vegetables and a lack of information on production practices are all likely reasons for the delayed development and poor status of underutilized vegetable crops. However, researchers may be able to commercialize these crops using different research methodologies.

Similarly, significant progress has been made in developing, standardizing, and managing technology in organic manure, mineral nutrients, bio-fertilizers, weedicides, irrigation, plant growth regulators, and plant protection measures such as bio-control, agents. Off-season cultivation in polyhouse/greenhouse and the vegetable-based cropping system can be utilized for getting higher yield and quality of underutilized vegetable crops. Internet of Things (IoT) is another recent technology used in agriculture nowadays in different countries to improve crop production. The applications of IoT in farming target conventional farming operations to meet the increasing demands and decrease production losses. IoT in agriculture uses robots, drones, remote sensors, and computer imaging combined with continuously progressing machine learning and analytical tools for monitoring crops, surveying, and mapping the fields and provides data to farmers for rational farm management plans to save both time and money.

The primary goal is to keep readers up to date and provide essential information on the fundamentals of commercially growing underutilized vegetable crops. This book is compiled in an easy language for easy understanding, and it contains very concise and precise information on the production aspects of all the underutilized vegetables. This book aims to provide every aspect of the production technology of underutilized vegetable crops and the knowledge regarding introduction, origin, and distribution. Further, nutritional importance and uses, medicinal uses, improved varieties/hybrids, climatic and soil requirements, seed rate and sowing, spacing, nutritional needs, irrigation, intercultural operations, and harvesting are included. The postharvest management, physiological disorders, insect pests, and disease management of different underutilized vegetables, like Alliaceae, Leguminosae, Cucurbitaceae, etc., have also been focused upon. It also includes information on the best possible solutions for increasing the production of underutilized vegetables using technology.

Furthermore, vegetable growers will find this book extremely useful and valuable in successfully and economically raising vegetable crops. The information presented in this book is based on the scientific records of scientists who work on vegetables for various organizations. Given the importance of vegetable production and the need to familiarize students with technical aspects of vegetable cultivation, the authors were compelled to write a book that would present vegetable cultivation by family in a sequential manner. The book would assist them in gaining a basic understanding of vegetable production. This book on production technology of underutilized vegetable crops, compiled for undergraduate, postgraduate students, and farmers, is one such attempt to help them learn and understand the subject of vegetable production more precisely. This book will motivate them to improve their knowledge in vegetable science to meet future needs and serve their best in the community. This book may also be helpful to others who want to broaden their basic knowledge in vegetable production and want to grow profitable vegetable crops on their farms. Obtaining scientific knowledge will undoubtedly benefit its users and, in the end, the country or world.

Dehradun, Uttarakhand, India — Savita
Phagwara, Punjab, India — Monisha Rawat
Dehradun, Uttarakhand, India — Vrince Vimal

Acknowledgment

Compiling a book is more complicated than we thought and more rewarding than imagined. None of this would have been possible without a great team of editors and contributors.

Our way in this world paves through the path curved by our teachers. Whom we wish to idolize throughout our lives and who taught the philosophy of nature and gave direction to our life. We are eternally grateful to our families, who have always supported and motivated us.

It gives us immense rejoice to owe my deep sense of gratitude to the Vice-Chancellor, Graphic Era Hill University. Nonetheless, to the Dean, School of Agriculture, Graphic Era Hill University for their sympathetic attitude, timely help, and providing other facilities during the writing of this book.

We thank God most of all because we couldn't do any of this without the almighty's wisdom.

Contents

Chapter 1
Underutilized Vegetables Introduction and Identification

Savita and Vrince Vimal

Introduction

Underutilized/underexploited vegetable crops are those that are not farmed commercially on a significant scale or traded widely. In other words "Underutilized crops/plant species are those species which have potential for contributing to food security, health (nutritional/medicinal), income generation, and environmental services" (Jaenicke and Höschle-Zeledon 2006). Leek, shallot, chive, winged bean, adzuki bean, sword bean, scarlet runner bean, Agathi, yard long bean, tannia, giant taro, swamp taro, chow-chow, sweet gourd, spine gourd, gherkin, ivy gourd, curry leaf, New Zealand spinach, Chinese potato, tree tomato, turnip-rooted parsley, rat tail radish, garden cress, pigweed, water spinach, Chekurmanis, drumstick, breadfruit Chinese water chestnut, purslane, West Indian arrowroot are the examples of underutilized vegetable crops. Some underutilized vegetables are listed below:

Family name	Name	Botanical name	Chromosome no. (2n)
Alliaceae/Amarylidaceae			
	Leek	*Allium ampeloprasum* L. var. *porrum*	32
	Shallot	*A. ascalonicum* L.	16
	Chive	*A. schoenoprasum* L.	16
	Multiplier onion	*A. cepa* var. *aggregatum* L.	16

(continued)

Savita (✉) · V. Vimal
Graphic Era Hill University, Dehradun, Uttarakhand, India

Savita et al. (eds.), *Production Technology of Underutilized Vegetable Crops*,
https://doi.org/10.1007/978-3-031-15385-3_1

Family name	Name	Botanical name	Chromosome no. (2n)
	Tree onion	*A. cepa* var. *viviparum* (Metz.) *Alef.*	16
	Welsh onion	*A. fistulosum* L.	16
	Kurrat	*A. kurrat*	32
	Great headed garlic	*A. ampeloprasum* L.	48
	Rakkyo	*A. chinense/chinensis*	16, 24, 32
	Chinese chive	*A. tuberosum* Rottler Ex Spreng	32
Leguminosae/Fabaceae			
	Winged bean	*Psophocarpus tetragonolobus* (L.) DC.	18
	Adzuki bean	*Phaseolus angularis* (Wild.) Wright	22
	Sword bean	*Canavalia gladiata* (Gucav.) DC.	22, 24
	Jack bean/Gotani bean	*Canavalia ensiformis* (L.) DC.	22
	Moth bean	*Phaseolus acontifolius*	22
	Scarlet runner bean/ Multiflora bean	*Phaseolus coccineus/Phaseolus multiflorus* Wild	22
	Yam bean	*Pachyrhizus erosus* Urbn.	22
	Potato bean	*Apios americana*	22
	Tepary bean	*Posphocarpus acutifolius*	22
	Catjung bean	*Vigna sesquipedalis* var. *cylindrical* or *biflora*	22
	Agathi	*Sesbania grandiflora* Poir.	24
	Thicket bean	*Phaseolus polystachyus*	22
	Velvet bean	*Mucuna deeringiana* (Bor.) Merr. (syn. *Stizolobium deeringianum* Bort.)	22
	Horse gram	*Dolichus uniflorus* Lam. (*Dolichus biflorus* Auct. L.)	22, 24
	Yard long bean or asparagus bean	*Vigna unguiculata* ssp. sesquipedalis	22
	African yam bean	*Sphenotylis stenocarpa*	22
	African locust bean	*Parkia biglobosa*	24
	Ground bean	*Vigna subterranea* (L.) Verdc.	22
	Marama bean	*Tylosema esculentum* (Burchell) Schreiber	44
Araceae			
	Tannia	*Xanthosoma sagittifolius* (L.) Scott.	26
	Giant taro	*Alocasia macrorrhiza* (L.) Scott.	26, 28
	Swamp taro	*Colocasia esculenta* var. stolonifera	26
Cucurbitaceae			
	Buffalo gourd	*Cucurbita ficifolia* Bouche	40

(continued)

Family name	Name	Botanical name	Chromosome no. (2n)
	Snap melon (phoot)	*Cucumis melo* L. var. momordica	24
	Chow-chow	*Sechium edule* (Jacq.) Swartz.	28
	Sweet gourd	*Momordica cochinchinensis* (Lour.) Spreng	28
	Spine gourd	*Momordica dioica* Roxb. Ex,,.	28
	Meetha karela	*Cyclanthera pedata* (L.) Schrader	32
	Gherkin	*Cucumis anguria*	24
	Ivy gourd	*Coccinia grandis* (L.) Voigt.	24
	Wild snake gourd	*Trichosanthes anguina* L.	22
	Balsam apple	*Momordica balsamina* L.	22
	Mango melon	*Cucumis melo* L. var. dudain	24
	Cassava melon		
	Pickling melon	*Cucumis melo* L. var. *conomon* Mak	24
Rutaceae			
	Curry leaf	*Murraya koenigii* (L.) Spreng	18
Dioscoreace			
	Potato yam	*Dioscorea bulbifera*	40
Aizoaceae			
	New Zealand spinach	*Tetragonia expansa* (Murr.)	32
Basellaceae			
	Malabar spinach	*Basella alba/rubra*	48
Labitae			
	Chinese potato	*Coleus parviflorus* Benth. (syn. *Coleus tuberosus*)	64, 84
	Chinese artichoke	*Stachys siebeldi* Miq.	16
	Hoary basil (Kalitulsi)	*Ocimum americanum* L.	24
Martynaceae			
	Martynia	*Proboscidea jussiecui* (Keller)	30
Solanaceae			
	Tree tomato	*Cyphomandria betacea*	24
Apiaceae			
	Turnip rooted parsley	*Petroselinum crispum* var. *tuberosum*	22
	Turnip rooted chervil	*Chaerophyllum bulbosum* L.	22
	Skirret	*Sium sisaarum* L.	40
Brassicaceae			
	Sprouting broccoli	*Brassica oleracea* var. italica	

(continued)

Family name	Name	Botanical name	Chromosome no. (2n)
	Knol-khol	*Brassica oleracea* var. gongylodes	
	Brussels sprout	*Brassica oleracea* var. *gemmifera*	
	Chinese cabbage		
	Rat tail radish	*Raphanus sativus* var. caudatus	18
	Garden cress	*Lepidium sativum* L.	16, 32
	Upland cress	*Barbarea vulgaris* R. Br.	16
	Sea kale	*Crambe maritime*	60
	Catran	*Crambe tatarica* Busch	60
Chenopodiaceae			
	French spinach (Orach)	*Artiplex hortensis* L.	12
	Pigweed	*Chenopodium album* L.	36
Compositae			
	Spanish salsify	*Scolymus hispanica* L.	14, 28
Convolvulaceae			
	Water spinach	*Ipomea aquatica* Forsk.	30
Euphorbiaceae			
	Chekurmanis	*Sauropus androgynous* (L.) Merr.	24
Moringaceae			
	Drumstick	*Moringa oleifera* Lamk.	28
Polygonaceae			
	Spinach dock/ sorrel	*Rumex acetosa*	
	Patience dock	*Rumex patientia* L.	80
	Bladder dock	*Rumex vesicarius* L.	18
Moraceae			
	Bread fruit	*Artocarpus altilis*	56
Cyperaceae			
	Chinese water chestnut	*Eleocharis dulcis*	64
Portulaceae			
	Purslane (Kulfa)	*Portulaca oleracea* L.	54
	Ceylon spinach	*Basella alba* L.	24
Cannaceae			
	Purple arrowroot	*Canna indica*	18, 48
Marantaceae			
	West Indian arrowroot	*Maranta arundinacea*	48

Importance of Underutilized Vegetables

Underutilized vegetables are the key source of vitamins, protein, and micronutrients; therefore, a valuable component of food safety (Jena et al. 2018) compared to significant vegetables such as tomato and cabbage (Keatinge et al. 2011). Mostly all the underutilized legume vegetables provide nutrients not only to the human beings (protein) but also to the soil (nitrogen) due to the well-established soil-enrichment property of symbiosis with nitrogen-fixing bacteria. They also used for the preparation of medicine, animal feed, natural manures, and environmental restorative commodities (Singh et al. 2007). Despite the fact that some crop species are prevalent around the world, they are only grown and consumed locally. Traditionally, many crops are grown for food, fiber, petroleum, and preparation of medicines – all play a very important role in the local people's lives. They are an essential part of our diet of important nutrients and often lack crucial crops due to their adjustment to often marginal land (Jain and Gupta 2013). In addition, underutilized vegetables play an important role in sustainable food production because of less environmental impact. Most of these plants are hardy and require minimum inputs for their cultivation, adapted to the specific marginal land under adverse environmental conditions (De la Peña et al. 2011; Hughes and Ebert 2011).

The Green Revolution has replaced many local, traditional crops and with high-yielding varieties developed through advanced breeding. Traditional methods do not usually match with the current uniformity standards, and other features as breeders in the private and public sectors have been neglected (Stamp et al. 2012). As a result, commercial cultivars are less competitive in the market. Genes that promote biotic/abiotic stress resistance, yield, quality, etc. are increasingly recognized and exploited in landraces and wild crops/species (Frison et al. 2011; Jackson et al. 2007; McCouch et al. 2013). However, if we develop more robust production systems, agricultural and biological variability should not be confined to the use of valuable genes for breeding programs. Presently, neglected food resources obtained from small grains and pulses, cultivated roots and tuber crops, fruits and vegetables in non-wood goods could contribute to food and nutrition security while protecting against internal and foreign market volatility and climate change (Keatinge et al. 2010). Extensive utilization of these crops would provide numerous possibilities for introducing temporal and geographical variation into a uniform cropping system, improving biotic and abiotic stresses resistance and, in the long run, leading to more sustainable food sources.

Futuristic Challenge

To feed the world's estimated 9 billion people by 2050, agriculture is increasingly pressurized on limited agricultural resources to produce more food, feed, and biofuel (Godfray et al. 2010). To cope with a projected 40% rise in global population by 2050, agricultural production will need to expand by 70% (Bruinsma 2009). Approximately 90% of this increase is expected to be achieved through increase in crop yield and intensity, the rest on the land that is not currently used for agriculture (Tilman et al. 2011). On the other hand, improper cultivation practices like excessive use of fertilizers directly affected the soil health, and maximum use of pesticides and fungicides on the vegetable crops affected human health due to their residual effects. Farmers or growers are following these practices to increase the yield and to get more profit.

Probable Solution

Increasing the production of underutilized vegetables could be one of the better options which meet the world's food demand because they are highly nutritious and give protection from several human diseases. Underutilized vegetables are important on a local or regional level, but they are rarely recognized or appreciated on a national or international level. The popularity of these vegetable crops can be increased to a greater extent through public awareness. Underutilized vegetable crops offer several advantages, such as being easy to grow, having a hardy nature, and providing a yield even in unfavorable soil and climatic conditions. Some underutilized vegetables improve soil health as well by increasing soil fertility (like underutilized legumes viz. adzuki bean,, African yam bean, Agathi, catjang bean, horse gram, jack bean, moth bean, potato bean, scarlet runner bean, sword bean, tepary bean, thicket bean, winged bean, yam bean, etc.). Mostly underutilized vegetables are resistant to insect pests and diseases; therefore, the use of pesticides and fungicides can be reduced to some extent. They require minimum inputs and hardy in nature and hence can adapt to the adverse environmental conditions on specific marginal land (De la Peña et al. 2011; Hughes and Ebert 2011).

Significance of Underutilized Vegetables

Rising consumer demand Many people, both young and old, have begun to choose underutilized vegetables in search of health benefits. Sprouting broccoli is one of the best examples because it has anticancerous property due to the presence of sulphoraphane.

Marketability It is still in the early phases of development, and the APMC (Agriculture Produce Market Committee) and the government must take various steps to improve the marketing and distribution of the product.

Buyers The migration of more wholesale clients toward growers of such underutilized vegetables indicates a bright future.

Storability of perishable underutilized vegetables In underutilized vegetable crops, the development of new varieties with improved shelf life is of prime importance, which hampers their market growth earlier. With suitable warehousing, seasonality and availability concerns associated with such crops can be eliminated.

Logistics and supply chain This has a substantial impact on agricultural crops, particularly vegetable crops. To satisfy increased demand of vegetables, their perishability and supply chain management must be addressed.

Protected cultivation Underutilized vegetable crops have high nutritive value, and uncontrollable environmental conditions necessitate their intensive care and maintenance. So, protected cultivation like polyhouse is the best alternative to grow the underutilized vegetable crops out of their normal season or during off season and make them available throughout the year. Broccoli, lettuce, celery, parsley, Chinese cabbage, and kale can be grown in off season under protected condition.

Constraints for Development of Underutilized Vegetable Crops

The following are some shortcomings that hamper the growth of underutilized vegetable crops. According to FAO, in 1998, the following limitations for the development of underutilized vegetable crops are listed (Williams and Haq 2002):

1. Availability of limited germplasm of underutilized vegetable crops limits the production.
2. Inadequate technical knowledge for the cultivation of underutilized vegetable crops.
3. National policy shortcomings for the production, distribution, and marketing of underutilized vegetable crops.
4. Researchers, agriculturists, and extension workers have shown little interest.
5. Lack of knowledge about postharvest handling of the produce.
6. Limited and insufficient marketing support, as well as transportation, storage, and processing infrastructure.
7. Lack of awareness about the nutritional and medicinal value of underutilized vegetable crops among the farmers.
8. There are a smaller number of researches on underutilized crops as they have a smaller number of seeds and planting material.
9. There is also a limitation in application of advanced techniques of agriculture.
10. Insufficient knowledge about the postharvest management practices of underutilized crops.
11. There is proper infrastructure facility for transporting, storing, and processing of these leafy vegetables.
12. Lack of identification of these underutilized crops in horticulture promotion program.

Development Strategies for Underutilized Vegetable Crops

Underutilized vegetable crops have a potential in the improvement of food security. Following measure can be taken into consideration to improve production of underutilized vegetable crops:

(i) Afforestation and forest rejuvenation could be done to augment and enrich the biodiversity of underutilized vegetable crops.
(ii) Domestication of potential wild species through household cultivation should be encouraged to avoid overexploitation of natural resources. Support is needed for the multiplication and distribution of planting materials, as well as market access for perishable vegetables.
(iii) More research and development is necessary to improve the production of underutilized vegetable crops because they are nutritionally rich and suited to low-input farming that helps to sustain human nutrition and food safety.
(iv) Underutilized vegetable crops are cultivated mainly by different ethnic communities under traditional farming systems. An increase in focus is required for the documentation of indigenous knowledge, such as ethno botany. This stress will help tap value add-ons as a multipurpose of indigenous diversity.
(v) In particular, strategies must be developed at national and regional levels to build and deliver promising selections/varieties, overcoming production restrictions on good seeds or planting materials, in vitro and tissue-cultured materials, etc. This would boost production, meet local needs, foster domestic markets, and improve smallholder farmers' incomes.
(vi) There should be systematic local special planning after agroclimatic appropriateness of the area.
(vii) The rapid expansion of infrastructure with a focus on market development, transport, and communication must be achieved.
(viii) Export-oriented production programs, border trade, with high-value products, such as off-season vegetables, should be emphasized in particular.
(ix) These crops' yields and quality are poor that interfere with productivity. The criteria such as high productivity, demand on the market, the absence of severe insect and disease insects, easier management after harvest, a high nutrient value, the availability of production technologies, etc. for the commercial exploitation of these crops need to be developed.
(x) The organization of micro and macro awareness camps/campaigns, exhibitions, multimedia applications such as radio, TV, the press, and other imprints helps the farming community to recognize the nutritional significance of underutilized vegetable crops (Sharma 2003). The organization should also emphasize the development of processing units in this area for appropriate exploitation and improved economic returns from underutilized vegetable plants. It would also employ the rural people.

(xi) Genetic erosion is a significant problem and will extinguish many land areas unless they are soon preserved. Commercial cultivation of nontraditional vegetable crops feasible by efficient production technologies and postharvest management is also required. The availability of nontraditional vegetable crops will help overcome the malnutrition of rural people.

Contribution of International Research Organization for the Development of Underutilized Vegetable Crops

International Plant Genetic Research Institute (IPGRI) In past years, IPGRI has led specific activities to improve the preservation and use of underutilized species both at the national and international levels. It is a very active global media in raising awareness and developing methods for sharing the advantages and uses of underutilized crops.

Global Facilitation Unit (GFU) It is a multi-initiative institute that works at global level to support and facilitate the work of other stakeholders to support the extensive use of unutilized species.

International Center of Underutilized Crops (ICUC) It was set up in 1948 to deal with ways to increase the use of underutilized crops in food, medicine, and the environment.

International Atomic Energy Agency (IAEA) The FAO's IAEA is organizing a project using irradiation technologies to improve production of underutilized species for low-income deficit nations.

Conclusion

Currently, underutilized plants play an essential role in a more diverse and sustainable food-producing system with great potential. However, to ensure competitiveness on the market, there must be more investment in long-term research and breeding programs and improved seed supply sources for these crops. Compared to the few major staple crops, there is insufficient funding for research and breeding underutilized vegetable crops. To achieve those goals and to generate interest in private sector breeders, considerable initial funding is required, if the significant market potential is reached, by the international donor community and the national state programs.

References

Bruinsma, J. (2009, June). The resource outlook to 2050: by how much do land, water and crop yields need to increase by 2050. In *Expert meeting on how to feed the world in* (Vol. 2050, pp. 24–26).

de la Peña, R. C., Ebert, A. W., Gniffke, P. A., Hanson, P., & Symonds, R. C. (2011). Genetic adjustment to changing climates: vegetables. In S. S. Yadav, R. J. Redden, J. L. Hatfield, H. Lotze-Campen, A. E. Hall (Eds.), *Crop adaptation to climate change* (pp. 396–410), 1st ed.; John Wiley & Sons, Ltd.: Chichester, UK.

Frison, E. A., Cherfas, J., & Hodgkin, T. (2011). Agricultural biodiversity is essential for a sustainable improvement in food and nutrition security. *Sustainability*, *3*(1), 238–253.

Godfray, H. C. J., Beddington, J. R., Crute, I. R., Haddad, L., Lawrence, D., Muir, J. F., Pretty, J., Robinson, S., Thomas, S.M. & Toulmin, C. (2010). Food security: the challenge of feeding 9 billion people. *science*, *327*(5967), 812–818.

Hughes, J. D. A., & Ebert, A. W. (2011, June). Research and development of underutilized plant species: the role of vegetables in assuring food and nutritional security. In *II International Symposium on Underutilized Plant Species: Crops for the Future-Beyond Food Security 979* (pp. 79–92).

Jackson, L. E., Pascual, U., & Hodgkin, T. (2007). Utilizing and conserving agrobiodiversity in agricultural landscapes. *Agriculture, ecosystems & environment*, *121*(3), 196–210.

Jaenicke, H., & Höschle-Zeledon, I. (2006). *Strategic framework for underutilized plant species research and development: With special reference to Asia and the Pacific, and to Sub-Saharan Africa.* Bioversity International.

Jain, S. M., & Gupta, S. D. (2013). *Biotechnology of neglected and underutilized crops.* Berlin, Germany:: Springer.

Jena, A. K., Deuri, R., Sharma, P., & Singh, S. P. (2018). Underutilized vegetable crops and their importance. *Journal of Pharmacognosy and Phytochemistry*, *7*(5), 402–407.

Keatinge, J. D., Waliyar, F., Jamnadas, R. H., Moustafa, A., Andrade, M., Drechsel, P. Hughes, J.D.A., Kadirvel, P., & Luther, K. (2010). Relearning old lessons for the future of food—by bread alone no longer: diversifying diets with fruit and vegetables. *Crop Science*, *50*, S-51.

Keatinge, J. D. H., Yang, R. Y., Hughes, J. D. A., Easdown, W. J., & Holmer, R. (2011). The importance of vegetables in ensuring both food and nutritional security in attainment of the Millennium Development Goals. *Food Security*, *3*(4), 491–501.

McCouch, S., Baute, G. J., Bradeen, J., Bramel, P., Bretting, P. K., Buckler, E., Burke, J.M., Charest, D., Cloutier, S., Cole, G. & Zamir, D. (2013). Feeding the future. *Nature*, *499*(7456), 23–24.

Sharma, D.V. (2003). Transfer of technology for increasing the scope of underexploited horticultural crops. Winter School on "Exploitation of Underutilized Horticultural Crops, 5–25th November, Department of Horticulture, College of Agriculture, MPUAT, Rajasthan, 313–320.

Singh, R. J., Chung, G. H., & Nelson, R. L. (2007). Landmark research in legumes. *Genome*, *50*(6), 525–537.

Stamp, P., Messmer, R., & Walter, A. (2012). Competitive underutilized crops will depend on the state funding of breeding programmes: an opinion on the example of Europe. *Plant breeding*, *131*(4), 461–464.

Tilman, D., Balzer, C., Hill, J., & Befort, B. L. (2011). Global food demand and the sustainable intensification of agriculture. *Proceedings of the national academy of sciences*, *108*(50), 20260–20264.

Williams, J. T., & Haq, N. (2002). *Global research on underutilized crops – an assessment of current activities and proposals for enhanced cooperation.* Southampton, UK: International Centre for Underutilized Crops. International Standard Book Number, 92-9043-545-3. Accessed on 21st August 2013.

Chapter 2
Production Technology of Underutilized Crops of Alliaceae Family

Kumari Shubha, Anirban Mukherjee, Nongmaithem R. Singh, Ramnath K. Ray, and A. K. Choudhary

Introduction

Regardless of huge achievements toward vegetable production globally, still food security is a big challenge for India. According to UNICEF (2019), every second child below 5 years is affected by some form of malnutrition. Underutilized *Allium* is rich in micronutrients such as zinc, iron, vitamin C, and provitamin A, which help protect people against lifestyle diseases like diabetes, heart disease, constipation, obesity, and cancer. *Allium* genus is one of the major genera in monocots, including more than 800 species (Fritsch et al. 2010) and popular vegetable crops, such as garlic and onions (Traub 1968). Additionally, minor and underutilized *Allium* viz. chive (*A. schoenoprasum*), bunching onion (*A. fistulosum*), Chinese chive (*A. tuberosum*), rakkyo (*A. chinense*), and leek (*A. porrum*) have also cultivated from the last few decades. The majority of the 850 species in the *Allium* genus have been identified as a rich source of micronutrients and secondary biological active metabolites (Khanum et al. 2004). *Allium* species' bulbs and leaves have been extensively proven to be antioxidants, antifungals, and antimicrobials (Bernaert et al. 2012; Mohammadi et al. 2012). Although some aqueous solvent aggregates contain antioxidants such as ascorbic acid and other organic acids, they lack non-growing *Allium* species in their profiles (Carocho and Ferreira 2013; Seabra et al. 2006). In several places (Picchi and Pieroni 2005; Tardío et al. 2006; Al-Qura'n 2010), and even in India, the bulb and leaf stalk with petioles are consumed as vegetables or condiments. As a vegetable, it is often eaten raw but more usually boiled, fried, sautéed, or combined with olive oil (Tardío et al. 2006; Dogan 2012). In South India, it is consumed as a special ingredient of Sambhar. In many restaurants, shallots are commonly offered in the condiments tray together with sauces and papads

K. Shubha (✉) · A. Mukherjee · N. R. Singh · R. K. Ray · A. K. Choudhary
Division of Crop Research, Indian Council of Agricultural Research (ICAR)-Research Complex for Eastern Region, Patna, India

Savita et al. (eds.), *Production Technology of Underutilized Vegetable Crops*,
https://doi.org/10.1007/978-3-031-15385-3_2

(Santosh et al. 2020). Underutilized *Allium* species have a very old and traditional heritage of treatment not only for various ailments (Osbaldeston 2000) but also for their presumed antihypertensive, diuretic, and anthelmintic characteristics, as well as ethnobotanical qualities (Guarrera and Savo 2013). As antagonistic bacteria and antifungal chemicals by *Allium* are produced, soil suppressiveness against fusarium wilting fungi also plays an important function (Nishioka et al. 2019).

Origin and Distribution

The genus comprises over 800 species, almost all distributed in the North. In Eurasia and a minor part of America, the majority of *Allium* species are present. The Mediterranean Basin covers Central Asia and neighboring nations with extraordinarily high species richness. A secondary center for a variety of *Allium* occurs in the western part of North America. Ecological diversification has followed the evolution of the genus. In sunny, open, and fairly humid regions, the majority of species grow. The details of the underutilized vegetables are given in Table 2.1.

Ploidy pressure in onion, shallot, garlic, and many other diploid species has not changed at home. Only seldom during selection was the introduction of additional

Table 2.1 Underutilized *Allium* crop species majorly grown worldwide

Common name	Scientific name	Area of cultivation
Shallot, potato, onion, multiplier onion	*A. ascalonicum* auct. hort.	Nearly worldwide
Ever-ready onion	*A. cepa* var. *perutile* Stearn	Great Britain
Onion	*A. cepa* ssp. cepa/var. *cepal*	Global
Garlic	*A. sativum* var. *typicum*, *A. sativum* var. *sativum*,	Global
Garlic	Regel *A. longicuspis* Regel	Central to the south and east Asia
Kurrat	*A. kurrat* Schweinf. ex Krause	Egypt and adjacent areas
Leek	*A. ampeloprasum* var. *porrum* (L.) J. Gay; *A. porrum* L.	North America and Europe
Great headed garlic	*A. ampeloprasum* var. *holmense* (Mill.) Aschers. et Graebn.	Eastern Mediterranean region
Bunching onion, Welsh onion	*A. fistulosum* L.	Europe, temperate Japan, East Asia, America
Rakkyo, Japanese scallions	*A. bakeri* Regel	Japan, China, Korea, South-East Asia
Chive	*A. sibiricum* L.	Worldwide in temperate areas
Top onion, tree onion, Egyptian onion, Catawissa onion	*A. cepa* var. *proliferum* (Moench) Alef.	North-East Asia, North America, Europe

Source: Rabinowitch and Currah (2002)

species of a function. However, *A. chinense* (Rakkyo) cultivated lines containing diploids, triploids, and tetraploids.

Nutritional Importance and Uses

As a basic component of human food, onions and other unused species, either raw or cooked in other vegetable recipes, have been widely used since ancient times. The health benefits of *Allium* sp. in treating many different major and minor disorders have been proven.

Allium porrum *L. (Leek)*

Leeks provide great vitamin C and folate quantities as well as valuable vitamin B, vitamin A, calcium, and iron portions (Table 2.2).

In addition, leeks were discovered to possess somewhat high specific carotenoids in the leaves (Heinonen et al. 1989). Two kaempferol glycosides in leek had been demonstrated to suppress blood-platelet aggregation (Fattorusso et al. 2000).

Allium ascalonicum *(Shallot)*

Shallot is a type of onion that grows in clusters and has a distinct tapered shape. Cloves are copper brown or reddish and have a sweet onion and garlic flavor. It has a morphology that is similar to both onion and garlic. The bulblets resemble garlic and their texture and color with onion. Further, it was also adopted in Ayurveda as a medicinal herb. Vitamin B6 is the primary core nutrient in shallots. Shallots contain a fair amount of vitamin C even if they are not present at levels as high as other *Allium* species. Shallots can include several other micronutrients in small but valuable quantities (Table 2.3). Shallots include saponins and high quercetin, isorhamnetin, and glycosides of quercetin (Fattorusso et al. 2002).

Table 2.2 Nutrient contents in leek

Nutrient	Content	Nutrient	Content
Energy (kcal)	77	Vitamin A (mg)	30
Moisture (g)	79	Calcium (mg)	50
Protein (g)	1.8	Iron (mg)	2.3
Fat (g)	0.1	Carbohydrates (g)	17.2

Source: USDA National Nutrient database

Table 2.3 Nutrient content in shallot

Nutrient	Content	Nutrient	Content
Energy (kcal)	72	Vitamin A (mg)	13
Moisture (g)	3.2	Calcium (mg)	3
Fat (g)	0.1	Carbohydrates (g)	17
Protein (g)	2.5	Iron (mg)	6

Source: USDA National Nutrient database

A. fistulosum *L. (Welsh onion)*

Vitamins B3 (284. 3 mg/100 g), B6 (5.4 mg/100 g), B2 (1.3 mg/100 g), B9 (2.2 mg/100 g), and iron (20. 8 mg/100 g) are all abundant in Welsh onion (Sung et al. 2014). Colds, influenza, abdominal pain, headaches, constipation, diarrhea, sores, and ulcers have all been treated with it in the past (Chen et al. 2000). *A. fistulosum* contains quercitrin, campesterol, isoquercitrin, *p*-Coumaric acid, stigmasterol, and allicin (Vlase et al. 2013).

Allium schoenoprasum *(Chive)*

Chives have medicinal properties similar to garlic but are weaker; the light effects may be the primary reason for their limited use as a medicinal herb when compared to garlic. Chive has a beneficial effect on the circulatory system due to the presence of allyl sulfides and alkyl sulfoxides. They can have moderate antiseptic, diuretic, and stimulant properties. Chive is rich in vitamins (A&C), minerals (calcium and iron), and a little amount of sulfur.

Agronomic Practices

In sandy, heavy clay, clay, or organic soil, *Alliums* are grown in all sorts of soils. Sandy loams are therefore the most suited soils for loams with decent organic content and a solid soil structure. Climatic requirement of underutilized *Allium* is similar to onion viz. open, dry, and sunny sites in humid and arid areas. Major cultivated *Allium* crops require low temperature for bolting including shallot (Krontal et al. 2000), chives (Poulsen 1990), bulb onion (Rabinowitch 1985), and Japanese bunching onion (Inden and Asahira 1990). In addition, for inflorescence initiation and further differentiation, some *Allium* crops need a long photoperiod; they include leek (Van der Meer and Hanelt 1990), Chinese chives (*A. tuberosum*) (Saito 1990), and rakkyo (*A. chinense*) (Toyama and Wakamiya 2020) (Table 2.4).

Table 2.4 Propagation method, varieties, and edible parts of major underutilized *Allium* species

Crop	Propagation method	Varieties	Edible part
Chive	Seeds or division of clumps	Grolau' Chives, Nelly' Chive, Staro' Chives, Kobold	Bundle of leaf sheaths (scapes)
Tree onion	Bulbs or sett Bulblets and also by seeds	MDU-1, CO-4, CO-3, CO-2, CO-1	Bulblet, bulbil
Welsh onion	Clumps and seeds		Bulbs, shoots, and leaves
Leek	Seeds	Palam Paushtik, PPL-1, American Flag, King Richard, Pandora, and Primor	Bulbs and lower parts of the shoots

Postharvest Management

Curing

Curing is required to store the bulbs for a long time by maintaining bulb quality. It removes extra moisture from the outer skin, neck tissue, and root of harvested *Allium* crop. This technique increases the quality and decreases the chances of disease infection during storage (Thompson et al. 1972; Petropoulos et al. 2017). The dry scales give mechanical protection (Maw and Mullinix 2005) and an aesthetic look of *Allium* (Maw and Mullinix 2005; Downes et al. 2009). Good ventilation and heat require during curing with low humidity, which dries out the neck and two to three outer scales of the bulb. The curing process is completed when the neck becomes tight and the outer scales also become dry enough to rustle (Hoyle 1948; Vaughn et al. 1964). During this process, *Alliums* have lost 3–5% of their weight (Anonymous 2016a). Immature bulbs have few outer layers and very high moisture; therefore, they require more extensive care during the curing than mature *Allium* (Anonymous 2016b). *Allium* curing can be done by natural convection of air or artificial hot air.

Storage

Allium can be best stored in well-ventilated rooms with lots of sunlight and aeration. Under a controlled atmosphere with a very low oxygen concentration of 0.5–2.0% and 3% CO_2, turning off the bulb and reducing rotten and sprouted bulbs is critical during storage (Adamicki 2004). *Allium* can be stored at a temperature of 30–35 °C with 65–70% of relative humidity (RH) for 6 months to avoid causing disease infection and rotting of bulbs (Saraswathi et al. 2017). Freezing injury occurs at temperatures below 2 °C, and rotting can occur at temperatures between 2 and 25 °C with more than 75% humidity (Saraswathi et al. 2017). Various pathogens, such as *Aspergillus*, *Alternaria*, *Botrytis*, *Rhizopus*, and others, attack *Allium* spp. during storage, causing economic losses as well as pathogens that are harmful to people's health (Fink-Grernmels 1999). Various chemicals such as mancozeb, carbendazim, salicylic acid, sulfur are used to control the pathogens (Kumar et al. 2015).

Grading

Generally, in India, *Allium* is performed manually grading as well as machine grading. *Alliums* are graded in three grade scales: A grade, B grade, and C grade scales. A grade scale should have more than 60 mm diameter of *Allium*, while B and C grade scales have 50–60 and 35–50 mm diameter of *Allium*, respectively (Tripathi and Lawande 2003). In addition, the price of *Allium* is differentiated significantly according to its graded size. Graded *Alliums* are more attractive to consumers and improve the graded *Allium* quality standard, showing uniformity in the size. Graded *Allium* will attract the consumer to buy the *Allium* at a given or higher than normal price (Aher et al. 2019).

Packaging

Allium crops can be packed in jute bags and wooden baskets for packaging. The nylon net bags can also be used for packaging due to the excellent ventilation provided to packed *Allium* (IIFPT). For easy handling of *Allium* during transportation, packaging should be made with small and suitable packaging materials that protect against physiological (mass), physical (firmness), and pathological (decay) deterioration (Saraswathi et al. 2017).

Marketing Strategies to Promote Underutilized *Allium* Species

Indonesia is the leading producer of leek, other alliaceous vegetables in the world producing 590.6 thousand tons of product in 2019, followed by Turkey and China, 234.1 and 167.0 thousand tons, respectively (Fig. 2.1). These countries also dominate the majority of the export. India has very little production of leek and other vegetables of the Alliaceae family.

In India, for *Allium* crops, majorly production and consumption data are available for onion and garlic. These two are the major crops largely produced and consumed worldwide. So the volume of production and marketing is also available in these two cops. If we consider the state-wise production of onion and garlic, we can see one or two-state dominance in production.

In India, Maharashtra is the highest producer of onion and produces 8854.09 thousand tons of onion every year and holds the share of 38.09% (Fig. 2.2). Madhya Pradesh is next to that, producing 3701.01 thousand tons and holding 15.92% of the

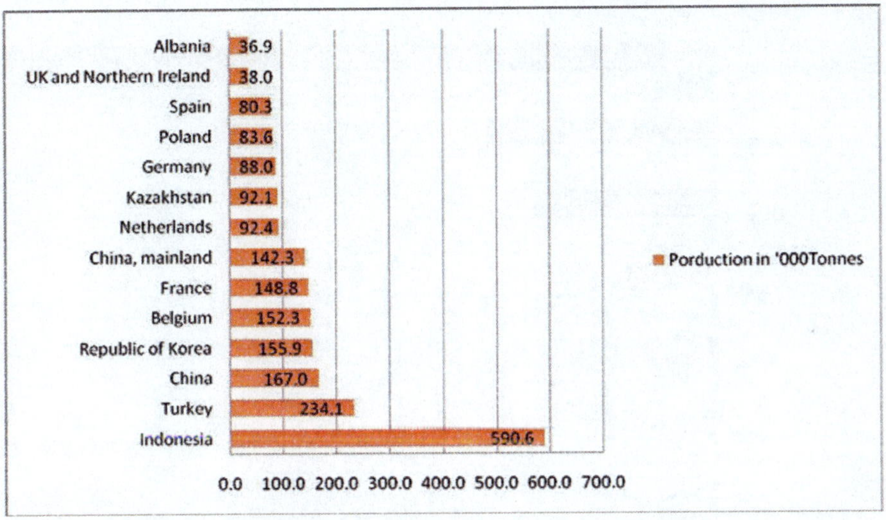

Fig. 2.1 Production of leeks and other alliaceous vegetables in the year 2019 in the top 14 countries in the world. (Data source: FAOSTAT (2021))

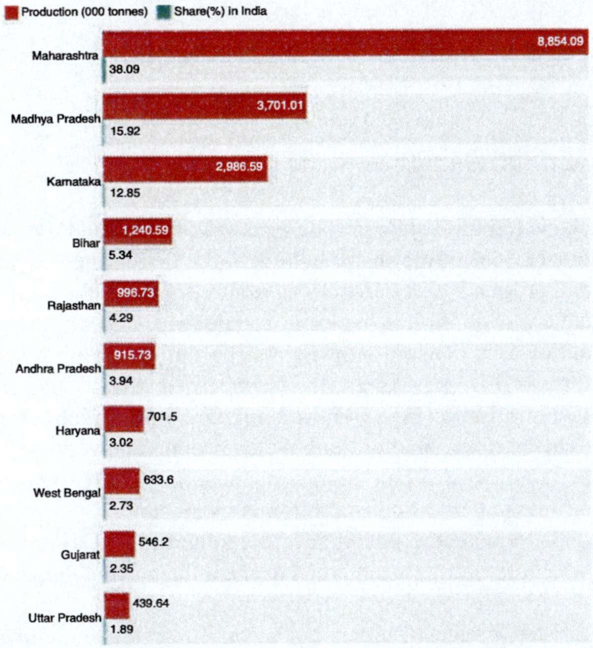

Fig. 2.2 Major onion producing states in India and their production share. (Data source: National Horticulture Board (NHB) Year 2017–18)

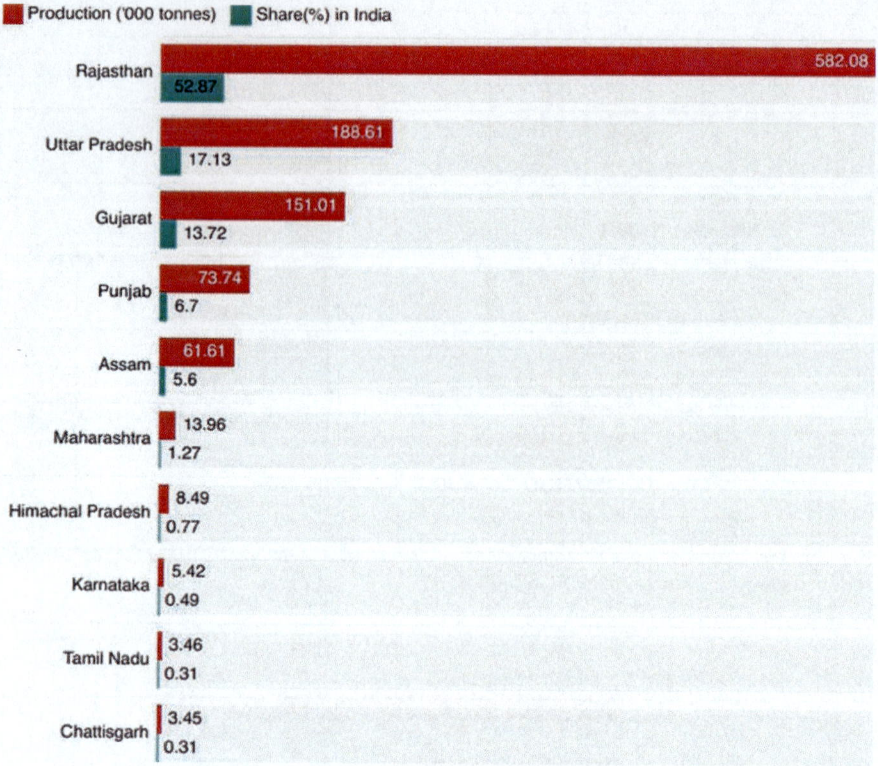

Fig. 2.3 Major garlic producing states in India and their production share. (Data source: National Horticulture Board (NHB) Year 2017–18)

share. Karnataka is the third state producing 2986.59 thousand tons of onion every year and holding 12.85% of the share. In total, these three states are producing two-thirds of the total production in India. These states dominate the entire marketing of onion. Interstate transportation is the way of trade (Fig. 2.3).

As consumption is a concern worldwide, the onion consumption in India per capita per annum is 14.7 kg. As garlic is concerned, the annual consumption is 5.86 kg per person in India (Fig. 2.4). Much data is not available for other *Allium* crops such as chives, leek, shallot, and Welsh onion. These crops are popular in China and other southwest Asian countries. In India, with increasing continental food preferences the pattern of consumption is increasing.

The above data indicate the potential of the underutilized *Allium*. There is massive potential for these nutrient-rich crops to meet the current demands of nutrient-rich foods.

Marketing strategy makes market chains more sustainable for minor *Allium* crops. They target campaigns for public awareness, select high-quality items, and promote certain features or attributes of species or variants. Marketing is essential for enhancing the profitability of farmers (Kumar et al. 2018). For onion and garlic,

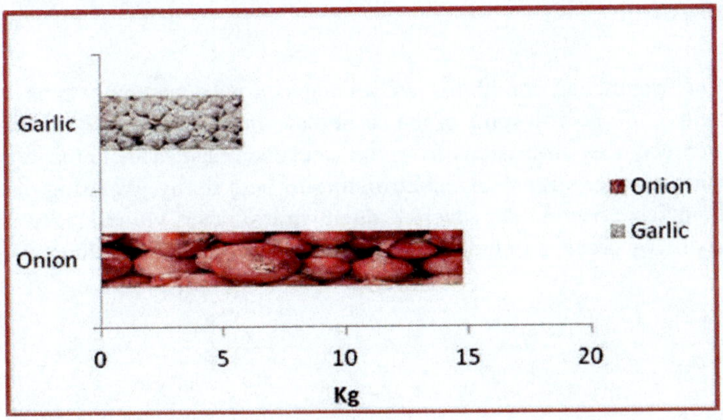

Fig. 2.4 Annual consumption of onion and garlic per person per annum in India. (Data source: Garlic, statista.com (2012); Onion, helgilibrary.com (2017))

the supply chain is well established but it is missing in the case of minor *Allium* as lack of sizable demand. Farmers can individually sell their low use of *Allium* crops through the local market and so even achieve a slightly higher price for their products (Kruijssen et al. 2008). However, growers indicate that they prefer to deliver to society as this is less time-consuming and secure. More demand for *Allium* crops in the market should make products based on *Allium* crops like pest powder flakes, etc. and such platforms like a store, malls, supermarkets, etc. can be more attractive to sell well-packed *Allium* and branded *Allium* products. In India, forming Farmers Producer Organization (FPO), particularly in these underutilized *Allium* crops, can be beneficial. It will be a win–win situation for farmers as well as the consumers. It has been reported from recent studies that FPO has been applicable in enhancing farmer's income (Mukherjee et al. 2018) and livelihood security in different sectors like dairy (Mukherjee et al. 2020a), poultry (Mukherjee et al. 2019), and traditional foods (Mukherjee et al. 2020b). For marketing of minor *Allium*, FPO can be made to market their product to processing industries and tie up with hotels and restaurants (IIFPT).

Physiological Disorders

Bolting

Most of the minor *Alliums* are biennials in nature. The emergence of the seed stalk before the formation of vegetative growth adversely affects the formation and development of bulbs. Bolting is premature flower head development, usually occurring in the first year of growth. Bolting can also result in small and poor vegetative development. It can be rectified by changing the transplanting time.

Bulb Splitting

Adverse environmental conditions and imbalanced nutrient supply cause the splitting of bulbs. *Allium* crop splits of the basal plate and the affected bulb's secondary growth can occur as protrusions from the weakened base. This damage can allow the invasion of microorganisms, contributing to bulb decay. Watering extensively, when plants have been under severe drought stress, raises vulnerability. Incidence can be severe by damage to the root base from other secondary pathogens.

Chimera

In this disorder, variegated leaf tissue usually appears longitudinally around the leaf in yellow and/or white strips. Extreme temperature causes this problem in some minor *Allium*.

Hail/Heavy Rain Damage

Some minor *Alliums* are sensitive to excessive rains or hail damage resulting in distinctive white marks, generally on one side of the leaf only. If serious, holes in the leaves may be punched or leaves may be torn off.

Thick Necks

This disorder occurs when immature bulbs are harvested, before proper bulbing. Phosphorous deficiency can increase incidence during development. It was common in low-population bulb crops or where bulb stimulation is weak, particularly in cold, wet summers.

Watery Scale and Leathery Skin

It is a very common disorder of bulb crops; it mainly occurs when harvested after heavy rain. Affected bulbs produce external water-soaked scales (thick scale). During storage, thick skins limit the absorption of oxygen and carbon dioxide into and out of the bulb scales resulting in reducing the postharvest life.

Conclusion

The better option for large-scale agriculture appears to be the cultivation of underutilized *Allium* crops on a huge scale. *Allium* species have the potential to provide food security, nutrition, health, and income generation. Due to the lack of competition from other crop species in the same agricultural environment, farmers and consumers are using these crops less. The underutilized *Allium* crops are a good option for the region because it is one of the richest reservoirs of genetic variability and diversity, with plant types, morphological and physiological variations, disease and pest reactions, adaptability, and distribution all present. Because these crops have the potential to directly alleviate hunger by increasing food production in difficult environments where major food crops are severely limited daily. The potential for agricultural–rural development, food and nutrition security, and the realization of the importance and uses of these underutilized *Allium* crops can be unlocked, enhancing nutrition, dietary and culinary diversification, health, and income generation.

References

Adamicki, F. (2004). Effects of pre-harvest treatments and storage conditions on quality and shelf-life of onions. *Acta Horticulturae* (ISHS), 688, 229–238.

Aher, D., Dabhade, A., Desai, P., Waghmare, D., & Suryawashi, K. (2019). The Onion Grading Machine, *International Journal of Scientific Research and Review, 7*(2), 631–638

Al-Qura'n, S.A. (2010). Ethnobotanical and ecological studies of wild edible plants in Jordan. *Libyan Agriculture Research Center Journal International, 1*(4), 231–243.

Anonymous. (2016a). Onion postharvest technology. http://nhb.gov.in.

Anonymous. (2016b). Onion storage guidelines for commercial growers.

Bernaert, N., De Paepe, D., Bouten, C., De Clercq, H., Stewart, D., Van Bockstaele, E., De Loose, M. & Van Droogenbroeck, B. (2012). Antioxidant capacity, total phenolic and ascorbate content as a function of the genetic diversity of leek (*Allium ampeloprasum* var. *porrum*). *Food chemistry, 134*(2), 669–677.

Carocho, M., & Ferreira, I. C. (2013). A review on antioxidants, prooxidants and related controversy: natural and synthetic compounds, screening and analysis methodologies and future perspectives. *Food and chemical toxicology, 51*, 15–25.

Chen, J. H., Chen, H. I., Wang, J. S., Tsai, S. J., & Jen, C. J. (2000). Effects of Welsh onion extracts on human platelet function in vitro. *Life sciences, 66*(17), 1571–1579.

Dogan, Y. (2012). Traditionally used wild edible greens in the Aegean Region of Turkey. *Acta Societatis Botanicorum Poloniae, 81*(4).

Downes, K., Chope, G. A., & Terry, L. A. (2009). Effect of curing at different temperatures on biochemical composition of onion (*Allium cepa* L.) skin from three freshly cured and cold stored UK-grown onion cultivars. *Postharvest biology and technology, 54*(2), 80–86.

FAOSTAT. (2021). Available at http://www.fao.org/faostat/en/#data/GN accessed on 16th January 2021.

Fattorusso, E., Iorizzi, M., Lanzotti, V., & Taglialatela-Scafati, O. (2002). Chemical composition of shallot (*Allium ascalonicum* Hort.). *Journal of agricultural and food chemistry, 50*(20), 5686–5690.

Fattorusso, E., Lanzotti, V., Taglialatela-Scafati, O., Di Rosa, M., & Ianaro, A. (2000). Cytotoxic saponins from bulbs of *Allium porrum* L. *Journal of agricultural and food chemistry*, *48*(8), 3455–3462.

Fink-Grernmels, J. (1999). Mycotoxins: Their implications for human and animal health. *Veterinary Quarterly,* 21, 115–120.

Fritsch, R. M., Blattner, F. R., & Gurushidze, M. (2010). New classification of *Allium* L. subg. Melanocrommyum (Webb & Berthel.) Rouy (Alliaceae) based on molecular and morphological characters. *Phyton (Horn)*, *49*(2), 145–220.

Guarrera, P. M., & Savo, V. (2013). Perceived health properties of wild and cultivated food plants in local and popular traditions of Italy: A review. *Journal of Ethnopharmacology*, *146*(3), 659-680.

Heinonen, M. I., Ollilainen, V., Linkola, E. K., Varo, P. T., & Koivistoinen, P. E. (1989). Carotenoids in Finnish foods: vegetables, fruits, and berries. *Journal of Agricultural and Food Chemistry*, *37*(3), 655–659.

Horticultural Statistics at a Glance 2018, National Horticulture Board (NHB) Year 2017–18 availableathttp://nhb.gov.in/statistics/Publication/Horticulture%20Statistics%20at%20a%20Glance-2018.pdf

Hoyle, B.J. (1948). Onion curing – a comparison of storage losses from artificial, field and non-cured onions. *Proceedings of the American Society for Horticultural Science, 52*, 407–414.

IIFPT-Indian Institute of Food Processing Technology, Tamil Nadu, *Model DPR on Integrated Onion Processing Unit*, 1–25. http://iifpt.edu.in/pmfme/dpr/DPR-Onion.pdf

Inden, H., & Asahira, T. (1990). Japanese bunching onion (*Allium fistulosum L.*). In: Brewster, J.L and Rabinowitch, H.D. *(eds) Onions and Allied Crops, III. Biochemistry, Food Science, and Minor Crops. CRC Press Boca Raton Florida,* 159–178.

Khanum, F., Anilakumar, K.R., & Viswanathan, K.R. (2004). Anticarcinogenic properties of garlic: A review. *Critical Review of Food Science Nutrition, 44*(6), 479–488.

Krontal, Y., Kamenetsky, R., Rabinowitch, H.D. (2000). Flowering physiology and some vegetative traits of short-day shallot–a comparison with bulb onion. *Journal of Horticultural Science and Biotechnology, 75,* 35–41.

Kruijssen, F., Giuliani, A., & Sudha, M. (2008). Marketing underutilized crops to sustain agrobiodiversity and improve livelihoods. *In International Symposium on Underutilized Plants for Food Security, Nutrition, Income and Sustainable Development, 806*, 415–422.

Kumar, S., Roy, M., & Mukherjee, A. (2018). Marketing behaviour of vegetable growers in Uttarakhand hills. *Journal of Community Mobilization and Sustainable Development, 13*(1), 68–74.

Kumar, V., Neeraj, Sharma, S., & Sagar, N.A. (2015). Postharvest management of fungal diseases in onion-A review. *International Journal of Current Microbiology and Applied Sciences, 4,* 737–752.

Maw, B.W., & Mullinix, B.G. (2005). Moisture loss of sweet onions during curing. *Postharvest Biology and Technology 35,* 223–227.

Mohammadi, P., Shoaie, N., Ghazanfari, T., & Mohammadi, S.R. (2012). Evaluation of antibiofilm activity of Allium sativum extract against biofilm formed by isolated Candida sp. from urinary catheters. *Mycoses, 55*, 95–96.

Mukherjee, A., Singh, P., Rakshit, S., Satyapriya, Burman, R.R., Shubha, K., Sinha, K., & Nikam, V. (2019). Effectiveness of poultry based Farmers' Producer Organization and its impact on livelihood enhancement of rural women. *Indian Journal of Animal Sciences*, *89*(10), 1152–1160.

Mukherjee, A., Singh, P., Ray, M., Satyapriya, & Burman, R.R. (2018). Enhancing farmers income through farmers' producers companies in India: Status and roadmap. *Indian Journal of Agricultural Sciences*, *88*(8), 1151–1161.

Mukherjee, A., Singh, P., Satyapriya, Maity, A. Shubha, K., & Burman, R.R. (2020a). Enhancing livelihood security of dairy farmers through farmers' producer company: a diagnostic study of Bundelkhand region. *Range Management and Agroforestry*, *41*(1), 156–167.

Mukherjee, A., Singh, P., Satyapriya, Rakshit, S., Burman, R.R., Shubha, K., & Kumar, S. (2020b). Assessment of livelihood wellbeing and empowerment of hill women through Farmers

Producer Organization: A case of women based Producer Company in Uttarakhand. *Indian Journal of Agricultural Sciences*, *90*(8), 1474–1481.

Nishioka, T., Marian, M., Kobayashi, I., Kobayashi, Y., Yamamoto, K., Tamaki, H., Suga, H., & Shimizu, M. (2019). Microbial basis of Fusarium wilt suppression by Allium cultivation. *Scientific reports*, *9*(1), 1–9.

Osbaldeston, T.A. (2000). The herbal of Dioscorides the Greek. Johannesburg, South Africa: IBIDIS Press.

Petropoulos, S. A., Ntatsi, G., & Ferreira, I. C. F. R. (2017). Long-term storage of onion and the factors that affect its quality: A critical review. *Food Reviews International*, *33*(1), 62–83.

Picchi, G., & Pieroni, A. (2005). Atlante dei prodotti tipici: Le erbe. Roma, Italy: Agra, RAI-Eri.

Poulsen, N. (1990). Chives Allium schoenoprasum L. In: Brewster, J.L. and Rabinowitch, H.D. (eds) Onions and Allied Crops, III. *Biochemistry, Food Science, and Minor Crops. CRC Press Boca Raton Florida,* 231–250.

Rabinowitch, H.D. (1985). Onions and other edible *Allium*. In: Halevy, A.H. (ed.) *Handbook of Flowering. CRC Press, Boca Raton, Florida*, 398–409.

Rabinowitch, H.D., & Currah, L. (Eds.) (2002). *Allium crop science: recent advances*. CABI.

Saito, S. (1990). Chinese chives Allium tuberosum Rottl. In: Brewster, J.L. and Rabinowitch, H.D. (eds) Onions and Allied Crops, III. *Biochemistry, Food Science, and Minor Crops. CRC Press Boca Raton Florida,* 219–230.

Santosh, K., Jotish, N., Dubey, R.K., Chaudhary, K.P., Leindah Devi, N., & Khangembam, J.K. (2020). Status, diversity and potential of indigenous and minor *Allium spp. Indian Horticulture,* 58–61.

Saraswathi, T., Sathiyamurthy, V.A., Tamilselvi, N.A., & Harish, S. (2017). Review on Aggregatum Onion (*Allium cepa* L. var. *aggregatum* Don.). *Int. J. Curr. Microbiol. App. Sci., 6*(4), 1649–1667.

Seabra, R. M., Andrade, P. B., Valentao, P., Fernandes, E., Carvalho, F., & Bastos, M. L. (2006). Anti-oxidant compounds extracted from several plant materials. *Biomaterials from aquatic and terrestrial organisms*, 115–174.

Sung, Y. Y., Kim, S. H., Kim, D. S., Park, S. H., Yoo, B. W., & Kim, H. K. (2014). Nutritional composition and anti-obesity effects of cereal bar containing *Allium fistulosum* (Welsh onion) extract. *Journal of functional foods*, *6*, 428–437.

Tardío, J., Pardo-de-Santayana, M., & Morales, R. (2006). Ethnobotanical review of wild edible plants in Spain. *Botanical journal of the Linnean society*, *152*(1), 27–71.

Thompson, A.K., Booth, R.H., & Proctor, F.J. (1972). Onion storage in the tropics. *Tropical Science, 14*, 19–34.

Toyama, M., & Wakamiya, I. (2020). Rakkyo *Allium chinense* G. Don. In *Onions and allied crops* (pp. 197–218). CRC Press.

Traub, H. (1968). The order of Allium. *Plant Life 24:*129–128.

Tripathi, P.C., & Lawande, K.E. (2003). Onion Graders, Bulletins, National Research Center for Onion & Garlic, Pune, Technical Bulletin, 6, 1–12.

UNICEF (2019). Malnutrition behind 69 per cent deaths among children below 5 years in India: UNICEF report Read more at: https://economictimes.indiatimes.com/news/politics-and-nation/malnutrition-behind-69-per-cent-deaths-among-children-below-5-years-in-india-unicef-report/articleshow/71618288.cms?utm_source=contentofinterest&utm_medium=text&utm_campaign=cppst, 71618288.

Van der Meer, Q.P., & Hanelt, P. (1990). Leek (*Allium ampeloprasum*). In: Brewster, J.L. and Rabinowitch, HD (eds) Onions and Allied Crops, III. *Biochemistry, Food Science, and Minor Crops. CRC Press, Boca Raton, Florida*, 179–196.

Vaughn, E. K., Cropley, M. G., & Hoffman, E. N. (1964). Effects of field-curing practices, artificial drying, and other factors in the control of neck rot in stored onions. *Oregon Agr. Exp. Sta. Tech. Bui.(77)*.

Vlase, L., Parvu, M., Parvu, E. A., & Toiu, A. (2013). Phytochemical analysis of *Allium fistulosum* L. and *A. ursinum* L. *Digest Journal of Nanomaterials & Biostructures (DJNB)*, *8*(1).

Chapter 3
Production Technology of Underutilized Vegetables of Leguminosae Family

Savita

Introduction

The *winged bean* (also called as Four-angled bean, Goa bean, ridged bean, Four-cornered bean, Manila bean, Cigarrillas, and Dragon bean) is a herbaceous legume plant with the potential to contribute significantly to global nutrition and food security. It is also known for its dense, nutrient-green pods, leaves, immature seeds, and tubers as "One Species Supermarket" (Tanzi et al. 2019). In Japan, the *adzuki bean* is the second most important legume crop (Hajika 2016), also called Azuki and small red beans (Gohara et al. 2016). It was brought from China to Japan some 1000 years ago and is now the country's sixth largest crop, as well as a prominent topic in Japanese scientific journals. *Sword bean* (*Canavalia gladiate*) is a very important underutilized legume crop (Ekanayake et al. 2000). It is known as "sword bean" because the green pods resemble the blade of a sword, and its Malay name is "kacang parang." It is also called the jack bean or beach bean (Black et al. 2006; Zhang and Yang 2007; Yang and Ma 2010; Luo et al. 2015). Vadivel et al. (2010) add that sword bean plants can convert solar light energy into chemical energy, so it has high biomass (Siddhuraju and Becker 2001; Niazi et al. 2017).

Jack bean (*Canavalia ensiformis*) is also named coffee bean, Pearson bean, wataka bean, Gotani bean, giant stock bean, horse bean, and wonder bean. It is also called *feijão-de-porco* ("pig bean") in Brazil, whereas in Jamaica, it first became well known and called the horse bean or the overlook bean. *Moth bean* (*Vigna aconitifolia* (Jacq.) Marechal) is named as dew bean, mat bean, moth bean, matki, or Turkish gram and is considered as potential future food sources due to its high protein. *Scarlet runner bean or Oregon runner bean* (*Phaseolus coccineus*) is the world's second most important species grown as a food as well as an ornamental plant (Łabuda 2010). It is an ancient vegetable crop that is grown in-home or market

Savita (✉)
Graphic Era Hill University, Dehradun, Uttarakhand, India

Savita et al. (eds.), *Production Technology of Underutilized Vegetable Crops*,
https://doi.org/10.1007/978-3-031-15385-3_3

gardens, as well as in truck gardens for canning. Green pods are utilized as green vegetables, and dried beans are used as a pulse. It is consumed as a vegetable in India, whereas in other countries, it is consumed in the off-season after processing when green pods are not available. The varieties having red flowers and multicolored seeds are often grown as ornamental plants.

Yam bean (*Pachyrhizus erosus* L.) is a plant native to the tropics and subtropics that has a long history of medicinal usage (Kumalasari et al. 2014), also known as Chinese potato, Mexican turnip, Mexican potato, Mexican yam bean, Mexican water chestnut, mishrikand, sakalu, or Jicama, and extensively cultivated both on small scale in kitchen gardens for its sweet and crisp tubers and on large scale for export purpose. There are five species out of which three are cultivated and two are wild types. The cultivated species include Mexican yam bean (*Pachyrhizus erosus*), Andean yam bean (*Pachyrhizus ahipa*), and Amazonian yam bean (*Pachyrhizus tuberosus*). *Pachyrhizus palmatilobus*, also known as Jicama de leche in the area, has a less pleasant flavor. In many parts of Central America, Southern Asia, the Caribbean, and South America, it is one of the most widely grown vegetables.

Potato bean (*Apios americana*) is a tuber-bearing leguminous plant also called groundnut, Wild Bean, Hopniss, and Indian potato eaten raw as well as cooked. It is mainly grown for its leaves and pods. The plant thrives on acidic soils in the tropical rainforests of South America. It is grown by the Amazon Rainforest's indigenous people. It is an ancient traditional crop from humid tropical South America. Tepary bean is primarily cultivated in Mexico and Arizona. It is widely grown in tropical Africa, particularly in Botswana, northeastern Kenya, Uganda, and Sudan, where most other legume crops are unable to thrive due to drought, and frequently required crop rotation, but its production is primarily for domestic consumption. *Agathi* is also known as hummingbird tree or august tree. It is mainly grown for its leaves, flowers, and as green manure and forage crop, valued. Its bark yields good fiber. It is grown as a standard for pepper and betel vines as a shading plant for coconut seedlings and as a windbreak for bananas. It is a multipurpose tree, which has food, fodder, and medicinal values.

Velvet bean is commonly known as Bengal velvet bean, buffalo bean, cowage, cowitch, donkey eye, Florida velvet bean, lacuna bean, Lyon bean, Mauritius velvet bean, monkey tamarind, and Yokohama velvet bean. It is produced in the tropics due to its tremendous food and feed potential. Its high protein content, combined with its digestibility, makes it an excellent source of dietary protein, comparable to other pulses such as soya, rice beans, and lima beans. It is used for weed-infested and degraded soils reclamation. Its plant is notorious, as its young foliage, pods, and seeds when coming in contact cause itching, which is known as mucunain. Scratching of the exposed area should be avoided, as it will transfer the chemical responsible for itching to all other touched areas. When the subject starts scratching tends to scratch vigorously and uncontrollably, which is why the local people in northern Mozambique refer to it as a mad bean. The seedpods are also known as Devil Beans in Nigeria. *Horse gram* is an important vegetable used to make "dal" and "rasam" for human use, as well as concentrated feed for bovine animals. It is cultivated in the vacant areas of the citrus fruit trees. *Yard long bean* is also called asparagus bean, Chinese long bean, cow bean, Daugok, Eeril, Kacangpanjang, long bean, long-podded cowpea, pea bean, snake bean, Thuafakyao, and Vali, mainly cultivated for its long green pods. Most of its common names refer to the length of the pod. Unlike cowpea, it bears delicious stringless pods, which are eaten like green beans.

African yam bean is also known as yam pea, tuber bean, haricot igname, pomme deterre dumossi, afrikanische yam bohne, or knollenbohne and several native lingual synonyms in Africa, namely akidi, ekulu, sunmunu, otili, kutreku, kulege, akitereku, apetreku, girigiri, kutonoso, roya, efik, nsama, ibibio, bitei, sesonge, gundosollo, sumpelegu, tschangilu, sese, sheshe, giliabande, pempo, or mpempo. The six botanical synonyms most often used for African yam beans are *Sphenostylis ornata*, *Sphenostylis congensis*, *Sphenostylis katangensis*, *Vigna katangensis*, *Vigna ornate*, and *Dolichos stenocarpus*. This is an important legume crop of tropical Africa, grown for both seeds and tubers. The *African locust bean* (*Parkia biglobosa*) is a tropical plant (Akande et al. 2010). The African continent has one of the world's richest biodiversities, with a wide range of plants used for herbal medicine, health food, and therapy (Farombi 2003). *Ground bean* (*Vigna subterranean*) is also known as Bambara bean, Bambara groundnut, Bambara nut, earth pea, Congo goober, or hog-peanut. It is valuable for the areas where the cultivation of other legumes is difficult because of the diseases, poor soils, and drought. *Marama bean* (*Tylosema esculentum*) is also called as gemsbok bean, Braaiboontjie Marumama, Gembok beans, Gemsbokboontjies, Muraki, Tamani berry, and Morama, a long-lived perennial legume.

The details of some underutilized legume crops are given below:

S. no.	Common name	Botanical name	Chromosome no.
1.	Winged bean	*Psophocarpus tetragonolobus* (L.) DC.	18
2.	Adzuki bean	*Phaseolus angularis* (Wild.) Wright	22
3.	Sword bean	*Canavalia gladiata* (Gucav.) DC.	22, 24
4.	Jack bean/Gotani bean	*Canavalia ensiformis* (L.) DC.	22
5.	Moth bean	*Phaseolus acontifolius*	22
6.	Scarlet runner bean/ Multiflora bean	*Phaseolus coccineus/Phaseolus multiflorus* Wild	22
7.	Yam bean	*Pachyrhizus erosus* Urbn.	22
8.	Potato bean	*Apios Americana*	22
9.	Tepary bean	*Posphocarpus acutifolius*	22
10.	Catjung bean	*Vigna sesquipedalis* var. *cylindrical* or *biflora*	22
11.	Agathi	*Sesbania grandiflora* Poir.	24
12.	Thicket bean	*Phaseolus polystachyus*	22
13.	Velvet bean	*Mucuna deeringiana* (Bor.) Merr. (syn. *Stizolobium deeringianum* Bort.)	22
14.	Horse gram	*Dolichus uniflorus* Lam. (*Dolichus biflorus* Auct. L.)	22, 24
15.	Yard long bean or asparagus bean	*Vigna unguiculata* ssp. *sesquipedalis*	22

(continued)

S. no.	Common name	Botanical name	Chromosome no.
16.	African yam bean	*Sphenotylis stenocarpa*	22
17.	African locust bean	*Parkia biglobosa*	24
18.	Ground bean	*Vigna subterranea* (L.) Verdc.	22
19.	Marama bean	*Tylosema esculentum* (Burchell) Schreiber	44

Origin and Distribution

Winged Bean

The center of origin of the winged bean is still uncertain. Some winged beans originated in Africa as five species belonging to Africa (Khan 1976; Pickersgill 1980; Harder 1992), namely *P. scandens*, *P. grandiforus*, *P. lancifolius*, *P. lecomtei*, and *P. palustris* (Harder and Smartt 1992). Burkill (1906) reported the Indian Ocean as the origin place of the winged bean, while Vavilov (1951) put them in his hypothesized India as center for crop domestications. The winged bean center of origin remains a subject that requires further research, despite the most recent phylogenetic analysis by Yang et al. (2018). It is widespread in humid tropics (South and Southeast Asia, particularly India, Bangladesh, Burma, Indonesia, Kampuchea, Laos, Malaysia, Philippines, Sri Lanka, Thailand, Vietnam, and Papua New Guinea) and subhumid tropics (NAS 1975; Khan 1982; Gross 1983; Lepcha et al. 2017).

Adzuki Bean

Vigna Angularis var. *nipponensis* is probably the ancestor of adzuki bean (Yamaguchi 1992) and is widely distributed in Bhutan, China, Japan, Korea, and Nepal (Kaga et al. 2008). It is originated from tropical Asia and Africa. Currently, erect plants are grown in the northern provinces of Japan, whereas in China, Manchuria, and other warmer areas the branching and vining types are cultivated. It is mainly distributed in India, Japan, Korea, New Zealand, South China, Taiwan, Thailand, and the Philippines (Tateishi 1983, 1984).

Sword Bean

Sword bean originated from the Indo-Malayan region and Tropical Africa is its secondary center of origin. It has also been consumed in the Asian tropics and subtropics (Moteetee 2016). It has a close relative, the jack bean, *Canavalia ensiformis*, which originated from South and Central America. Both species are easily confused with one another. The probable progenitor species *Canavalia virosa* is found to occur in some parts of the northeastern region. It is also spread in northern and peninsular India (Vadivel and Janardhanan 1998). The indigenous people of Arunachal Pradesh, Nagaland, Manipur, Mizoram, Tripura, and Meghalaya consume the fruits of sword beans (Borthakur 1996). It is grown in South and Southeast Asia, particularly in India, Burma, and Sri Lanka. It is used for local consumption in humid tropics.

Jack Bean

The West Indies and the surrounding continent are native place of jack bean. Jack bean is similar to *Canavalia gladiata*, a species of beans cultivated in the Old World and the prevailing wild species of Africa (*Canavalia virosa*). In South Africa, it has been called the sunshine bean. Because of the close similarity between the three species, some authors consider both jacks and sword beans as derivatives of wild Africans. It is primarily found in tropical, subtropical, western Africa, Asia, Latin America, South America, India, and the Southern Pacific (Tiamiyu et al. 2016).

Moth Bean

Moth bean is originated from India, Pakistan, and Myanmar. This is extensively grown in India, Thailand, Australia, Sri Lanka, and China, as well as in the southwest (Texas and California), Canada, and also in the United States (Munro and Small 1997; Brink and Jansen 2006; Kochhar 2016).

Scarlet Runner Bean

Scarlet runner bean is originated from Mexico and Central America's tropical highlands and lush valleys. South and Central America are the principal centers of origin, with Peru, Ecuador, and Bolivia serving as secondary centers of origin. For decades, this species was grown in Mesoamerica's highlands. It is extensively cultivated and distributed in pre-Colombian Mexico by the Anahuac people. The introduction in Southern Colombia and Europe, where it was referred to as "scarlet runner bean," might occur in the seventeenth century before it reaches the Ethiopian mountains in other parts of the world and, subsequently, it was transported to India from Europe, where it was distributed to Andhra Pradesh, Karnataka, Maharashtra, Tamil Nadu, Gujarat, and West Bangkok. It is popularly grown mostly in America and Europe. It is also grown in the United Kingdom and Asia, especially in China and Sri Lanka (Suttie 1969; Kay 1979; Caiger 1995; Brink 2006).

Yam Bean

Yam bean is native to the hot moist region of Mexico and northern South America, in the head-water region of the river Amazon. Tropical America is considered the origin place of yam bean. In the sixteenth century, the Spaniards most likely introduced the yam bean to Southeast Asia via the Philippines. The cultivation has since spread to Indonesia and the rest of the Far East and the Pacific. It is widely cultivated in Burma, China, Hawaii, India, Indonesia, Mexico, Nepal, Philippines, and Singapore. In India, it is cultivated in parts of Bihar, Odisha, West Bengal, Tripura, Assam, and eastern Uttar Pradesh.

Potato Bean

The potato bean is originated from eastern North America. There are three species of potato bean native to Asia, namely *A. fortunei*, *A. carnea*, and *A. delavayi*, while two species belong to North America, i.e., *A. americana* and *A. priceana* (Li et al. 2014; Woods 2005). During the 1845 Irish famine of potato, an attempt was made in Europe to plant *Apios*, but this was abandoned when the potential of cultivating potato became possible (NRC 1979). Hoshikawa (1994) tried to cultivate this plant for about 10 years and named the new crop plant *Apios*. Through his efforts, recently, cultivation of the crop has become widespread in Japan, especially in Tottori, Akita, and Miyagi prefectures.

Tepary Bean

Northern Mexico and the United States are the origin place of the tepary bean. It was domesticated in Tehuacan Valley, Mexico, 2300 years ago. Its cultivation dwindled strongly after the Second World War, but nowadays the crop is regaining the interest of people for its cultivation. It was introduced into Africa, Asia, Australia, and Madagascar. Its cultivation is also recorded in Morocco, Algeria, South Africa, Swaziland, and Lesotho.

Catjang Bean

Catjang bean is originated in India from the West African-based cowpea (cultigroup Unguiculata) (Ng and Marechal 1985). Catjang is grown extensively in India, China, Cambodia, Japan, Korea, Laos, Vietnam, Africa, and the United States.

Agathi

Its actual origin of agathi is unknown (India or Indonesia have been cited), although it is thought to be endemic to many Southeast Asian countries. *Sesbania formosa*, an endemic Australian plant, is extremely closely related to it. This evidence backs up the theory that agathi originated in Indonesia. It is mostly spread from the southern part of Mexico to South America and also cultivated in Punjab, Assam, Tamil Nadu, and Kerala in India.

Velvet Bean

The western hemisphere is covered by Mauritius with velvet bean, originally from South Asia. It is cultivated in Asia, America, Africa, and the Pacific Islands, where its pods are consumed by human beings and fed to animals. It was broadly cultivated in the easterly Himalayas and Mauritius in the 18th and 19th centuries. It was brought to the southern regions of America as an ornamental species in the late nineteenth century.

Horse Gram

It is native to India (Purseglove 1974; Smartt 1985; Vavilov 1951; Zohary 1970; Arora and Chandel 1972) and grown for food purposes in Australia, Africa, India, Myanmar, Nepal, Malaysia, Mauritius, and Sri Lanka (Asha et al. 2006).

Yard Long Bean

The yard long bean is originated from Southeast Asia and is a conventional crop of Africa, which is capable of improving food, improving food security, promoting rural development, and promoting sustainable land management. It is also extensively cultivated in India for its pods, containing the number of pea-like seeds known as chowlee by Hindus and forms a considerable article of food.

African Yam Bean

Its center of origin of African yam bean is Africa and is cultivated particularly in Nigeria, Ivory County, Ghana, Togo, Gabon, Zail, Central Africa and Western Africa. The cultivation outside the African continent is relatively unknown.

African Locust Bean

It is primarily cultivated in the savannah region of West Africa (Hopkins 1983; Akande et al. 2010).

Ground Bean

It is native to West Africa. It is widely distributed in tropical Africa, particularly Nigeria, Mali, Burkina Faso, Upper Volta, Niger, Benin, and Togo (Kay 1979; Obasi and Agbatse 2003).

Marama Bean

Marama bean is originated from South Africa's Kalahari Desert, also known as the Green Gold (Uzabakiriho 2016). Namibia and Botswana are two nations where it grows naturally. It is being grown in Kenya, Australia, Israel, +and the United States.

Widely accepted center of origin of different minor legume vegetables are given as follows:

S. no.	Vegetable name	Center of origin
1.	Winged bean	India (Burkill 1906)
2.	Adzuki bean	China (Tomooka 2009)
3.	Sword bean	Indo-Malayan region
4.	Jack bean	West Indies
5.	Moth bean	India and Pakistan (Adsule 1996; Gupta et al. 2016)
6.	Scarlet runner bean	Mexico, Guatemala, Honduras (Debouck and Smartt 1995; Freytag and Debouck 2002), and Mexico (Guerra-García et al. 2017)
7.	Yam bean	River Amazon (Potter and Doyle 1992)
8.	Potato bean	North America
9.	Tepary bean	The Southwestern United States and northern Mexico
10.	Catjang bean	West Africa (Ng and Marechal 1985).
11.	Agathi	India or Indonesia
12.	Velvet bean	Southern Asia
13.	Horse gram	India (Arora and Chandel 1972; Purseglove 1974; Smartt 1985; Vavilov 1951; Zohary 1970)
14.	Yard long bean	Southeast Asia
15.	African yam bean	Africa
16.	African locust bean	Unknown
17.	Ground bean	West Africa
18.	Marama bean	South Africa (Uzabakiriho 2016)

Nutritional Importance and Uses

Winged Bean

Generally, immature and tender pods are used as green vegetables, while leaves, flowers, seeds, stem, and tuberous roots are also eaten in a variety of preparations. Owing to its high nutritional value and the fact that all plant parts are edible, the winged bean is known as the "vegetable of the twentieth century." The seeds are boiled in water for 30 minutes to destroy the toxic substances present in the seeds such as trypsin inhibitors and hemagglutinating activities (Kadam and Smithard 1987). Young leaves are consumed as either cooked or salad in Asia. The breeding objectives for winged bean improvement are shown in Fig. 3.1.

Adzuki Bean

Adzuki bean has a sweet, nutty taste (Kumar et al. 2012; Reddy et al. 2017; Tazawa et al. 2018). "An" is a mixture of adzuki, sugar, and water used as a filling for bread, steamed bread or dumplings, sweet cakes, and desserts (HDC 2017). It has significant global potential for health improvement and illness prevention (Shahrajabian et al. 2019). Dietary fiber, saponins, and polyphenols are all found in azuki beans.

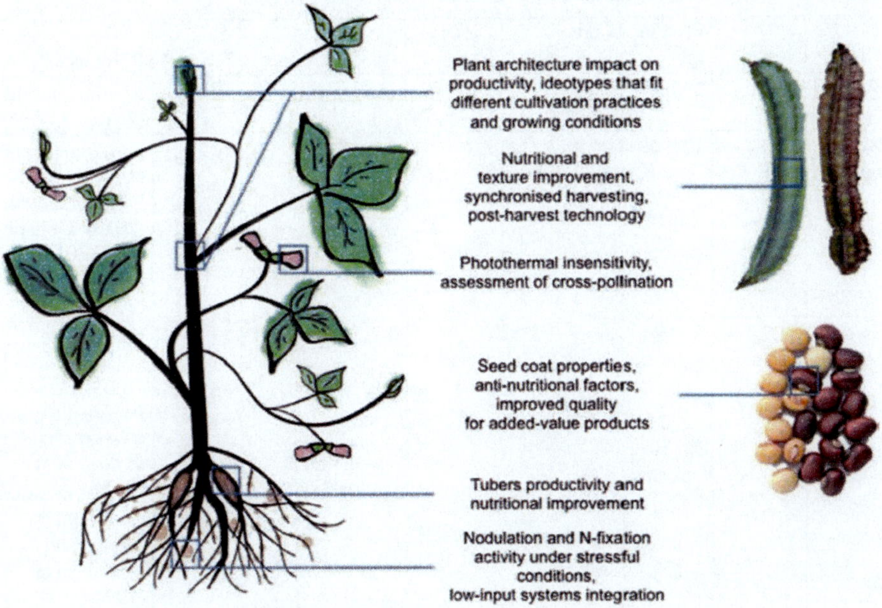

Fig. 3.1 Breeding objectives for winged bean improvement (Tanzi et al. 2019)

Furthermore, the adzuki bean also has antioxidant (Yao et al. 2012; Agarwal and Chauhan 2019), anti-inflammatory (Mukai and Sato 2011; Zha et al. 2011), antihypertensive (Mukai and Sato 2009; Takahashi et al. 2008), anticancer (Zhang and Popovich 2010), anti-obesity (Liu et al. 2017), and antidiabetic properties (Yao et al. 2012; Itoh et al. 2009). The adzuki bean extract is a popular dietary ingredient to reduce Alzheimer's disease progress (Miyazaki et al. 2019). Adzuki beans with high polyphenol concentrations may have therapeutic effects on oxidative stress-related disorders (Lee et al. 2018). Dainagon starch is a health-promoting ingredient extracted from *Vigna angularis* (Honda et al. 2020) (Table 3.1).

Sword Bean

Sword bean is a good source of polyphenol content (Koley et al. 2019). As a coffee substitute, roasted and ground beans have been utilized (Bressani et al. 1987; Vadivel et al. 1998; Eknayake et al. 1999). The seeds and green portions of the plant are used to make animal feed. The mature seeds contain a high amount of protein but immature green pods contain canavanine and toxic amino acid, and antimetabolite of arginine should never be eaten raw. Human pancreatic cancer cells are cytotoxic to canavanine. The canavanine content of seeds can be greatly reduced by soaking them overnight and boiling them in excess water.

Table 3.1 Nutritional composition of underutilized legume vegetables

Crop	Water	Protein (g)	Carbo-hydrates (g)	Dietary fiber	Fat (g)	Ca (mg)	Fe (mg)	References
Winged bean	–	29.65	41.7	25.9	16.3	440	13.44	Martín-Cabrejas (2019)
Adzuki bean	66%	12	24.8	7.3	0.1	28	2	Lam Sanchez (1990), Palombini et al. (2013), Gohara et al. (2013), Durak et al. (2013)
Sword bean	11.2 (dry)–88.6 (fresh)	3	8	1	0	60	2	Vadivel and Janardhanan (2005), Pugalenthi and Vadivel (2005), Vadivel et al. (1998), Bressani et al. (1987), Mohan and Janardhanan (1994), Purseglove (1968)

(continued)

Table 3.1 (continued)

Crop	Water	Protein (g)	Carbo-hydrates (g)	Dietary fiber	Fat (g)	Ca (mg)	Fe (mg)	References
Jack bean	10–11	10.85	12.15	3.98	1.59	380.5	631	Patel et al. (2016), Osuigwe et al. (2002), Osuigwe et al. (2006), Ologhobo (1992), Adebowale and Lawal (2004)
Moth bean	9.7	22.9	61.5	–	1.6	150	10.9	USDA (2005)
Scarlet runner bean	93.5 g	8	21	9	0	4	10	–
Yam bean	108.08 g	29.2 g	31.3 g	7.5 g	24.1 g	12 mg	0.6 mg	Kisambira et al. (2015)
Potato bean	–	Seed: 25–30% Tuber: 11–14%	–	–	–	–	–	Carlisi and Wollard (2005), Neacsu et al. (2021)
Tepary bean	73 g	8 g	12	37.0%	0.5%	–	10.7	Bhardwaj and Hamama (2004)
Agathi	73.1 g	8.4 g	12 g	2	1 g	1130 mg	4 mg	Gopalan et al. (2004)
Velvet bean	75 g	15 g Seed: 314.4 g	5.2 g	27 g	1 g	10 mg	15.9 mg	Siddhuraju et al. (1996)
Yard long bean	–	2.8 g	8.35 g	5 g	0.4 g	60 mg	0.47 mg	Suma et al. (2021)
African yam bean	81 g	0.6 g	85.3 g	1.1 g	8.25 g	–	12.6 mg	Adeyeye et al. (1994), Abioye et al. (2015), Ade-Omowaye et al. (2015), Ikpa (2017), Baiyeri et al. (2018), Anya and Ozung (2019), George et al. (2020)
Catjang bean	11.05 g	23.85 g	59.64 g	10.7 g		85 mg	9.95 mg	USDA (2010)
Horse gram	–	16.9–30.4%	57.2%	–	0.5%	–	–	Patel et al. (1995)
African locust bean	–	30–40%	30–40%	8.82–94%	–	–	–	Aliero et al. (2004), Campbell-platt (1980)
Ground bean	–	21.5 g	73.9 g	6.1 g	1.2 g	21.4 mg	–	Obasi and Agbatse (2003)

Excessive use of sword bean causes drowsiness and vomiting. The seeds are used to cure nausea, abdominal dropsy, lower soreness, injury pain, cure hernia, coughs, constipation, headache, pain relief around kidneys and waist in Korea (Zhu and Wang 2002; Duranti 2006; Li et al. 2007). Athletes' feet and acne are cured using soap made from bean extracts.

Jack Bean

Its young green pods are consumed as vegetables (Stanton 1966). The young leaves may be cooked and the whole plant is used for fodder. In Indonesia, steamed flowers and leaves are used as flavorings. It is rich in protein (29%) and helps in alleviating protein malnutrition (Adebowale and Lawal 2004; Abitogun and Olasehinde 2012). It is also rich in vitamin C and carbohydrates (Patel et al. 2016). Hemagglutinin, a toxic substance, is used to cure abdominal swelling, hiccups, vomiting, lower back pain (lumbago) due to kidney failure, asthma with sputum, etc. The seeds are also used as an antibacterial and antiseptic in Nigeria (Olowokudejo et al. 2008).

Moth Bean

The immature pods and seeds of moth bean are an important edible parts of the plant. The seeds are used medicinally in the healing of paralysis, weight loss, cough, fever, liver ailments, and rheumatism (Kushwah et al. 2002) and roots have narcotic properties (Brink and Jansen 2006). Herbs contain some bioactive compounds that are being successfully used for neurological disorders treatment (Phukan et al. 2015). The seeds contain antinutritional factors such as phenolics, protease inhibitors, trypsin inhibitors (Singh and Ansari 2018; Gupta et al. 2016; Franks et al. 2002; Kadam et al. 1985), polyphenols, phytic acid (Gupta et al. 2016), albumin, globulin (Gupta et al. 2016; Siddhuraju 2006), saponins, and oxalic acid and amylase inhibitors. It also has antioxidant, antidiabetic, and hypocholesterolemic properties (Kaleem et al. 2020).

Scarlet Runner Bean

Scarlet runner bean helps in the proper development of our body due to its nutrient content. The pods contain a good amount of carbohydrates, protein, minerals (calcium, iron, phosphorus, manganese, and magnesium), carotene, vitamin C, and dietary fiber. It is also a very good source of niacin, thiamine, and riboflavin. Pods are rich in antioxidants but very low in fat and energy. It also contains traces of the poisonous lectin and phytohemagglutinin, which are also found in common beans. Seeds contain substantial levels of protein and energy, as well as nutritionally essential minerals, equivalent to groundnut and soybean (Aremu et al. 2010). Seeds can be added to salads, soups, stews, etc. Its flowers having a bean-like flavor are also used in salads. The pods and seeds are used for garnishing many dishes and adding flavor and color at the same time. It is used as fodder to feed cattle (Asch and Hart 2004; Makri and Doxastakis 2006; Contreras-Govea et al. 2009). It is also grown as an ornamental crop because of its attractive flowers. It is a diuretic that can help with kidney disorders. It is considered good for cardiac problems, diarrhea, itching, and rheumatism. Its pods are good for individuals with diabetes, provide protection against breast cancer, and help in lowering blood cholesterol levels. The presence of fibers also helps in stabilizing glucose metabolism and preventing constipation.

Yam Bean

Yam bean is rich in carbohydrates (present in the form of starch, sugar, and dietary fiber) and ascorbic acid. The higher sugar content in tubers imparts a sweet taste when eaten raw. Young, sensitive, crisp, succulent, and sweet tubers are eaten as a fruit in India, and the flavor is similar to a Chinese water chestnut. Young tender pods of *P. erosus* are boiled and eaten as a vegetable but *P. tuberosus* pods are not eaten because of the presence of irritant hairs. The seeds contain a high amount of proteins (Sørensen 1996), lipids, Fe, and Ca in comparison to other legumes. It also possesses antioxidant and anti-melanogenic properties (Lee et al. 2017). The starch extracted from tubers is used in confectionery. In some areas, the tubers are used in pickles and useful addition to stir-fries, salads, soups, and stews. In Latin America, it is eaten raw or cooked as a vegetable or sometimes made into pickles and chutney, whereas in China, mature dried roots are used as a cooling agent for people suffering from high fever. In Fiji, its tough and fibrous stem is used for making fishnets. It is cooked the same way as the potato. Its tubers help in reducing weight, increasing immunity, and improving digestive, circulatory, and nervous systems. The roots are used to treat insomnia (Abid et al. 2006), osteoporosis, and skin rashes (Ong and Nordiana 1999; Nurrochmad et al. 2010), whereas the tubers are used to treat tumors (Kardono et al. 1990; Park and Han 2015; Thaptimthong et al. 2016).

Potato Bean

Potato bean is rich in carbohydrates, protein, dietary fiber, and iron-rich tubers (Kikuta et al. 2011) and is used as food by the North American Indians and early European settlers (Seabrook and Dionne 1976). A low level of reducing sugar in tubers (Carlisi and Wollard 2005) helps in preventing browning when they are fried (Reynolds et al. 1990) and thus makes excellent chips. The powder from the dried tubers is sold in markets as "Apios powder," which is used as an ingredient in cookies, doughnuts, dumplings, and bread (Kikuta et al. 2011). It is used for postnatal medication and to relieve fatigue, and it was also used to keep children healthy. It has anticancerous properties (breast and prostate cancer) due to the presence of genistein (Peterson and Barnes 1993; Barnes et al. 1994; Agullo et al. 1997).

Tepary Bean

Tepary bean is rich in carbohydrates, protein, vitamins, and minerals (phosphorus and calcium). It contains anti-nutritional factors but only the lectin activity is exceptionally high, which can easily be denatured by cooking. The seed coat of white cultivar seeds is more permeable than the seed coat of black-seeded cultivars, resulting in a shorter cooking time. Dry seeds are consumed after boiling and frying. Intermittently, it can be eaten raw or cooked as a green bean or bean sprouts. The remaining pods and stems are used as cattle feed after the seeds are removed.

Catjang Bean

In Japan, the catjang bean is used to make "sekihan" (steamed glutinous rice colored red by boiled cowpea or azuki) for traditional ceremonies and celebrations. It is used as a fodder for livestock (Lim 2012). The seed has diuretic properties and is used as a stomach-strengthening supplement. When boiled and consumed as food, it is thought to kill stomach worms. Broken bones are treated with a poultice made from crushed leaves.

Agathi

Agathi leaves and flowers are a good source of protein, fat, vitamins (vitamin A, thiamine, riboflavin, and niacin), minerals (calcium, phosphorus, and iron), fiber, and ash. The leaves, flowers, and green fruits are eaten as a vegetable or salad. Its tender portions serve as cattle feed; however, its overeating causes diarrhea. In Asian buildings, wood is used similarly to bamboo. All the plant parts have medicinal uses in Southeastern Asia and India. The leaves are often used as a tonic, diuretic, laxative, antipyretic, and mouth and throat disinfectant. Flowers are used to relieving headaches, improve appetite, visual problems, and catarrh. Its astringent and bitter bark is used for cooling, as a tonic, anthelmintic, febrifuge, and in diarrhea and smallpox. Fever, discomfort, bronchitis, anemia, tumors, colic, jaundice, and poisoning are treated with its bitter and caustic fruits, which act as a laxative. Rheumatism, severe swelling, and catarrh are treated with roots as a demulcent. It can be grown as an ornamental plant and used as windbreaks, and living fences. It is also used for the production of gum and tannin in the manufacturing of pulp and paper.

Velvet Bean

It is a good source of protein, fat, starch, and amino acids. It is used as food, feed, and environmental safety source; thus, it has a great value in agricultural and horticultural use. Certain ethnic groups in many countries used it traditionally as a food source. In Central America, Guatemala, and Mexico, it has been used after roasting and grinding as a coffee substitute for at least several decades and its seeds are widely known in these regions as Nescafe. In South India, the ethnic group (the Kanikkars) conventionally consumes its dry seeds after repeated boiling. Sometimes, its dry leaves are used for smoking. Growing velvet bean reduces soil erosion and controls nematodes. The plant exhibits low susceptibility to insect pests due to the presence of toxic compounds. It can also be used as a biological control for problematic weeds, as the plant has an allelopathic effect, which may suppress competing plants. It is a popular medicinal plant, used to treat Parkinson's disease. Its constituents have analgesic, aphrodisiac, anti-inflammatory, neuroprotective, antiepileptic, antineoplastic, antidiabetic, and antihelminthic properties. Its seeds also contain a kind of substance that counteracts the harmful effects of snake venom. The methanolic extracts of its leaves have antimicrobial and antioxidant properties, and are used to treat the skin diseases, nervous disorders, and arthritis. Its paste is applied to the scorpion bites and absorbs the poison.

Horse Gram

Horse gram is rich in minerals (phosphorus and iron), vitamins (carotene, thiamine, riboflavin, niacin, and vitamin C) (Sodani et al. 2006), and lysine content (Gopalan et al. 1978). It has antiurolithiatic activity (Chaitanya et al. 2010) and lipids used to treat peptic ulcers (Jayaraj et al. 2000). It is used for treating cough and bronchitis, kidney problem, and irregular periodicity of the menstrual cycle.

Yard Long Bean

It is rich in carbohydrates, protein, vitamins (vitamin B, vitamin C, and vitamin K), minerals (calcium, iron, magnesium, manganese, phosphorus, and potassium), and dietary fibers. The young and slender pods have an excellent gourmet flavor. Its tender pods are sometimes cut into small pieces and used in omelets. They can also be cooked with rice, added to soup, or used fresh in salads. Its pods can be pickled in vinegar and added to dals, curries, and cooked salads in Malaysian and Indonesian cuisine. In West Indian dishes, the pods are often stir-fried with potatoes and shrimp. In the Philippines, they are widely eaten stir-fried with soy, garlic, or hot pepper sauce. Relatively, the smaller seeds are bitter in taste. It is regarded as a low-calorie food, and for this reason, it is consumed by obese individuals to lose weight. Its constituents have an anti-inflammatory property and relieve inflammation during arthritis and asthma. Since it is a green vegetable, it contains phytochemicals and antioxidants, which help in combating cancer and holding up aging. Its consumption provides cardiovascular defense and lowers blood pressure.

African Yam Bean

African yam bean is regarded as a nonfatty food and a good source of protein and starch. The shelled beans from mature green pods are eaten after boiling. The tubers are eaten both raw and cooked.

African Locust Bean

It is high in protein, and the seed pulp, also known as dozim, is a high-energy food (Audu et al. 2004). Soumbala balls made from fermented seeds are offered at the market as a condiment (credit: Barbara Vinceti, Bioversity International) (Termote et al. 2020). Leprosy can be treated with the bark. It is used to treat malaria, jaundice, stomach ache, diarrhea, and hypertension (Ojewumi et al. 2014, 2016). It also has antioxidants, hepatoprotective, and antiparasitic properties (Lawal et al. 2015, 2017; Bashir et al. 2015). Farmers should be encouraged to plant *Parkia biglobosa* on their farms for food production, shade, and environmental preservation (Oluwalana et al. 2019) (Fig. 3.2).

Ground Bean

It is rich in proteins and minerals (Minka et al. 1999; Minka and Bruneteau 2000). The seeds contain a low amount of tannin, which helps in improving human health and nutrition (Akindahunsi and Salawu 2005).

Marama Bean

Marama bean is rich in calcium (241 mg per 100 g) (Museler and Schonfeldt 2006) and produces large nutritious tubers (Hamunyela et al. 2020). Its mature and dried seeds are used for boiling, grinding, or roasting (Phuthego 2014). Marama bean milk is similar to soymilk or dairy milk in terms of nutrition and health (Jackson 2017; Jackson et al. 2010).

Fig. 3.2 Balls of soumbala from the fermented seeds of *Parkia biglobosa*. (Credit: Barbara Vinceti, Bioversity International)

Botany and Taxonomy

Winged Bean

It is a twining perennial herb grown as annuals and reaches up to 3–4 m (9.8–13.1 ft) in height, with trifoliate leaves (Khan 1982) and tuberous roots. The waxy pods are 15–22 cm long and have four wings with frilly edges. Pods may be colored cream, green, pink, or purple. Seeds are of cream (NAS 1975; Anonymous 1982; Mohanty et al. 2013), black (Klu and Kumaga 1999), brown (NAS 1975; Klu and Kumaga 1999; Mohanty et al. 2013) to purple and mottled colors (NAS 1975) depending on growing and storage conditions (Khan 1982). It is normally a self-pollinated crop, and the cross-pollination occurs to the extent of 7.6%. The cleistogamous floral system implies autogamy (self-pollination). Large flowers are pale blue and open (anthesis) in the morning (Karikari 1972; Erskine and Bala 1976; Erskine 1980). The flowering starts between 50 and 90 days after planting and the pods mature in 3–5 months. The pollen is viable for 24 hours (Koshy et al. 2013) and stigma receptivity for 33 hours (Senayake and Sumanasinghe 1978) after anthesis, ensuring the possibility of cross-pollination.

Adzuki Bean

It is a fast-growing annual crop and exhibits indeterminate growth (Duke 2012). The leaves are pinnate, trifoliate, and somewhat hairy; the leaflets are ovate, sometimes slightly cleft with the odd leaflet supported on a long petiole with yellow flowers. Anthesis occurs in the morning hour and continues for up to 40 days. Raceme inflorescence has 6–20 flowers. The pods are smooth, cylindrical, thin-walled and change from green to white after attaining maturity. Seeds are often dark red, although they can also be green, straw-colored, black-orange, or mottled (Duke 2012; Kramer et al. 2012). Adzuki bean is cross-compatible with *Vigna angularis* var. *nipponensis*, *V. hirtella*, *V. minima*, and *V. tenuicaulis* (section Angulares) and gives multiple resistance genes to Fusarium wilt and brown stem rot (Kondo and Tomooka 2012).

Sword Bean

It is a short-day, fast-growing, shade-tolerant, deep-rooted, perennial climber (4.5–10 m in height), often cultivated as an annual (Knoblauch et al. 2001). The leaves are trifoliate and pubescent. The inflorescence type is an axillary raceme that bears numerous white or pink flowers. The pods are curved with 8–16 seeds on average (Purseglove 1968). The seed germination is epigeal.

Plant parts of sword bean (Xianzong et al. 2017)

Jack Bean

Jack bean is a climbing or bushy annual or weak perennial legume that grows up to 1 m (3.3 feet) tall. It has a taproot system, making it tolerant to drought. The plant can spread via long runners with pink-purple flowers. The pods are sword-shaped with 20 large smooth white seeds per pod (Purseglove 1968; Bressani et al. 1987; Smartt 1976).

Heuzé et al. (2015)

Moth Bean

It is a short day, annual herbaceous trailing legume crop (Sathe and Venkatachalam 2007). The leaves are compound and trifoliolate. The inflorescence is an axillary, dense false raceme on a peduncle. The flowers are bisexual, papilionaceous, and yellow. The fruit is a hairy, brown, or pale gray cylindrical pod that measures 2.5–5 cm in length and 0.5 cm in width. The pods contain 4–10 seeds. The seed has a test weight of 10–35 g (NARO 2020; Brink and Jansen 2006).

Heuzé et al. (2020)

Scarlet Runner Bean

Scarlet runner bean is a long day (Kaloo 1995; Mulanya et al. 2019), perennial plant but grown as an annual. The stem, which emerges from fleshy edible tuberous roots, is vigorous and several meters (2.5–4.5 m) in height. It has a taproot system (Galadzhun 2015). Its leaves are typical alternate dark green with purple-tinged veins, trifoliolate, and triangular stipules. The inflorescence is a terminal raceme, with bisexual flowers that are borne in clusters and are papilionaceous and campanulate. The corolla is pink or white, with a conventional hood-shaped circular or broadly obovate keel. Pods are slender and long. Runner bean is a self-compatible and cross-pollinated species assisted by its exerted stigma and nectaries. Pollination is carried out by insects (Quagliotti and Marletto 1987) and birds, with the honeybee serving as the main pollinator (Kendall and Smith 1976; Anasiewicz 1994), and the bumblebee (*Bombus terrertris* L.) is undesirable (Quagliotti and Marletto 1987). The pods are knife-shaped, linear, lanceolate, and normally green in color. The pale green pods are 10–20 cm long depending upon the variety. Seeds are large, deep crimson color speckled with brilliant black to violet-black color and shape similar to lima bean seeds. It carries out hypogeal germination (Borowy 1996).

Yam Bean

Yam bean is a trailing vine that reaches 6 m in height. The leaves are dentate to the palate with bluish violet or white flowers. The inflorescence is complex racemes and 8–45 cm in length. Its tubers are broad spindle-shaped or turbinate with easily peelable skin. The pods are curved with 8–11 seeds. The seed shape can range from flat and square to round.

Potato Bean

It is a climbing perennial vine that reaches up to 10–20 feet (Kalberer et al. 2020). The pinnately arranged alternating compound leaves comprise a single terminal leaflet and variable pubescence along the rachis (Reynolds et al. 1990). The inflorescence is axillary racemes with fragrant pink and purple flowers. It is known for its stem tubers, which usually develop as thickenings along stolons. The stolon length varies from 2 to 4 feet (Belamkar et al. 2015). Several buds are formed at one side of the upper part (shoulder) of the tuber, but only one bud developed into the above-ground shoot.

Tepary Bean

It is a climbing vine with an erect stem that reaches 4 m in length. The root system of the plant is adventitious. The leaves are alternate and 2–3 mm long. It has three foliolate and lanceolate stipules appressed to the stem. The petiole length is 2–10 cm and the leaflets are ovate to oval in shape. Inflorescence is an axillary raceme. Flowers are bisexual, consisting of the campanulate calyx; white, pink, or pale purple corolla; and superior ovary. The pods are curved with 2–9 seeds. The seeds are globose to oblong, with white, yellow, brown, purple, and black with epigeal germination.

Catjang Bean

It is a 0.5–3 m long annual or perennial climbing vine. The leaf is trifoliolate and ovate or rhombic in shape with pink, purple, or white flowers. Flowers and seeds are smaller than cowpea. Fruits are linear cylindrical, glabrous, scabrous legumes that can be straight or curved, pale or speckled, and can be erect, suberect, or horizontal. The seeds are either black, red, or brown.

Agathi

Agathi is a 15-m tall perennial, deciduous, or evergreen plant with 20 years of life span. Flowers are borne on an unbranched pendulous inflorescence with pinnately compound leaves. There are two varieties of flowers: (i) variety "grandiflora" with white flowers and (ii) variety "coccinea" with rose-pink or red flowers. In shape and arrangement, the flowers are similar to pea flowers. It bears raceme axillary 2–4 flowered which are white, yellow, rose-pink, or red in the color. The pods are 20–60 cm long with broad sutures, 15–50 seeded, glabrous, vertically hanging, and indehiscent. The seeds are dark brown and subreniform.

Thicket Bean

It is a perennial climbing vine with coarsely pubescent to glabrate stems. The leaves are trifoliolate, alternating, and pinnately complex. The perfect, zygomorphic flowers are pink-purple flowers (occasionally white) in color. The fruit is glabrous, flat, 3–8 cm long, and slightly curved, with 4–6 seeds inside (Femald 1950). The seeds are small, smooth, kidney-shaped, and black or gray. The seeds have hypogeal germination.

Velvet Bean

It is a summer season climbing shrub, with a long vine that grows to a length of 6–18 m. It has a taproot system with slender stems. The leaves are slightly pubescent, alternate, and trifoliolate. The inflorescence type is axillary raceme and white to purple flowers borne in clusters. Sepals are longer or of the same length as the shuttles. Corollas are butterfly shaped. The pods are 10–13 cm long, stout and curved, containing 2–6 seeds. The seeds are elongated ellipsoid in shape and weigh 55–85 g per 100 seeds.

Heuzé et al. (2015)

Horse Gram

Horse gram is a slender, twining annual herb up to 0.6 m long with cylindrical tomentose stems. It has trifoliate leaves with persistent stipules and petiole. Flowers are short, sessile, or subsessile and 10–12 mm long, with 2–4 flowered axillary racemes. The pods are 30–55 cm long, upcurved toward the apex, usually containing 6–7 seeds per pod (Purseglove 1974). It is a self-pollinated crop that matures in 3–4.5 months.

Yard Long Bean

It is a climbing annual legume that reaches up to 2.7–3.6 m in height. Its stem is square, usually smooth, violet in color, and often twined about the nodes. Its leaves are trifoliate, green, oval, and smooth-edged. The flowers with five petals are large and pale pink to violet-blue. Its plant continues to grow after flowering and fruiting. Anthesis starts after 5 weeks of seed sowing and bears pods in pairs 60 days after sowing, which lasts for a long time. Anthesis occurs in the morning hour from 6:30 to 9:00 a.m. Anther dehiscence occurs from 10:00 to 12:45 a.m. The stigma becomes receptive 12 hours before and 6 hours after anthesis. Although it is a self-pollinated crop, cross-pollination is also done mainly by yellow jackets and ants. The pods formation occurs after 10–12 days of anthesis. Its pods are pendulous, long, thin, fleshy, stringless, and 30–70 cm long, depending on the variety. Seeds are elongated and kidney-shaped.

African Yam Bean

It is a perennial climbing vine and reaches 1–3 m in height. Its red stems are often strongly branched, bearing trifoliate leaves. Some roots become thick to form the storage organ, known as tubers. These tubers have a similar appearance to sweet potato and the taste resembles a potato. The butterfly-shaped flowers borne on racemes may be of greenish-white, yellowish-white, pink, purple, red, magenta, lilac, and blue color. The crop is pollinated by insects. Pods are 20–30 cm long, 1 cm wide, and become brown at maturity. Each pod contains 20–30 seeds, which range in color from white to brown to black.

African Locust Bean

The multipurpose perennial tree (Akande et al. 2010) reaches up to 7–20 m high. The plant produces bipinnate leaves and hermaphrodite flowers (Adewumi and Igbeka 1993). The plant produces 120–300 mm long and 15 mm in diameter pods in huge bunches (Oni 1990), and the seeds are brown-blackish in color (Fig. 3.3).

Fig. 3.3 Fruits (pods) of *Parkia biglobosa* from a Parkland in Burkina Faso. Pulp and seeds are consumed as edible products. (Credit: Barbara Vinceti, Bioversity International)

Ground Bean

It is an annual herbaceous legume and has a tap-root system. The flowers are yellow and the leaves are trifoliate. The petiole is tall and rigidly grooved. The peduncle elongates during pollination and fertilization. The pods formed below the ground because pedicles enter the soil after fertilization contaminating 1–2 seeds.

Marama Bean

It is a perennial crop with a stem that reaches 3 m long at maturity. When the leaves are young, they are delicate and succulent, but as they mature, they become thick, leathery, and grayish. The yellow blooms are borne in a raceme. It produces hardy pods with 1–3 reddish-brown-black seeds per pod (Omotayo and Aremu 2021).

Varieties

Crop	Varieties
Winged bean	Indian varieties are No. 21, 60, 71 (Indian Institute of Horticultural Research), WBC-2 (Meghalaya), and JCV44 (West Bengal). Certain varieties like Bogor, Ribbon, Butterfly, Maripusa, Lunita, Always, Alipasto, Dual, Chimbu, and Tinge are available in markets of Puerto Rico and Southeast Asia
Adzuki bean	Japanese Red, Chinese Red, Adzuki Express (Johnny's selected seeds), Takara (Japanese import), Minoka (Minnesota Agricultural Experiment Station), Erimo, Bloodwood, and Dainagon
Sword bean	SBS 1 (India), Jingpin Jiadouwang and Hong Daodou (China)
Jack bean	No variety
Moth bean	Early cultivars: RMO-40, RMO-257, FMM-96, RMO-225, RMO-435 Mid cultivars: Jadia, Jawala, Maru Moth, IPCMO-800, 912, CAZRI Moth 1 Late cultivars: Type-1,3, MG-l, Baleshwar-12
Scarlet runner bean	Butler (stringless), Painted Lady (red with white flowers), Kelvedon Wonder (early with long pods), Sunset (pink flowers), and Scarlet Emperor (scarlet flowers with good pods). Nyeri, KIN 2 local varieties of Kenya (Kimani et al. 2019)
Yam bean	Rajinder Mishrikand-1
Potato bean	Ashipa (South America), Yushpe, and Chuin (Peru)
Tepary bean	Cultivars based on the seed color (i) White (Redfield) and (ii) Dark-yellow-seeded developed through mass selection. Cultivars based on growth habit (i) Indeterminate shrubby varieties with short leaves and (ii) Indeterminate creepers with long leaves, which climb if they find support. Based on the shape of leaflet and seeds, there are three botanical varieties, viz. (i) *Phaseolus acutifolius* var. *acutifolius*, (ii) *Phaseolus acutifolius* var. *tenuifolius*, and (iii) *Phaseolus acutifolius* var. *latifolius*.
Agathi	There are two varieties: (i) variety bears white flowers and (ii) variety bears red flowers called red august tree leaves.
Horse gram	BR 5, BR 10, Madhu Bihar, HPK-4, VLG 1, PDM 1, VZM 1, K82, Birsa Kulthi, S27, S8, S39, S1264, Co-1, Hebbal Hurali 2, PHG 9, KBH 1, Maru Kulthi, KS 2, AK 21, AK 42, and VLG 1
Velvet bean	Arka Dhanvanthari, Arka Daksha, and Arka Shukla – Indian varieties Yokohama, Georgia, and Alabama – USA varieties
Yard long bean	Arka Mangala (pole type, vigorous, photo-insensitive, high yielding), Stickless Wonder (bush type), Liana (day-neutral), Chinese Red Noodle, and Purple Podded
Ground bean	BBP (black), WBP (white)
No variety is available in African yam bean, African locust bean, and Marama bean.	

Climate and Soil

Winged bean	
Climate	Temperature plays an important role in photoperiod for flowering (Herath and Ormrod 1979). It thrives best in high temperature and humidity (Basamba and Ruegg 1982). The ideal growing temperature is 25 °C. The optimum temperature requirement for vegetative growth is 22–30 °C, while 18–32 °C temperature and short days require for flowering and pod development. A temperature of 13–24 °C is best for root and tuber enlargement. Lower temperatures suppress germination but increase tuber production, and extremely high temperatures inhibit yield (Khan 1982)
Soil	It is susceptible to frost, moisture stress, and waterlogging. It can be grown on sandy loam to loam soil except for sodic and alkaline soil. It prefers sandy to clay soil with 4.3–8.5 pH
Adzuki bean	
Climate	It is adapted to warm temperate climate and frost-free periods with cool nights. It is moderately drought tolerant. It requires 16 °C temperature for seed germination and 15–30 °C throughout the growing season (Hardman et al. 1989; Duke 2012). A temperature of 10–14 °C is ideal for its cultivation but below 10 °C the growth of the plant slows down. It can grow well in regions having an annual rainfall ranging from 530 to 1730 mm (Myers 1998; Duke 2012). Due to climate change, there is an increase in atmospheric temperature, and a decrease in precipitation intensified during the red bean growing period (June–August) (Lee et al. 2020). It is shallow-rooted and does not nodulate well; therefore, they grow best on alluvial river soils and sandy loams
Soil	It can be grown on sandy and silt loam soil but not on heavy soil. It cannot withstand waterlogging conditions. Its pH requirement is 5–7.5
Sword bean	
Climate	It prefers a warm climate (Li et al. 2007; Huang 2008) and grows well in lowland tropical areas at elevations up to 1500 m. It grows best in areas where annual daytime temperatures are within the range of 15–30 °C but can tolerate 12–36 °C. Low temperature during flowering results in low growth and flowering (Yang et al. 2014). The plants are sensitive to frost, but mature beans remain unaffected. It prefers a mean annual rainfall in the range of 800–1800 mm
Soil	It can be cultivated in a range of soils, but it favors organically rich soil. It can withstand saline soil, drought (Siddhuraju and Becker 2001), and is slightly tolerant to waterlogging (Duke 1981). It prefers a pH in the range of 5–6
Jack bean	
Climate	It prefers a warm and humid climate but leaf fall occurs due to extremely high temperature and slight frost (Florentin et al. 2004; Rodino et al. 2007). It grows well at a moderate temperature of 20–30 °C. Optimum growth occurs in full sunlight but can tolerate shade to some extent. It can grow at 600 m above mean sea level and can serve to shade the soil during hot summer to prevent loss of organic matter. Below 500 m of mean sea level, it often stops growing without rain but can be used as a green manure crop up to 1600–1800 m above mean sea level. Its annual rainfall requirement ranges from 650 to 2000 mm
Soil	It can be grown in soils with high lead concentrations and has the potential to be used for the restoration of lead-contaminated soils (Pereira et al. 2010). It is highly drought-resistant and its roots penetrate deep into the soil but do not grow well in excessively wet soil. It is more tolerant to waterlogged, saline, and acid soil (pH 4.3–6.8) conditions

(continued)

Winged bean	
Moth bean	
Climate	It is a warm-season crop that grows better at 24–32 °C but can tolerate temperatures up to 45 °C (Brink and Jansen 2006). It is grown from sea level up to an altitude of 1300 m (NARO 2020; Kochhar 2016; Brink and Jansen 2006). It thrives well with an annual rainfall of 500–750 mm but has grown successfully in areas with low annual rainfall (200–300 mm). The plant provides yield even with as little as 50–60 mm in 3–4 showers during the growing period (Brink and Jansen 2006). Plants remain green and succulent even at the end of the hot season (Sherasia et al. 2017)
Soil	It can withstand extreme drought and help to prevent soil erosion (Roy et al. 2010). It can be grown on dry light sandy soil with 3.5–10 pH. It is sensitive to waterlogging but slightly tolerant to salinity (Brink and Jansen 2006)
Scarlet runner bean	
Climate	It is a warm-season crop that requires 12–22 °C temperature for its successful cultivation but above 25 °C inhibits its growth and pod development. It requires a 15–20 °C temperature for seed germination, and 30 °C during flowering may cause a high percentage of flower drops and abortion of ovules, resulting in poor yield. It is more tolerant to cool conditions (Hardwick 1972; Holubowicz and Dickson 1989; Hołubowicz and Khan 1989; Otubo et al. 1996; Revilla et al. 2005; Rodino et al. 2007) but temperature below 5 °C causes damage to the plants. It grows widely at an altitude of 1500–2800 m. It requires 400–2600 mm annual rainfall. It needs high relative humidity for setting seeds
Soil	It can be grown on well-drained medium-textured fertile sandy loam soil. Soil containing heavy clay should be avoided, as it hampers root penetration in soil but light soil is preferred for early yield and heavy soils for higher yield. Slightly acidic soils are most preferable but the extremely acidic and alkaline soil is not good for its cultivation. It is tolerant to salt (Al Hassan et al. 2016) and drought (Schwember et al. 2017)
Yam bean	
Climate	It grows well in hot, humid areas ranging from subtropical to tropical, although it is frost-sensitive, it requires a 9-month frost-free period for a healthy tuber yield. The crop performs better under the day and night temperatures of 30 °C and 20 °C, respectively. It can grow up to an altitude of 1000 m. It requires 14–15 hours of photoperiod for good vegetative growth, but hot days and cooler nights are suitable for tuberization. LED exposure has the potential to enhance the growth characteristics, metabolite accumulation, and antioxidant properties of yam beans (Chung et al. 2020)
Soil	It can be grown on well-drained sandy loam with good humus content. It can withstand drought but is extremely susceptible to waterlogging. The pH requirement for its successful cultivation is 5.2–7.0
Potato bean	
Climate	It is very sensitive to frost and does not survive the winter. It is cultivated from sea level to 1900 m elevation (Balbin et al. 2007)
Soil	It can be grown on sandy or gravelly soils (Thayer 2002) with neutral pH (Reynolds et al. 1990). It can also be grown on acidic soil (Reed 1985) and tolerate severe flooding and moderate drought (Thayer 2002; Reynolds et al. 1990; Belamkar et al. 2015)

(continued)

Winged bean	
Tepary bean	
Climate	It can withstand hot and dry conditions, making it ideal for growing in arid areas. It requires 17–26 °C day temperature and >8 °C night temperature for its proper growth and development. It can successfully be grown in regions with an annual rainfall of 500–1700 mm. In subtropical areas, the crop-growing season is very short, while in humid regions, it may be cultivated round the year. Some of its cultivars are photosensitive as they require either short or long days for flowering, whereas some of the cultivars seem day-neutral. It can be cultivated from 50 to 1920 m above sea level. However, the best crop can be taken at medium altitudes
Soil	Although it can be grown on light sandy soil with pH 6.7–7.1, the high organic matter content results in a better yield. Heavy soil is not appropriate for cultivation, but it can be grown on saline and alkaline soil because it is slightly tolerant of these soils. It is sensitive to waterlogging conditions
Agathi	
Climate	It needs warm and humid weather with an average temperature ranging from 24.3 to 26.7 °C. It is best adapted to regions with an annual rainfall of 2000–4000 mm but has been grown successfully in semi-arid areas with 800 mm annual rainfall and up to 9 months of the dry season. It is adapted only to the lowland tropics up to 800 m, occasionally to 1000 m. Agathi is sensitive to frost, shade, and extended periods of cool temperature. Its rapid early growth and erect growth habit usually enable it to access sunlight by overtopping neighboring plants
Soil	Agathi is tolerant to a wide range of soils including alkaline soils, poorly drained, saline, or low fertility. It can tolerate drought, heavy soils, poor soil, and waterlogging conditions. When flooded, the plants initiate floating adventitious roots and protect their stems, roots, and nodules with spongy aerenchyma tissue. It tolerates acidic soil (pH 4.5) also. However, its tolerance to high acid and aluminum saturated soils is unknown. It requires a soil pH of 6.6–8.5 for good growth
Velvet bean	
Climate	It requires a warm and humid climate coupled with high light intensity and annual rainfall of 650–2500 mm. It can be grown up to an altitude of 2100 m above mean sea level. A temperature of 20–30 °C is optimum throughout its growing period and a night temperature of about 21 °C has been reported to stimulate flowering. A low night temperature of about 10 °C reduces the plant growth, as they are dependent on *Rhizobium* for nitrogen fixation, which increases with increasing night temperature and is poor below 18 °C. The plants are susceptible to frost and require a frost-free period of 180–240 days
Soil	It may grow on sandy to clay soil, but it grows best in well-drained light sandy loam soil with a pH of 5–6.5. It can tolerate drought, soil acidity, and poor fertility but cannot survive in wet or waterlogged soils
Yard long bean	
Climate	It prefers a warm, humid climate with frequent rains. Its plant survives in extreme humidity and heat but is very sensitive to cold and frost. It needs full sunlight for its better growth and development, but it can also be grown in partial shade. A soil temperature of about 15 °C or higher is ideal to ensure good germination. A temperature of 25–30 °C is optimum throughout its growing period

(continued)

Winged bean	
Soil	Yard long bean can be grown on a wide range of soils from sandy to clay soils but it thrives best in well-drained sandy loam soils with a pH range of 5.5–6.8, enriched with organic matter. It can also be grown in weak acid or alkaline soils but is sensitive to waterlogging as it causes root rot
African yam bean	
Climate	It needs a warm and dry climate but is sensitive to frost. It grows better in regions where annual rainfall ranges from 800 to 1400 mm and where the temperature comprises between 19 and 27 °C. It can be grown at elevations from sea level to 1800 m
Soil	It can be cultivated on well-drained sandy loam or loam soils rich in organic matter content. It prefers a soil pH in the range of 5–6
Catjang bean	
Climate	It prefers a warm and humid climate and can be grown successfully at an elevation of 2000 m from MSL (mean sea level). It requires 25–35 °C temperature but can also tolerate up to 40 °C temperature. It cannot tolerate −1 °C or lower temperature. An annual rainfall of 500–1500 mm is the requirement of the crop
Soil	The plants can grow on well-drained sandy loam soils. It prefers 5.5–7.5 pH. It can tolerate moderate drought
Thicket bean	
Climate	It is a very frost-hardy climber with perennial tuberous roots. It can tolerate −35 °C temperature when well established
Soil	It can be grown on sandy and clay soil
Horse gram	
Climate	It prefers a warm and dry climate but is sensitive to frost. It can be cultivated up to an altitude of 1000 m above sea level. It requires 25–30 °C temperature and 50–80% relative humidity for the plant growth. Its annual rainfall requirement is 800 mm
Soil	It can be grown as a rainfed crop (Jansen 1989) but does not tolerate waterlogging. The crop can be produced on a variety of soils, from light to heavy soil
African locust bean	
Climate	The annual rainfall requirement of the crop is less than 400 mm (Gbadamosi 2005)
Soil	It is grown on well-drained sandy loam soil rich in organic matter and it prefers 6–7 pH
Ground bean	
Climate	It is a warm-season crop that requires bright sunshine and an average temperature between 20 and 28 °C. It is typically a short-day plant. It is usually cultivated up to 1600 m elevation from MSL (mean sea level). Its annual rainfall requirement is in the range of 600–750 mm but the optimum yield requires an annual rainfall of 900–1200 mm
Soil	It may grow in a variety of soils, but it prefers light sandy loam soil with a pH of 5–6.5. Calcareous soils are unsuitable for their cultivation
Marama bean	
Climate	It is a warm-season crop that needs temperatures of 28–37 °C. It thrives in dry climates with a yearly rainfall of less than 100 mm and ideal rainfall of 250–600 mm
Soil	It is a drought-resistant plant. Marama bean grows best in soil with a pH of 7 (neutral) and is susceptible to waterlogging

Agronomic Practices

Winged Bean

Seed Rate and Sowing

It is propagated from seeds as an annual crop but tubers are also used as a perennial crop. The weight of 1000 seeds is about 250 g. It requires a 10–35 kg/ha seed rate depending upon spacing, genotypes, and purpose of the crop. Scarification and seed soaking increases the speed and percent of germination. Germination occurs within 5 or 15 days after the sowing of seeds. Two to three seeds are placed on each hill by keeping 90–100 cm × 40–60 cm spacing between rows and plant, 2.5–3.8 cm deep (HDC 2014; Hang et al. 1993; Hardman et al. 1989). Sowing is done from the end of July to October.

Nutrient Management

Winged bean can be grown without the addition of fertilizers in the soil to their nitrogen-fixing capacity. This crop should be provided with nitrogen 40 kg, phosphorus 100 kg, and potash 40 kg per ha. The half quantity of N along with full P and K should be given during seed sowing, and the remaining half amount of N is applied at the time of flowering as top-dressing. The activity of nitrogenase is stimulated by the addition of potassium after 60 days of sowing and also enhances root growth: the number of nodules and the dry weight of nodules (Parthipan and Kulasooriya 1989).

Irrigation

Generally, rain destroys most other beans, but winged bean flourish in rain. If rains are not frequent, irrigation will be necessary. Irrigation is required weekly until the plants are large and have roots deep enough to resist short droughts.

Intercultural Operations

Winged bean can grow very easily with very little care (Karikari 1978). Its biomass can hold a significant amount of nitrogen, enabling it to be used in a variety of cropping systems such as crop rotation, intercropping, as a cover crop (Mupangwa et al. 2017), and as a cash crop in commercial plantations alongside fruit and oil trees (Raai et al. 2020). It has the potential to be advantageous to low-input farming due to its nitrogen-fixing capacity (Tanzi et al. 2019). Manual weeding is done as the slow plant growth at the early stage. The vines are of twining growth habit; hence, just after a few days of germination, they essentially require proper support from the beginning itself. If the vines are allowed to creep on the ground, the yield is reduced to even less than half when compared to properly staked plants. The staking can be provided by thick bamboo poles or any wooden sticks to get a higher yield.

Harvesting and Yield

It is a perennial vegetable crop that flowers after 40–140 days of sowing. After pollination, pods take around 2 weeks to attain a maximum length, while 3 weeks later they become fibrous, and mature seeds are ready to harvest 6 weeks later. Tubers are also developed but some varieties do not form tuberous roots. The pod turns from green to ash-brown and splits open when they become fully mature. It has a prolific pod-bearing capacity. A yield of 10–15 tons/ha of green pod yield, 1–1.5 tons/ha of dry seed yield, and 5–10 tons/ha of tubers can be obtained.

Adzuki Bean

Seed Rate and Sowing

It is propagated by seeds and requires a 27–37 kg/ha seed rate. The seeds are sown in May–June. The seeds take 10–14 days to germinate when sown in late May although earlier sowing results in late germination, i.e., 20 days after sowing (Hardman et al. 1989; Yoshida et al. 1995). The seedling emergence is hypogeal. The recommended space between rows varies between 30–45 cm and 45–75 cm; the sowing depth of the seed should be 3–5 cm.

Nutrient Management

The ability of legume crops in symbiosis with rhizobia to fix atmospheric nitrogen into the soil contributes significantly to their economic and environmental importance (Shahrajabian et al. 2017; Yong and Shahrajabian 2017; Yong et al. 2017, 2018). If sufficient inoculation has been provided to the seed before planting, some nitrogen is required to achieve good nodulation in legumes that can fix nitrogen. *Bradyrhizobium japonicum* strains help in the root nodule formation and fixation of nitrogen (Delić et al. 2010). At the time of field preparation, 25–30 tons/ha of farmyard manure (FYM) is applied. It requires 40:30–40:20–30 kg/ha N:P:K. Maintaining phosphorus and potassium at immature seedling and early blooming phases, which are quite similar to other edible beans, may boost crop yields, especially on nutrient-deficient soils. A fertilizer ratio of N:P:K = 1:0.82:0.55 resulted in a high yield (Yin et al. 2019). Organic farming with integrated pest control boosts yields, whereas organic farming without it exposes plants to pests and diseases, which can boost antioxidant activity (Hou et al. 2018).

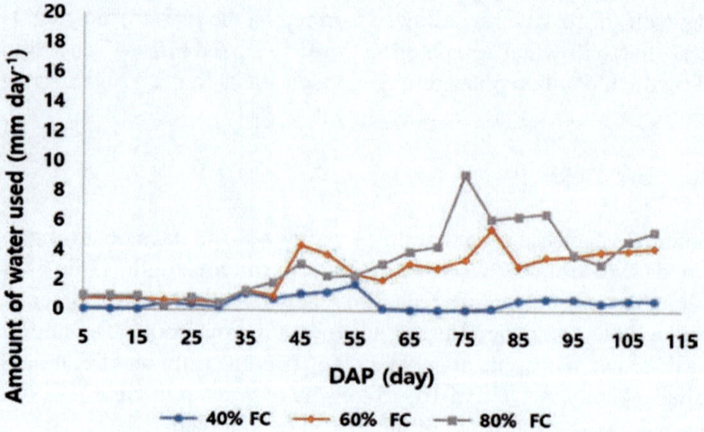

Graph 3.1 DAP represents days after planting and FC refers to field capacity

Irrigation

It is sensitive to both water excess and stress due to the shallow root system. Adequate moisture is essential on the topsoil layer at the time of sowing to ensure good germination. About 6–7 irrigations are needed during the growing season depending upon varieties, soil type, and environmental conditions. The plants are susceptible to moisture stress, so irrigation must be given during the pre-flowering, flowering, and pod filling stage.

Water is used daily from adzuki beans grown in soils with 40, 60, and 80% of field capacity. Graph 3.1 shows that the adzuki bean was extremely vulnerable to drought stress during the reproductive stage, requiring at least 60% of field capacity in soil moisture to avoid drought damage (Chun et al. 2020).

Intercultural Operations

The field is kept weed-free at the initial stage of plant growth because of its poor weed competitors and slow growth. Two to three weedings are required where the first weeding is done after 7–10 days of sowing and subsequent weeding is done at 10–20 days intervals. Pre-sowing application of Fluchloralin 2 l/ha or Alachlor 1.5 kg/ha as preemergence weedicide, or Pendimethalin 2 kg/ha effectively controls the weeds. It is sensitive to chemical-based weed control (OMAFRA 2016). The use of new herbicide strategies for weed management is required to provide a significant margin of crop safety (Belfry and Sikkema 2018). Adzuki bean is suitable for relay intercropping, intercropping (maize + adzuki bean), and multiple cropping (rice + adzuki bean) systems (Li et al. 2017).

Harvesting and Yield

The seeds in the adzuki bean pods for fresh market are harvested when they are visible in the pod. Picking is usually done at 5- to 6-day intervals to avoid toughening of pods due to high fiber and wax content. For grain purposes, the crop becomes ready for harvesting 115–120 days after sowing when the pods turn from green to yellowish-brown and seeds contain 12–15% moisture. Delay in harvesting results in the shattering of seeds in the field. The average dry seed yield varies from 5 to 6 q/ha.

Postharvest Handling and Storage

The entire plant with dry pods is pulled and stacked in a well-ventilated place to dry the seeds completely, which occurs 1–2 weeks after harvesting. Thereafter, the seeds are separated from the pods by threshing and winnowing and stored in refrigerated airtight containers for the next planting season.

Sword Bean

Seed Rate and Sowing

Sword bean is propagated by seed and requires about 25–40 kg seeds per hectare. In bush-type varieties, 60 cm spacing is required between the rows and plants, while pole-type varieties require 4 × 3 m spacing. The seeds are sown 5–7.5 cm deep. Sowing is done either in May–June or September–October.

Irrigation

Adequate moisture should be available in the soil during seed sowing for quick and good germination. A single irrigation 2–3 days before sowing and light irrigation 2–3 days after sowing help in better germination and crop stand. It grows well under rain-fed conditions but still responds well to irrigation. Irrigation should be given to the crop every week. Irrigation is very important during flowering and pod development.

Nutrient Management

During field preparation, 25 tons/ha of well-decomposed farmyard manure is applied. Generally, 63:100:75 kg N:P:K is required for a one-hectare area. Half quantity of N, along with full P and K, is given during seed sowing, with the remaining half N applied 3 weeks later. After harvesting, roots are left in the soil, where they decompose and release nitrogen. There was no change in dry matter or calcium absorption between VAM (Vesicular Arbuscular Mycorrhiza)-inoculated and non-VAM-inoculated plants; there was a substantial difference in VAM-inoculated plants reported by Indriani et al. (2019). Organic manures provide a good medium for microorganism growth and help soil ecosystems maintain a healthy nutritional balance (Shariff et al. 2017). Phosphorus is essential for legume crops, particularly for root growth and development (Rathore et al. 2015). Rock phosphate (300 kg/ha) is the best alternative phosphate fertilizer for seed and forage production (Indriani et al. 2019; Chaudhary et al. 2015).

Intercultural Operations

One or two shallow hoeings are necessary to make the environment suitable for root development. A hand weeding at 20–25 days after sowing should be done at the early stage of crop growth. In pole-type varieties, staking of vines promotes growth, flowering, and pod development. Trailing of the vines over pandals is also in practice. In homestead gardens, the vines are allowed to trail over the roof. It can grow as cover, green manure, forage crop, border crop, intercrop, and shade crop.

Harvesting and Yield

The harvesting of immature pods starts after 3–5 months of sowing and mature seeds after 6–10 months. Dry seed yields range from 0.7 to 1.5 tons/ha on average, with 4.6 tons/ha being ideal. It yields green manure/green vegetation in the range of 40–50 tons/ha.

Jack Bean

Seed Rate and Sowing

The seed rate depends on the germination percent, growing season, method of sowing, and soil properties. Usually, 40–60 kg seeds are sufficient for a one-hectare area. May–June and September–October are the sowing time. If mild climatic conditions prevail round the year, it can be grown throughout the year. The ridge and furrow method of planting is ideal for its cultivation. The broadcasting method of sowing is practiced if cultivated as a green manure crop, and the seeds are properly spaced if it is cultivated as a pulse crop. The seeds of pole-type varieties are sown at a spacing of 40 × 30 cm and bush-type varieties at 60 × 60 cm. Pruning with a 50 cm × 50 cm × 100 cm spacing in a double-row pattern produces better growth and yield (Sarijan et al. 2020).

Irrigation

Excessive watering and waterlogging conditions should be avoided. Pre-sowing irrigation is essential for proper seed germination. The critical stages for irrigation are flowering and pod setting. Further irrigation is given as per the need of the crop. The quantity of water required by the crop depends on the growing season, the kind of soil, and the amount of organic matter in the growing medium.

Nutrient Management

During field preparation, approximately 5 tons/ha of well-decomposed farmyard manure is used. As a base dose, a 7:10:5 ratio of N:P:K is applied. Half of the N is applied as a basal dose, along with a full P and K, and the remaining half of N is given 1 month after germination is completed. Around 187–229 kg N/ha can be fixed by jack bean (Bayorbor et al. 2006; Benjawan et al. 2007). The foliar application of boron, copper, molybdenum, zinc, manganese, and magnesium is effective in enhancing quality and pod yield. Seeds are inoculated with *Rhizobium* culture before sowing as it helps in quick root nodulation, which in turn fixes atmospheric nitrogen.

Intercultural Operations

Generally, two to three hand weeding and hoeings are needed to control the weeds and to get a higher yield. The first hoeing is done after 2–3 weeks of sowing and second is done at flowering. The hoeing operation should be followed by light earthing-up to strengthen the plants and encourage root growth. Alachlor at 2–2.5 kg/ha as a preemergence weedicide is recommended for controlling weeds. Another preemergence application of weedicides like Stomp at 3 l/ha or Goal at 750 ml/ha can control the weeds most effectively for about 45 days. Wooden or iron poles are used for the support of twining varieties for obtaining higher yields with the best quality pods. It does not require staking but staking helps in making the harvesting operation convenient. It is used as a cover crop (Echo 2006; Maceno 2019) because it produces photochemical that acts as a fungicide, bactericide, pesticide (Morris 1999), and control weeds under fruit trees (Rugare et al. 2020; Bunch and Staff 1985).

Harvesting and Yield

The seeds should be harvested by hand to prevent shelling. Harvesting of green immature pods is done after 90–120 days of seed sowing and the pods for the extraction of mature seeds are harvested 180–300 days after sowing. A single peduncle bears one to three pods and a plant has about four to seven pods. The bush-type cultivars with early flowering give 9–10 seeds per pod, whereas bush-type cultivars with mid- and late-season flowering give 11–12 seeds per pod. The average yield of green tender pods is 3–4 tons/ha (Okonkwo and Udedibie 1991). The average yield of mature seeds is 10–15 kg per plant and about 7–9 q/ha. The yield of green herbage is 40–50 tons/ha.

Moth Bean

Seed Rate and Sowing

Seeds are sown by broadcasting method and require 10–20 kg seeds per ha for seed production, while forage requires a 7–34 kg/ha seed rate. The seeds are planted at a 30–90 cm spacing and 2.5–4 cm depth.

Irrigation

Light irrigation should be given immediately after the sowing of seeds. The second and third irrigation is given 14 days after the first and second irrigation. The remaining six irrigations should be applied 7–10 days apart (Yadav and Patel 2019).

Nutrient Management

During field preparation, 20–25 tons/ha of well-decomposed FYM is applied, combined with 10 kg/ha N and 40 kg/ha P.

Intercultural Operations

Weed should be removed by hand weeding or hoeings. It can be planted in rotation with cotton as an intercrop and green manure. To minimize soil erosion, it is used as a cover crop (Munro and Small 1998).

Harvesting and Yield

The harvesting starts after 70–90 days of planting. When the leaves become dry and pod color changes from green to yellow, the whole plant is uprooted. The harvested plants are then dried in the sun for 3–5 days. After completion of drying, seeds are separated from the pods by threshing with the help of sticks. The seeds are dried again until the moisture content reaches 8–10%. It gives 6–8 q/ha grain yield and 12–25 q/ha fodder yield.

Scarlet Runner Bean

Seed Rate and Sowing

Generally, it is propagated by seeds and the direct seed sowing method is used. It requires a 20–35 kg seed rate per hectare. Before sowing, the seeds are treated with Captan or Thiram at 3–4 g/kg of seed. The spacing is kept at 15–30 cm × 90 cm for bush types and 15–30 cm × 150 cm for pole-type varieties. Sowing of seeds is done at 1.5–2.5 cm depth in heavy soils and 2.5–3.5 cm in light soils. May–June is the sowing time for scarlet runner bean.

Irrigation

Adequate soil moisture is essential during seed sowing to ensure good germination. It requires frequent irrigation because of its shallow root system. The crop requires 100–450 mm of irrigation on average. About 6–7 irrigations at intervals of 5–10 days are required during the growing season. Blossoming and pod set are the critical stages for irrigation. Hot and dry soil conditions at these stages cause severe flower drops with no pod set.

Intercultural Operations

After germination, the first 30–40 days are critical for crop–weed competition since it is a short-duration crop, requiring only 50–60 days from sowing to maturity. During the early stage of plant growth, two light hoeings and weedings should be done to keep the field weed-free. Fluchloralin 2 l/ha as pre-sowing or Oxadiazon 0.5 kg/ha, Alachlor 1.5 kg/ha, Pendimethalin 2 kg/ha, or Diuron 1.5 kg/ha as pre-emergence weedicides are used to control weeds. In summer under a condition of low-moisture stress, mulching with straw is very effective for good plant growth since it conserves soil moisture and suppresses weeds growth, and regulates soil temperature around the plant roots. Staking of pole-type varieties is done with bamboo poles or trellis.

Nutrient Management

During field preparation, 25–30 tons/ha of well-decomposed farmyard manure, 30–40 kg nitrogen, 40–60 kg phosphorus, and 40–50 kg potash are applied per hectare. A half quantity of N along with a full P and K is given during seed sowing, and the remaining N is given during flowering.

Harvesting and Yield

The harvesting of the crop starts when the pods and seeds are bright green, which usually occurs approximately 7–10 days after flowering. Picking of maturing beans prolongs the vines' life. The immature tender pods are harvested before they become fibrous and waxy. For seed production, harvesting is done when the pod color changed from green to yellow. On average, the green pod yield varies from 2 to 9 q/ha depending upon varieties, spacing, and rain. If the crop is grown with the recommended dose of fertilizers, the yield varies from 10 to 20 q/ha.

Postharvest Management

Pre-cooling of fresh pods soon after harvesting is an essential operation to remove field heat. The pods can be stored at 0 °C and 90% RH (relative humidity) for 10–15 days. After cleaning of pods, they are crossed cut into pieces of 1.0–1.5 cm in length, and then the pods are sorted, packed, and frozen or canned.

Yam Bean

Seed Rate and Sowing

It is propagated by seeds but sprouted roots from the previous crop are also used. Seeds are soaked in lukewarm water for 12 hours before sowing, which helps in softening of seed coat and speeding up germination. The weight of 1000 seeds is about 200 g. It requires about 20 to 45–50 kg/ha seed rate. The field is prepared thoroughly by ploughing with a harrow to a fine tilth for making smooth seedbeds. Two to three seeds per site are placed in holes at 5 cm depth on ridges and thinning is done as necessary. Spacing varies and can be 15–20 cm apart with 15–75 cm between rows but best results are obtained at 15 × 15 cm spacing. For vegetable production, the seeds are sown from June to September, although seeds can be sown in June–July for seed production and August–September for tuber production. For vegetable and seed production, 30 × 30 cm is used between rows and plants, and 30 × 15 cm or 15 × 15 cm for tuber production.

Irrigation

In dry soil, the first irrigation should be done immediately after seed sowing. Do not over-irrigate the field, as excessive soil moisture can damage the crop by increasing the chance of fungal infections. If the rain distribution is normal, crops sown during the rainy season may not require irrigation; however, crops sown in September require irrigation at regular intervals to maintain sufficient soil moisture for proper tuber development.

Intercultural Operations

Two hand weedings are required during the first few months. Mounting up of soil on both sides of the ridges is also essential to provide enough space for the growing tubers and also to cover them, as the exposed tubers are damaged by rodents and insect pests. Yam bean climbing or trailing vines (except for *P. ahipa*) require trellises and good support of the above-ground parts increases yields. Plants are traditionally pruned to stimulate better vegetative growth, and flowers are frequently removed to boost tuber yield and quality. In some regions, not only the flower removal is done but also the top portion of the aerial part of the plant is removed. Two prunings are sufficient to get a high yield in most varieties. Foliar application of 2–4 D 50 ppm at flower initiation also leads to dehiscence of the flowers, which in turn will result in better tuber yield.

Nutrient Management

After ploughing and planking, compost or 15–20 tons/ha of FYM along with half dose of nitrogen (40 kg/ha), full P (60 kg/ha), and K (80 kg/ha) are usually incorporated into the soil as basal dose to boost production. The remaining 40 kg/ha N is applied at 40–50 DAS (days after sowing) along with earthing-up (Ramaswamy et al. 1980; Sen and Mukhopadhyay 1989), whereas high potassium application reduced the cracking of tubers (Mishra et al. 1993).

Harvesting and Yield

It can be ready for harvesting within 150 days of sowing but can be harvested a little earlier. Tubers are dug by hand or with the aid of ploughing. After the tops are removed completely the tubers are washed and sorted to remove diseased and unsalable before the tubers are taken to market. The tubers can be kept at 5 °C for at least 1–2 months (Paull and Jung Chen 1988) 10 °C for 14 days, and less than 10 °C for a shorter period (Barile and Esguerra 1984). The crop gives an average yield of 18–20 tons/ha and the variety Rajendra Mishrikand gives 36–40 tons/ha.

Physiological Disorder

The *cracking* of tubers in yam beans is a major problem. Some cultivars like Mexican types are prone to cracking. Studies have shown that the availability of proper soil moisture during 45–90 days checks to crack. Delay in harvesting also causes cracking. The application of potassium reduces the cracking of tubers to a great extent.

Postharvest Management

The postharvest operations followed in yam beans are cleaning, sorting, and chlorine treatment for disinfection and bleaching. The tubers after cleaning adhered soil particles and curing in shade are packed in jute sacks. The tubers can be stored for 1–2 months at 12–16 °C temperature. When tubers are kept in storage for a long time, the starch to sugar ratio changes, with starch content decreasing and sugar content increasing as storage time passes.

Potato Bean

Seed Rate and Sowing

It is propagated through seed or tuber. Seeds require 10–30 days for germination (Reynolds et al. 1990). Tubers are used for planting in the spring season when the frost is over and they take a few days to several weeks for sprouting. Around 30 cm spacing is maintained between the plant to get a high yield.

Irrigation

It is a drought-tolerant crop but regular watering is required in the first week. Irrigation is essential during the dry months; however, there is no need to water the field during the rainy season.

Intercultural Operations

Apios has a weak vine that makes it difficult to compete with weeds. Weeds can be controlled by hand weeing, using plastic mulch, and following a vertical farming system.

Nutrient Management

Although a symbiotic relationship occurs between *Rhizobium* bacteria and *Apios*, which results in atmospheric nitrogen fixation in the soil (Putnam et al. 1991; Parker 1999), *Apios* and *Bradyrhizobium japonicum*, bacteria have a symbiotic association. In a study, tuber weight is improved when the *Apios* is inoculated with *B. japonicum* and nitrogen application of more than 100 ppm (Krishnan 1998).

Harvesting and Yield

When tubers are collected in the winter and when frost is a typical occurrence, they become sweet (Dean and Houston 2007). It gives 2.4 kg tubers per plant (Foster and Duke 1990).

Postharvest Handling

After harvesting, tubers are kept in cold storage, such as a refrigerator or a root cellar, or in the soil outside to avoid sprouting and storage infections (Thayer 2002). After peeling and drying tubers in the sun, they can be stored on racks (Beardsley 1939). Dried tubers have higher relative protein content than fresh ones (Beardsley 1939).

Tepary Bean

Seed Rate and Sowing

In general, the seed rate varies from 15 to 20 kg/ha. Direct seed sowing is done at 2.5–10 cm depth on the flat seedbeds. However, in low-lying areas, sowing on raised beds is preferred over sowing on flat beds, as it facilitates drainage. Seeds are sown either by broadcasting or line sowing prepared at a spacing of 60–90 cm × 10–45 cm between seeds/plants. However, in Kenya, it is sown at 60 × 30 cm spacing. The seeds must be treated before sowing with Thiram or Captan to disinfect the seed for better germination and with *Rhizobium* culture for better nodulation. The seeds are sown from September to November in the tropical climate, while February–March in the subtropical climate and April to October in the temperate climate.

Irrigation

Generally, the crop requires about six or seven irrigations at regular intervals for the successful cultivation of the crop. There should not be any water stress during flowering and pod formation. If moisture is not enough during sowing, light irrigation is given soon after sowing; however, excessive watering is harmful to the crop.

Nutrient Management

The crop requires about 25–30 tons/ha of well-decomposed FYM during field preparation. A general fertilizer dose of 40–50 kg N, 60 kg P, and 50 kg K should be applied per hectare. A half dose of nitrogen should be applied at seed sowing, along with a full dose of P and K, and the remaining half dose of nitrogen should be applied at flowering.

Intercultural Operations

Two hoeings should be done until the crop plants are well established to prevent weed competition with the crop. It is also beneficial for providing good aeration into the soil, which is very essential for the survival and efficiency of *Rhizobium* bacteria to fix atmospheric nitrogen in the soil. Benatazone at 2 kg/ha along with Thiobencarb at 5 kg/ha is effective for controlling weeds as preemergence weedicides. It can be grown as green manure, cover crop, fodder crop, and an intercrop in low-density orchards and the agroforestry system. It can be intercropped with cereals like maize, pearl millet, and sorghum, as well as vegetable crops like *Capsicum* spp., *Allium* spp., *Brassicas* spp., and *Cucurbita* spp. and other legumes.

Harvesting and Yield

The yield depends upon cultivars, soil types, sowing density, rainfall, and cultural practices adopted. It gives a 5–8 q/ha yield when grown as a rainfed crop and a 9–17 q/ha yield as an irrigated crop in irrigated conditions. The average seed yield of tepary beans is 4.5–6.7 q/ha.

Catjang Bean

Seed Rate and Sowing

It is propagated by seeds, and presoaking of seeds for 12 hours in warm water improves germination with 1–3 cm sowing depth. The crop requires 30 cm spacing between the rows and 15 cm between the plants for busy-type varieties, while pole-type varieties require 45 and 30 cm spacing between the rows and plants. Pole-type varieties are supported by growing on the trellis.

Irrigation

It needs light, frequent irrigations but is susceptible to waterlogging. The timing of irrigation is determined by the growing season. If the crop is cultivated during the rainy season, irrigation should only be done when there is no rain; however, in the summer, irrigation should be done in 8–10 days intervals.

Intercultural Operations

The entire field should be kept weed-free from sowing to 1 month after seed sowing. Generally, two hand weedings and hoeings are required to suppress the weeds. Weedicide like Fluchloralin at the rate of 1 kg/ha is applied as PPI (pre-plant incorporation) or Alachlor/Nitrofen at the rate of 2.0 kg/ha as the preemergence spray for weed control. To improve pod set and yield, maleic hydrazide at the rate of 50–200 ppm is also used before the commencement of flowering.

Nutrient Management

The plants fix atmospheric nitrogen into the soil because of the nodule formation in the roots where some amount of nitrogen is consumed by the plants and can also be utilized by other nearby growing plants. After harvesting, the plants are removed from the field while the roots remain in the soil because decomposition of the roots adds nitrogen to the soil. The crop requires about 25–30 tons/ha of well-decomposed FYM during field preparation. A general fertilizer dose of 40–50 kg N, 60 kg P, and 50 kg K should be applied per hectare. A half dose of nitrogen should be applied at seed sowing, along with a full dose of P and K, and the remaining half dose of nitrogen should be applied at flowering.

Harvesting and Yield

The harvesting of pods starts 45–50 days after flowering in early varieties and continues up to 100 days. For vegetable purposes, picking of pods is done after 15 days of pod setting. Generally, crop matures in 75–125 days depending on the variety and season. It gives a 5–8 q/ha yield when grown as a rainfed crop and a 9–17 q/ha yield as an irrigated crop in irrigated conditions. The average seed yield of tepary bean is 4.5–6.7 q/ha.

Agathi

Seed Rate and Sowing

Generally, the seeds are sown directly in the field but sometimes the seedlings are raised in the nursery. During nursery raising, the seeds are either sown on polythene bags or containers for seedling preparation. The seedlings are ready for transplanting after 1 month of seed sowing. During the rainy season, the seeds can be sown directly onto well-prepared land. A total of 7.5 kg seed is sufficient for a one-hectare area. The spacing requirement between the rows and plants is 100 cm. Agathi can also be propagated through stem cutting, though this is not a commercial method of propagation. Agathi can be planted quite intensively for wood production, with over 3000 plants per hectare grown in Australia and India.

Irrigation

Agathi is irrigated every 4–5 weeks with one to two buckets of water per plant, which means when the soil has been dry for days. If the season is very hot and arid, the number of irrigations can be intensified although leaving the substratum soaked with water must be avoided.

Intercultural Operations

Weeding should usually be done in an early crop stage, as this is more practical and more effective when the weed plants are smaller. Weeds can be controlled mechanically or by using chemicals (herbicides). Thinning is also required to maintain the plant-to-plant spacing for their optimum growth by reducing the competition among plants for space, light, nutrients, and moisture.

Nutrient Management

Agathi can be grown on soil with low fertility because of its less nutrient requirement. Lime is used in strongly acidic soils and 20 kg/ha phosphorus is used in soil with poor fertility for agathi cultivation. When agathi seeds are treated with the *Azorhizobium caulinodans* (ORS571), they can fix a considerable amount of nitrogen up to 50–200 kg/ha within 6–8 weeks. During field preparation, 12.5 tons/ha of FYM is applied. Nitrogen, phosphorus, and potash fertilizers (30, 15, and 33 kg/ha, respectively) are applied in the field.

Harvesting and Yield

Agathi yield depends on soil fertility and growing condition. When side branches are trimmed regularly, a tree can produce up to 27 kg of fresh green leaves every year. A green fodder yield of 100 tons/ha per year is obtained.

Velvet Bean

Seed Rate and Sowing

For sowing, about 22 kg/ha seed is required during the summer season and 12–15 kg/ha seed during the rainy season. The test weight of velvet bean seed is 1 kg. Seeds are sown either by broadcasting, dibbling, or in furrows at a spacing of 1 × 1 m with two seeds per hill. One or two interrow cultivations improve the early development of the crop. In subtropical climate, sowing is done from February to March as a summer crop and June to July as a rainy season crop, while in the hills, sowing is done from April to July.

Irrigation

The summer crop is irrigated once in 7–10 days. The crop is sensitive to waterlogging conditions; thus, water stagnation should be avoided. In the rainy season, irrigation depends on the rain.

Intercultural Operations

Light hoeing and weeding are required in the initial stages of crop growth to keep the field weed-free. Being a legume crop, it needs proper soil aeration for the normal activities of *Rhizobium* bacteria present in root nodules, which can be improved by hoeing. Two to three weedings are sufficient to control the weeds in the entire growing season.

Nutrient Management

It does not require much nitrogen but responds well to phosphorus. Farmyard manure about 15–20 tons/ha should be applied during field preparation. At the time of sowing, 20–25 kg/ha of nitrogen along with 50–70 kg/ha each phosphorus and potassium is applied. It can be grown as green manure and cover crop due to its heavy root and vine growth that significantly increases soil fertility by adding 10 tons of organic matter and 331 kg/ha of nitrogen.

Harvesting and Yield

The harvesting of pods starts as soon as the color changes from green to dark brown. In tropical regions, many of its cultivars grown for seed purposes become ready for harvesting in 180–270 days after sowing, and for forage, it is harvested 90–120 days after sowing when the pods are still young. The Yokohama variety generally matures within 110–120 days; however, harvesting after 120 days of planting results in high biomass yield and nutritive value. On average, velvet bean yields 20–35 tons/ha green material and 2.5–33 q/ha seed, depending on variety, climatic conditions, soil fertility, and cultivation practices.

Horse Gram

Seed Rate and Sowing

Generally, two methods of seed sowing are followed, i.e., broadcasting and line sowing method. A total of 40 kg/ha seeds are broadcasted for grain and fodder crops, while 25–30 kg/ha seed is sufficient for a grain crop following the line sowing method. Prepare the land to a fine tilth. The seeds are treated with Carbendazim or Thiram at the rate of 2 g/kg seed. They are treated with Rhizobial culture and Phosphobacteria at the rate of 200 g/ha. Biofertilizers can also be used to improve germination, crop growth, and yield like *Trichoderma viridi* at the rate of 4 g/kg seed. During the kharif and rabi seasons, row spacing is kept at 40–45 cm and 25–30 cm, respectively, with 5 cm between plants. Seeds are sown from late August to November as a fodder crop and from June to August as a fodder crop.

Irrigation

Irrigation must be done on time to maximize crop growth and yield. Pre-flowering and pod formation stages are the two critical stages of irrigation of the crop.

Intercultural Operations

During the early stage of crop growth, one hand weeding and hoeing are enough for weed control and are done 25–30 days after sowing. As a preemergence weedicide, Pendimethalin is used at a rate of 0.75–1 kg a.i./ha.

Nutrient Management

Generally, 12.5 tons/ha FYM should be applied during field preparation. At the time of planting, 20 kg/ha N and 30 kg/ha P are also applied.

Harvesting and Yield

The pods are ready for harvesting when the color changes from green to yellow. Whole plants are cut and spread on the threshing floor for 1–2 days for drying. The seeds are then removed from the pods by beating them with sticks. The average seed yield is about 700–900 kg/ha and 1000 kg/ha fodder yield can be achieved.

Yard Long Bean

Seed Rate and Sowing

The yard long bean seed rate varies between 20 and 25 kg/ha. Seeds are directly sown on well-prepared fine seedbeds, for which the field should be ploughed 2–3 times and leveled properly by planking. It takes 6–10 days for the seed to germinate, although soaking the seeds in water before sowing improves germination. One to two seeds per hill are sown in rows at a depth of 2–5 cm, with 60–90 cm row and 40 cm plant spacing for dwarf- and pole-type cultivars. Flat and raised seedbeds are used for the sowing in the spring-summer and rainy season, respectively. It can also be grown by transplanting without disturbing the root system.

Irrigation

It is sensitive to moisture excess and stress; thus, it requires light irrigation at regular intervals as compared to other legume vegetables. Heavy irrigation and waterlogging lead to the decaying of roots, whereas low soil moisture causes yield loss and produces short and fibrous pods. During the rainy season, the crop does not require any irrigation. However, the summer crop needs irrigations at 4–6 days intervals or once a week depending upon the weather conditions.

Nutrient Management

During field preparation, 15–20 tons/ha of well-decomposed farmyard manure is applied in the soil. As it is a legume crop, its nitrogen requirement is low but phosphorus requirement is high for nodulation and potash for the utilization of nitrogen.

A general fertilizer dose of 25 kg N, 75 kg P, and 60 kg K should be applied in one hectare area. A half dose of the N along with the entire dose of P and K should be applied during final land preparation and the remaining half dose of N should be applied 15–20 days after sowing along with weeding and earthing-up.

Intercultural Operations

In the initial stage of crop growth, shallow hoeing is required to suppress weed growth and to improve aeration in the soil. In addition to earthing-up, two hoeings are required to keep the field weed free. Light earthing-up, combined with fertilizer application, is also beneficial for greater root growth and protects young seedlings from lodging. Nipping the apical part is also a typical method to stimulate branching, flowering, and fruiting, especially when rain occurs during the flowering and fruit set. The vines should be encouraged to trail on trellises or a bower system. The long bamboo strips are fixed into the ground next to bean plants to support them. A trellis can also be made using two stakes and a length of wire. Both the stakes are hammered lengthwise into the ground along the bean plant rows. The bean seeds are sown along the bottom of the wire mesh. The bean vines will grow up and into the trellis. In addition, a simple trellis is formed using staking strings.

Harvesting and Yield

The yield of yard long bean is affected by a variety of factors such as growing season, soil type and fertility, irrigation facilities, and cultural practices used during cultivation. On average, yard long bean yields 150–180 q/ha pods and a seed yield of 4.5–6 q/ha.

Postharvest Management

The pods can stay fresh for a maximum of 3 days at room temperature but they start shriveling due to loss of moisture, and as a result, they can be kept in a refrigerator for 5–6 days at 4–8 °C and 80–90% relative humidity, or for 7–10 days if packaged in low-density polyethylene film bags, which prevent pod shriveling. They can be kept at 2–4 °C and high humidity for up to 4 weeks. However, pitting and russet symptoms develop at a temperature below 2 °C. The pods can be stored longer if blanched and frozen.

African Yam Bean

Seed Rate and Sowing

African yam bean can be propagated by seeds, small tubers, or root pieces, but commercially, the seed is used for planting. The seed rate depends on the season, region, and cultivation system. Generally, a 20–60 kg seed rate is required for a one-hectare area; however, the optimum seed rate is 35–40 kg/ha. Soaking of seeds before sowing in lukewarm water overnight helps in softening of seed coat and speeding up the germination. The field is prepared thoroughly by ploughing repeatedly with a harrow to a fine tilth for making smooth seedbeds and ensuring even germination. In ridges that are 60 cm apart and 15 cm height, seeds are sown at a depth of 5 cm, with plants spaced 30 cm apart. In the rainy season, seeds are sown from May to July. It can be sown throughout the year in tropical areas, while in subtropical areas, it is sown in late spring. In warm temperate areas, the crop can be sown at the end of spring frost. In cooler regions, the crop can be grown where it has a frost-free period of at least 5 months. The seeds are sown in June–July and in September as a late crop in India.

Irrigation

When the seeds are sown in dry soil, the first irrigation is applied immediately after sowing and when the crop is fully established, it is very important to make a moisture balance in the soil. The crop grown in the rainy season may not require irrigation if the distribution of rain is normal, but the crop sown in September requires irrigation at frequent intervals to keep the soil moist enough for the development of tubers.

Intercultural Operations

Mounting up of soil on both sides of the ridges is also essential to provide enough space for the growing tubers and also to cover them, as the exposed tubers are damaged by rodents and insect pests. This operation also indirectly helps in controlling weeds in a yam bean field. An important cultural operation in African yam bean is reproductive pruning, which helps in obtaining a maximum yield, as excessive flowering reduces the tuber size and yield. The plants are normally supported by trellises or stakes, but as far as yield is considered, no significant difference between supported and unsupported crops has been reported. Two reproductive prunings are usually enough to ensure a better yield in most varieties. When grown on the same field for two seasons, the second season produces a larger yield than the first. Crop rotation must be followed with maize, beans, and onions for sustaining land productivity.

Nutrient Management

At the time of land preparation, 10–20 tons/ha of well-decomposed FYM should be incorporated into the soil. The crop requires 60 kg N, 60 kg P, and 60 kg K per hectare. Before sowing, the full dose of P and K is applied along with a half-dose of nitrogen into the soil and the remaining half dose of the nitrogen is applied 40–45 DAS (days after sowing).

Harvesting and Yield

The harvesting of pods starts at 5 months after planting and tubers take 5–8 months. Tubers are picked up manually after loosening the soil around the plant. After collecting tubers, the vegetative tops are removed and left in the field to be used as pasture or organic manure. The yield also varies with the cultivar, soil type, climatic conditions, and cultural management. African yam bean yields an average of 13.69 q/ha.

Postharvest Management

After harvesting the cleaning of tubers is done by washing, trimming (removing roots and stem parts), sterilization, and bleaching by dipping in concentrated chlorinated water. The tubers, after cleaning adhered soil particles and curing in shade, are packed in sacks made of plastic wire or jute. The tubers can be stored for 1–2 months at 12.5–17.5 °C temperature. Prolonged storage may alter the starch to sugar ratio where starch content decreases and sugar content increases with the storage period. After removing the vegetative tops, the smallholders tend to leave the tubers underground until eaten or sold.

African Locust Bean

Seed Rate and Sowing

It has an exogenous dormancy due to hard seed coats (Hall et al. 1997). Seed dormancy can be broken by peeling off the coat or treating the seeds with H_2SO_4, which also improves the seedling growth (Muhammed et al. 2019).

Nutrient Management

Considering the economic and environmental implications of utilizing inorganic fertilizers, 5 g of poultry dung can be used to raise seedlings to maximize yield.

Intercultural Operations

Weeding and thinning are performed as needed.

Harvesting and Yield

The tree can start bearing after 5–7 years of planting Musa (1991) and a 20–30-year-old tree can produce a ton or more fruits. A light hooked pole is used to harvest the pods. To reach every bunch, the grower climbs up a locust bean tree, leans on larger branches, and pulls out the hooked pole (Akande et al. 2010). Each locust bean tree produces roughly 25–52 kg of pods, from which 6–14 kg of bean can be harvested (Odunfa 1982).

Ground Bean

Seed Rate and Sowing

It is propagated by seeds and requires about 75–90 kg/ha seed rate (Cumberland 1978). The weight of 1000 seeds is 500–750 g (Kay 1979). One seed is sown per hill as sole crop and four seeds per hill as intercrop. Seeds are treated with Captan or Thiram at 2–3 g/kg seed before sowing. The field is prepared by one deep ploughing followed by planking. Direct seed sowing is done on the ridges to reduce the risk of erosion and waterlogging (Stanton et al. 1966; Johnson 1968). Sowing is usually done in March–April, July–August, and November–December months; however, late sowing results in high yield. Sowing is done at a spacing of 45–90 cm between rows and 10–15 cm between plants.

Irrigation

Sufficient moisture should be available in the soil at the time of sowing until the seeds are germinated.

Intercultural Operations

Generally, two weedings are required. The first weeding is done 15 days after germination and the second weeding is done 30–45 days (Rassel 1960). Earthing-up at the time of flowering encourages underground pod formation (Holm 1940). It is cultivated as a sole crop or intercropped with cereals like pearl millet, maize, or sorghum (Ezueh 1977), root crops, tuber crops, and other legumes (Doku 1967, 1977).

Nutrient Management

At the time of field preparation, 10–20 tons/ha of well-decomposed FYM is applied. NPK fertilizers are applied before sowing, whereas nitrogen and potash can be applied 3 weeks later. It requires 100 kg N, 40 kg P, and 80 kg K per hectare (Uchehara et al. 2013).

Harvesting and Yield

The harvesting starts when the leaves turn green to yellow and wither, and the pods become mature to avoid the losses by rotting and premature germination. The plants are uprooted with extreme care because 65% of losses occur during the lifting of the pods. After that, the pods are dried under the sun and stored or shelled by hand with flails or by modified groundnut shellers, and then stored (Kay 1979). Shelling percentage is about 75% by weight (Johnson 1968). Seed yield may vary from 56 to 3870 kg/ha in different countries (Stanton 1966; Johnson 1968).

Marama Bean

Seed Rate and Sowing

The crop is propagated through seed and the sowing of seeds is done from October to November. Before sowing, scarification of seeds is done by exposing them to low temperatures to improve germination (Chingwaru et al. 2015). To facilitate quick germination, scarification can also be accomplished by layering seeds in the sand in shallow boxes (Kains 2020). Seeds are placed at 5–15 cm depth, whereas 20 cm is used in wet soil.

Nutrient Management and Irrigation

The crop does not require fertilizer or irrigation after the seeds get germinated.

Intercultural Operations

Hand weeding or hoeing is done to remove weeds.

Harvesting and Yield

Pods become ready for harvesting when the pod color changes from green to brown (April–June).

Plant Protection

Diseases

Common diseases of minor legumes are discussed below:

Disease	Symptoms	Control measures
White mould (*Sclerotinia sclerotium*)	Small, circular, dark green, water-soaked lesions on pods, leaves, and branches enlarge and become slimy	Follow crop rotation; Use disease-free seed; Sowing at wider spacing; Seed soaking in Bavistin or Benomyl 0.2% for 30 minutes; Spray of Bavistin 0.1% or Mancozeb 0.25% at 10–15 days intervals
Altenaria leaf spot (*Altenaria aiternata*)	Small irregular-shaped brown lesions on leaves, which later turn large oval gray-brown with concentric rings; small water-soaked, reddish-brown lesions develop that merge into long streaks develop on pods	Follow crop rotation; Use disease-free seed; Spray of Mancozeb 0.25% or copper oxychloride 0.3% at 10–15 days intervals
Powdery mildew (*Erysiphe polygoni, Plasmopara halstedii*)	Initially, yellow spots with powdery growth appear on the upper surface of leaves and spread to the underside of the leaves and other parts of the plant. Finally, leaves turn completely yellow, curl, die, and fall	Grow resistant varieties; Spray the crop with Karathane, Calyxin Bavistin, Benlate 0.2%, or Sulfex 0.5% at 10–15 days intervals
Downy mildew (*Peronospora parasitica*)	Yellow spots on the upper surface of the leaves, while bluish-white fluffy growth develops on the lower surface of the leaves. The infected leaves and branches may be distorted and die	Use disease-free seed; Follow crop rotation; Plant at wider spacing; Remove and destroy infected plants; Spray with Dithane M-45 0.2%, Chlorothalonil 0.2%, or Nitrophenol 0.05% twice at 10–15 days intervals

(continued)

Disease	Symptoms	Control measures
Common blight (*Xanthomonas campestris* pv. *phaseoli* and *fuscans*)	Initially, small translucent water-soaked spots on leaves, later spots enlarge, the tissue dies, leaving brown spots with yellow margins. Water-soaked sunken lesions on pods, which later turn brownish-red. The main stem and branches are also killed by bacteria	Follow crop rotation; Use disease-free seed; Remove and destroy infected plants; Seed treatment with Streptomycin 0.01% and Captan 0.25%; Spray the crop with Blitox 0.35%
Anthracnose (*Colletotrichum lindemuthianum*)	Seed-borne disease, necrosis of leaves, become thin, papery, and produce pinkish slimy mass. Dark brown sunken lesions with raised, reddish, or yellowish margins appear on pods	Use disease-free seed; Follow clean cultivation; Seeds treatment with Cerasan 0.15% for 30 minutes; Spray with 0.1% Bavistin at 10 days intervals
Stem blight/charcoal rot (*Macrophomina phaseolina*)	Black cankerous lesions appear on the cotyledons, later the collar region becomes brown and vascular bundles of the roots turn brown, resulting in dry root rot	Use disease-free seed; Follow crop rotation; Seed treatment with Captan 2–3 g/kg seed; Fumigate the soil with Chloropicrin or methyl iodide
Fusarium wilt (*Fusarium oxysporum* f. sp. *phaseoli*)	Yellowing and wilting of leaves, resulting in wilting and dying of entire plant; reddish discoloration throughout the root, stem, petioles, and peduncle; water-soaked lesions on pods	Grow resistant variety; Soil solarization; Deep ploughing; Removal and destruction of the infected plant; Seed treatment with Cerasan, Thiram, or Benomyl 2 g/kg of seeds; Soil drenching with Bavistin 0.1% or Thiram 0.2%
Fusarium root rot (*Fusarium solani* f. sp. *phaseoli*)	Lesions on primary roots which later become brown and extend up to soil surface; death of the primary roots and stem become hollow	Grow resistant variety; Follow crop rotation; Soil solarization; Deep ploughing; Removal and destruction of the infected plant; Seed treatment with Captan or Thiram 2 g/kg of seeds; Soil drenching with Carbendazim 0.1%

(continued)

Disease	Symptoms	Control measures
Angular leaf spot (*Phaseoriopsis griseola*)	Brown spots with tan or silvery center on the leaves; elongated and dark brown lesions on stems and petioles. Infected pods may develop shriveled or discolored seeds. A temperature of 20–25 °C and humid climate favor the disease	Grow resistant variety; Follow crop rotation; Deep ploughing; Removal and destruction of the infected plant; Seed treatment with Captan or Thiram 2 g/kg of seeds; Spray with streptomycin sulfate 100 ppm + copper oxychloride 0.25% at 15 days intervals
Ascochyta leaf spot (*Ascochyta phaseolorum*)	Large dark gray to black spots on leaves, blackening of nodes on stems. Flower infection leads to end rot of pods and causes extensive cankers	Grow resistant variety; Follow crop rotation; Use disease-free seed; Removal and destruction of the infected plant; Seed treatment with Captan or Thiram 2 g/kg of seeds
Leaf rust (*Uromyces appendiculatus*, *Phakopsora pachyrhizi*)	Rusty brown pustules develop on the lower surface of the leaves and give a rusty appearance. Later leaves turn brown, curl upwards, dry, and drop prematurely. Pod setting, pod filling, yield, and seed size are reduced in severe infection	Grow resistant variety; Follow crop rotation; Plant at wider spacing; Spray the crop at bloom stage with chlorothalonil, sulfur, or Maneb 0.5% at weekly interval
Halo blight (*Pseudomonas syringae* pv. *phaseolicola*)	Small water-soaked pinprick spots on the undersides of the leaves, later become yellow-green, resembling a halo, greasy spots on branches and petioles, turn reddish-brown and water-soaked lesions appear on infected pods turn into reddish discoloration. Seeds may become shriveled, rotten, or discolored	Grow resistant variety; Use disease-free seed; Follow crop rotation; Removal and destruction of the infected plant; Spray with copper-based fungicides Blitox 50 WP 0.2% at 10 days intervals
Common mosaic virus or bean mosaic virus	It is transmitted by aphids and white flies. Light to green or yellow to dark green mosaic patterns on leaves followed by puckering, rolling, distortion of the affected leaves, mottling, downward curling, and malformation of leaves	Use disease-free seed; Removal and destruction of the infected plant; Spray of Rogar or Metasystox 2 ml/l at 10–12 days intervals; Spray mineral Krishi oil 1.5% at 10 days intervals

(continued)

Disease	Symptoms	Control measures
Sincama mosaic virus	Irregular chlorosis on leaves, young shoots becoming brittle and the seed set is reduced due to reduced pollen fertility. Aphids and spider mites transmit the disease	Removal and destruction of the infected plant; Soil application of Disulfotan or Phorate 10G granules 1.5 kg/ha; Spray with Monocrotophos 0.05% or dimethoate 0.05% at 10–12 days intervals
Witches' broom disease	Excessive branching dwarfed leaves and deformation of the flowers. Transmitted by whitefly, aphids, and mealy bugs	Remove and destroy infected plant; Soil application of Carbofuran, Fensulfothion, or Disulfotan 1.5 kg/ha; Spray with Monocrotophos 0.05% or dimethoate 0.05% at 10–12 days intervals

Insect Pests

Common insects pests of minor legumes are discussed below:

Insect name	Damaging symptoms	Control measures
Aphid (*Aphis craccivora*)	Nymphs and adults both attack the young plants and suck the cell sap from growing tips and later from the whole plant	Spray the crop with 0.03% dimethoate or 0.05% phosphamidon and soil application of carbofuran granules 1.0 kg a.i./ha
Whitefly (*Bemisia tabaci*)	Nymphs suck the cell sap from tender parts of the plant and transmit viral diseases. The affected leaves become shriveled and curled gradually	Spray the crop with Metasystox or Rogor 0.25%
Bean pod borer (*Maruca vitrata, Muruca testulalis, Lampides boeticus, Helicoverpa armigera*)	Larvae cause damage to the corolla and caterpillars to distort the pods by making round holes and eating seeds	Removal and destruction of infested pods; Spray 0.03% dimethoate, or 0.04% monocrotophos, or 0.5% neem seed kernel extract; Apply nuclear polyhedrosis virus (NPV)
Red spider mite (*Tetranychus urticae*)	Adults and nymphs cause damage to the plants by feeding on the undersides of the leaves	Foliar spray of 0.5% neem seed kernel extract, wettable sulfur 0.3%, Malathion 0.05%, Dicofol 0.035%, or Ethion 0.005%

(continued)

Insect name	Damaging symptoms	Control measures
Stem fly (*Ophiomyia phaseoli*)	Maggots mine the leaf, bore inside the petiole, and make tunnels inside the tender stem, which ultimately cause a girdling effect and the death of the plant	Removal and destruction of the infected plant; Spray with Rogar 1 ml/l of water at 10–12 days intervals or neem seed kernel extract 5%
Leaf miner (*Spilasoma oblique Euproctis, Liriomyza trifolii*)	Larvae feed by making tunnels in the leaves and later entire leaflet becomes brown, rolls shrivels, and dries up	Follow crop rotation; Use resistant varieties; Spray with 0.03% dimethoate at 10–12 days intervals; Use larva parasites, i.e., *Bracon gelechidae* Ashm. and *Elasmus brevicornis* Gah
Root-knot nematodes	Gall formation on roots, stunted plant growth, and yellowing of leaves	Follow crop rotation; Deep summer ploughing; Soil application of Furadon 3 kg/ha granules at the time of sowing
Bean weevil (*Acanthoscelide obtectus*)	Larvae bore into the seeds and feed until maturity; adults chew their way out of the larval chambers, exposing holes in the seeds	Fumigate the seeds with Celphos or Phosfume 1–2 tablets per ton of seed
Hairy caterpillar (*Ascotis imparata, Spilosoma obliqua*)	Caterpillars feed on the leaves, leaving them skeletonized. In severe cases, defoliation occurs	Collection and destruction of the larvae; Spray with Quinolphos 0.05%, Monocrotophos 0.15%, or Fenvalerate 0.05% at 10–12 days intervals
Drosophilid fly (*Protostegana lateralis*)	Maggots bore into the tender shoots of mature plants, causing wilting of affected parts; dropping of infested pods	Use traps for monitoring the levels of male flies; Follow crop rotation
Tree hopper (*Otinotus oneratus*)	Insects cut the bark to lay their eggs inside	Use of biopesticides like azadirachtin or pyrethrin; Spray with chlorpyrifos or imidacloprid
Stem borer (*Azygophleps scalaris*)	Larvae make tunnels on the main stem, also feed on roots, and eat the pith region without damaging the epidermis	Uprooting the stumps immediately after harvesting and burning them may prove effective means of control
Seed chalcid (*Bruchophagus mellipes*)	Larvae make small round holes in mature pods and damage the seeds	Fumigation with carbon disulfide; Spray with Fenvalerate 0.05%
Red spider mites (*Tetranychus urticae*)	Mites make thick silky webs and suck the cell sap from the undersurface of leaves, resulting in chlorotic specks, a silvery appearance on leaves, and reduced yield	Use neem oil like Econeem or Neemazal 0.5% under the surface of leaves; Spray with Dicofol 0.25% or Ethion 0.05%

(continued)

Conclusion

All the underutilized legume vegetables are very important in improving soil health by fixing atmospheric nitrogen into the soil as well as improving human health because of nutrient content, especially protein. Apart from that, they do not require extra care and inputs during cultivation, which makes them easy to grow. Therefore, growers can gain more profit by producing these crops at a commercial level.

References

Abid, M., Hrishikeshavan, H. J., & Asad, M. (2006). Pharmacological evaluation of *Pachyrrhizus erosus* (L) seeds for central nervous system depressant activity. *Indian journal of physiology and pharmacology*, *50*(2), 143.

Abioye, V. F., Olanipekun, B. F., & Omotosho, O. T. (2015). Effect of varieties on the proximate, nutritional and anti-nutritional composition of nine variants of African yam bean seeds (Sphenostylis Stenocarpa). *Don. J. Food Sci. Technol*, *1*(2), 17–21.

Abitogun, A. S., & Olasehinde, E. F. (2012). Nutritional evaluation of seed and characterization of crude jack bean (*Canavalia ensiformis*) oil. *Journal of Applied Chemistry*, *1*(6), 36–40.

Adebowale, K. O., & Lawal, O. S. (2004). Comparative study of the functional properties of bambarra groundnut (*Voandzeia subterranean*), jack bean (*Canavalia ensiformis*) and mucuna bean (*Mucuna pruriens*) flours. *Food Research International*, *37*(4), 355–365.

Ade-Omowaye, B. I. O., Tucker, G. A., & Smetanska, I. (2015). Nutritional potential of nine underexploited legumes in Southwest Nigeria. *International Food Research Journal*, *22*(2), 798.

Adewumi, B. A., & Igbeka, J. C. (1993). The effect of steaming on the physical and dehulling characteristics of locust bean (*Parkia biglobosa*). *Tropical Agriculture*, *70*(3), 380–382.

Adeyeye, E. I., Oshodi, A. A., & Ipinmoroti, K. O. (1994). Functional properties of some varieties of African yam bean (Sphenostylis stenocarpa) flour II. *International journal of food sciences and nutrition*, *45*(2), 115–126.

Adsule, R. N. (1996). Moth bean (*Vigna aconitifolia* (Jacq.) Marechal). In *Food and Feed from Legumes and Oilseeds* (pp. 203–205). Springer, Boston, MA.

Agarwal, S., & Chauhan, E. S. (2019). Adzuki Beans-Physical and Nutritional Characteristics of Beans and Its Health Benefits. *International Journal of Health Sciences & Research*, *9*(4), 304–310.

Agullo, G., Gamet-Payrastre, L., Manenti, S., Viala, C., Rémésy, C., Chap, H., & Payrastre, B. (1997). Relationship between flavonoid structure and inhibition of phosphatidylinositol 3-kinase: a comparison with tyrosine kinase and protein kinase C inhibition. *Biochemical pharmacology*, *53*(11), 1649–1657.

Akande, F. B., Adejumo, O. A., Adamade, C. A., & Bodunde, J. (2010). Processing of locust bean fruits: Challenges and prospects. *African Journal of agricultural research*, *5*(17), 2268–2271.

Akindahunsi, A. A., & Salawu, S. O. (2005). Phytochemical screening and nutrient-antinutrient composition of selected tropical green leafy vegetables. *African Journal of Biotechnology*, *4*(6).

Al Hassan, M., Morosan, M., López-Gresa, M. D. P., Prohens, J., Vicente, O., & Boscaiu, M. (2016). Salinity-induced variation in biochemical markers provides insight into the mechanisms of salt tolerance in common (*Phaseolus vulgaris*) and runner (*P. coccineus*) beans. *International journal of molecular sciences*, *17*(9), 1582.

Aliero, B. L., Umaru, M. A., Suberu, H. A. & Abubaka, A. (2004). A handbook of common plants in Northwestern Nigeria, Sokoto university press, Nigeria, Pp. 130.

Anasiewicz, A. (1994). Insects pollinating runner bean (*Phaseolus coccineus* L.). Mat. General Conf. Science. Legumes protein plants. I. Beans. Lublin, 114–117.

Anonymous. (1982). The winged bean Flyer (4 No.1). International Documentation Centre for the Winged Bean Agricultural Information Bank of Asia, Laguna.
Anya, M. I., & Ozung, P. O. (2019). Proximate, mineral and anti-nutritional compositions of raw and processed African Yam bean (Sphenostylis stenocarpa) seeds in Cross River State, Nigeria. *Global Journal of Agricultural Sciences, 18*(1), 19–29.
Aremu, M. O., Olaofe, O., Basu, S. K., Abdulazeez, G., & Acharya, S. N. (2010). Processed cranberry bean (Phaseolus coccineus L.) seed flour for the African diet. *Canadian Journal of Plant Science, 90*(5), 719–728.
Arora, R. K., & Chandel, K. P. S. (1972). Botanical source areas of wild herbage legumes in India. *Tropical Grasslands, 6*(3), 213–221.
Asch, D. L., & Hart, J. P. (2004). Crop domestication in prehistoric eastern North America. *Encyclopedia of plant and crop science*, 314–319.
Asha, K. I., Latha, M., Abraham, Z., Jayan, P. K., Nair, M. C., & Mishra, S. K. (2006). Genetic resources. In D. Kumar (Ed.), *Horsegram in India (pp. 11–28)*. Jodhpur, India: Scientific Publisher.
Audu, I., Oloso, A., & Umar, B. (2004). Development of a concentric cylinder locust bean dehuller.
Baiyeri, S. O., Uguru, M. I., Ogbonna, P. E., Samuel-Baiyeri, C. C. A., Okechukwu, R., Kumaga, F. K., & Amoatey, C. (2018). Evaluation of the nutritional composition of the seeds of some selected African yam bean (Sphenostylis stenocarpa Hochst Ex. A. Rich (Harms)) accessions. *Agro-Science, 17*(2), 37–44.
Balbin, I. O., Sorensen, M., Kvist, L. P., & Vasquez, O. D. (2007). Review of the *Pachyrhizus tuberosus* (Lam.) Spreng. cultivar groups in Peru. *Plant Genetic Resources Newsletter, 151*, 2.
Barile, T. V., & Esguerra, E. B. (1984). Low temperature storage of yam beans. *Postharvest Research Notes, 1*(2), 23–25.
Barnes, S., Peterson, G., Grubbs, C., & Setchell, K. (1994). Potential role of dietary isoflavones in the prevention of cancer. In *Diet and Cancer* (pp. 135–147). Springer, Boston, MA.
Basamba, S. & Ruegg, J. (1982). The performance of winged bean in the field and in controlled environment – a comparative study. In: Stephenson RA (ed) The winged bean flyer, Int Doc Centre for winged bean, Philippines vol. *4*(1), P 15 (Abstr).
Bashir, L., Shittu, O. K., Sani, S., Busari, M. B., & Adeniyi, K. A. (2015). African natural products with potential antitrypanosoma properties: A review. *Int J Biochem Res Rev, 7*(2), 45–79.
Bayorbor, T. B., Addai, I. K., Lawson, I. Y. D., Dogbe, W., & Djabletey, D. (2006). Evaluation of some herbaceous legumes for use as green manure crops in the rainfed rice based cropping system in Northern Ghana. *Journal of Agronomy*.
Beardsley, G. (1939). The groundnut as used by the Indians of eastern North America. *Michigan Academy of Science, Arts, and Letters, 25*, 507–515.
Belamkar, V., Wenger, A., Kalberer, S. R., Bhattacharya, V. G., Blackmon, W. J., & Cannon, S. B. (2015). Evaluation of phenotypic variation in a collection of *Apios americana*: an edible tuberous legume. *Crop Science, 55*(2), 712–726.
Belfry, K. & Sikkema, P. H. (2018). Weed management in adzuki bean: a review. *Canadian Journal of Plant Science, 98*(6), 1221–1233.
Benjawan, C., Chutichudet, P., & Kaewsit, S. (2007). Effects of green manures on growth, yield and quality of green okra (*Abelmoschus esculentus* L.) Har Lium Cultivar. *Pakistan journal of biological sciences: PJBS, 10*(7), 1028–1035.
Bhardwaj, H. L., & Hamama, A. A. (2004). Protein and mineral composition of tepary bean seed. *HortScience, 39*(6), 1363–1365.
Black, M., Bewley, J.D., & Halmer, P. (2006). The Encyclopedia of Seeds. Science, Technology and Uses. CAB International, London, UK 696.
Borowy, A. (1996). Reaction of *Phaseolus coccineus* L. to competition of weeds and to the action of several herbicides. Ed. AR in Lublin, Rozpr. Science. 196.
Borthakur, S.K. (1996). Wild edible plants in markets of Assam, India – an ethnobotanical investigation. In: Ethnobiology in Human Welfare (edited by S.K. Jain). Pp. 31–34. New Delhi, India: Deep Publications.

Bressani, R., Brenes, R. G., García, A., & Elías, L. G. (1987). Chemical composition, amino acid content and protein quality of *Canavalia spp.* seeds. *Journal of the Science of Food and Agriculture, 40*(1), 17–23.

Brink, M. (2006). *Phaseolus coccineus* L. In M. Brink, & G. Belay (Eds.), *PROTA 1: Cereals and pulses/Cereales et legumes secs.[CD-Rom]*. Wageningen, the Netherlands: PROTA.

Brink, M., & Jansen, P.C.M. (2006). *Vigna aconitifolia* (Jacq.) Maréchal. [Internet] Record from PROTA4U. Brink, M. & Belay, G. (Editors). PROTA (Plant Resources of Tropical Africa/ Ressources végétales de l'Afrique tropicale), Wageningen, Netherlands. <"Archived copy". Archived from the original on November 8, 2013. Retrieved October 20, 2013.>. Accessed 15 November 2013–2006.

Bunch, R., & Staff, E. C. H. O. (1985). Green manure crops. *ECHO Technical Note*, 8–9.

Burkill, I.H. (1906). Goa beans in India. Agric Ledger 4:101–114.

Caiger, S. (1995). *Runner bean profile for Sri-Lanka.* The agro-enterprise development project, Colombo, Sri-Lanka: Miscellaneous reports. (p. 43).

Campbell-Platt, G. (1980). African locust bean (*Parkia species*) and its West African fermented food product, Dawadawa. *Ecology of food and nutrition, 9*(2), 123–132.

Carlisi, J., & Wollard, D. (2005). History, culture, and nutrition of Apios americana. *Journal of nutraceuticals, functional & medical foods, 4*(3–4), 85–92.

Chaitanya, D. A. K., Kumar, M. S., Reddy, A. M., Mukherjee, A. M., Sumanth, M. H., & Ramesh, A. (2010). Anti urolithiatic activity of *Macrotyloma uniflorum* seed extract on ethylene glycol induced urolithiasis in albino rats. *J Innov Trends Pharm Sci, 1*, 216–26.

Chaudhary, S. K., Kumar, R., Singh, A. K., & Kumar, R. (2015). Effect of acidulated rock phosphate on growth yield attributes and yield of wheat (*Triticum aestivum* L.). *Indian Journal of Agricultural Research, 49*(6). https://doi.org/10.18805/ijare,v49i6.6690.

Chingwaru, W., Vidmar, J., Kapewangolo, P. T., Mazimba, O., & Jackson, J. (2015). Therapeutic and prophylactic potential of Morama (*Tylosema esculentum*): a review. *Phytotherapy Research, 29*(10), 1423–1438.

Chun, H. C., Lee, S., Gong, D. H., Choi, Y. D. & Jung, K. Y. (2020). Characterizations of Growth, Yield and Water Use Efficiency of Adzuki Bean as Affected by Drought Stress. *Korean J. Soil Sci. Fert., 53*(4), 471–480.

Chung, I. M., Paudel, N., Kim, S. H., Yu, C. Y., & Ghimire, B. K. (2020). The influence of light wavelength on growth and antioxidant capacity in *Pachyrhizus erosus* (L.) Urban. *Journal of Plant Growth Regulation, 39*(1), 296–312.

Contreras-Govea, F. E., Muck, R. E., Armstrong, K. L., & Albrecht, K. A. (2009). Nutritive value of corn silage in mixture with climbing beans. *Animal Feed Science and Technology, 150*(1–2), 1–8.

Cumberland, J. (1978). Jugo beans: Research Report No. 19. *Agriculture Research Div, Ministry of Agriculture*; Malkerns Research Station, Swaziland. 11pp.

Dean, T., & Houston, P. B. J. (2007). Stalking the wild Groundnut. *Orion Magazine*.

Debouck, D. G., & Smartt, J. (1995). Beans, *Phaseolus spp.* (Leguminosae Papilionoideae). *Evolution of crop plants. 2a edición, Longman Scientific & Technical, Harlow, United Kingdom.*

Delić, D., Stajković, O., Rasulić, N., Kuzmanović, D., Jošić, D., & Miličić, B. (2010). Nodulation and N2 fixation effectiveness of Bradyrhizobium strains in symbiosis with adzuki bean, Vigna angularis. *Brazilian Archives of Biology and Technology, 53*, 293–299.

Doku, E. V. (1967). Are there any alternatives to the traditional bush fallow system of maintaining fertility?. *Ghana Farmer, 11*(1), 27–3038.

Doku, E. V. (1977). Grain legume production in Ghana. In *Proceedings of the Joint University of Ghana Council for Scientific and Industrial Research Symposium on Grain Legumes in Ghana, The Institute of Adult Education, University of Ghana, Legon* (pp. 1–7).

Duke, J. A. (1981). *Handbook of legumes of world economic importance*. New York, NY: Plenum Press.

Duke, J. (2012). *Handbook of legumes of world economic importance*. Springer Science & Business Media.

Durak, A., Baraniak, B., Jakubczyk, A., & Świeca, M. (2013). Biologically active peptides obtained by enzymatic hydrolysis of Adzuki bean seeds. *Food chemistry*, *141*(3), 2177–2183.

Duranti, M. (2006). Grain legume proteins and nutraceutical properties. *Fitoterapia*, *77*(2), 67–82.

Echo 2006 (Roland Bunch, 1985). Green manure crops, Echo technical note; http://echonet.org/ Accessed on 09/06/2013.

Ekanayake, S., Jansz, E. R., & Nair, B. M. (2000). Literature review of an underutilized legume: *Canavalia gladiata* L. *Plant foods for human nutrition*, *55*(4), 305–321.

Eknayake, S., Jansz, E. R., & Nair, B. M. (1999). Proximate composition, mineral and amino acid content of mature *Canavalia gladiata* seeds. *Food chemistry*, *66*(1), 115–119.

Erskine, W. (1980). Measurements of the cross-pollination of winged bean in Papua New Guinea. SABRAO J 12:11–14.

Erskine, W. and Bala, A.A. (1976). Crossing technique in winged bean. Trop Grain Legum Bull 6:32–35.

Ezueh, M. I. (1977). Cultivation and utilization of minor food legumes in Nigeria. *Tropical Grain Legume Bulletin*, *10*, 28–32.

Farombi, E. O. (2003). African indigenous plants with chemotherapeutic potentials and biotechnological approach to the production of bioactive prophylactic agents. *African Journal of biotechnology*, *2*(12), 662–671.

Femald, M. L. (1950). Grays Manual of Botany. 8th (Centennial) Ed. *Illustrated. New York: American Book Company*, 1179–1180.

Florentin, M.A., Penalva, M., Calegari, A. & Derpsch, R. (2004). Green manure/cover crops and crop rotation in the no-tillage system on small farms. Ministry of Agriculture and Livestock of the Republic of Paraguay and the German Technical Cooperation. http://www.fidafrique.net/IMG/pdf/NoTill_SmProp_Chptr_1AF912.pdf (accessed 7 Jan. 2013).

Foster, S., & Duke, J. A. (1990). A field guide to medicinal plants: eastern and central North America. *The Peterson field guide series (USA)*.

Franks, P. W., Wong, M. Y., Mitchell, J., Hennings, S., & Wareham, N. J. (2002). Non-esterified fatty acid levels and physical inactivity: the relative importance of low habitual energy expenditure and cardio-respiratory fitness. *British journal of nutrition*, *88*(3), 307–313.

Freytag, G. F., & Debouck, D. G. (2002). Taxonomy, Distribution, and Ecology of the Genus Phaseolus (Leguminosae-Papilionodeae) in North America, Mexico and Central America. Taxonomía, distribución y ecología del género Phaseolus (Leguminosae-Papilionodeae).

Galadzhun, I. (2015). *Effect of grafting runner beans on crop yield and fruit quality* (Master's thesis).

Gbadamosi, A. E. (2005). Genetic variation in Enantia chlorantha (Oliv.)–a medicinal plant.

George, T. T., Obilana, A. O., & Oyeyinka, S. A. (2020). The prospects of African yam bean: past and future importance. *Heliyon*, *6*(11), e05458.

Gohara, A. K., Souza, A. H. P. D., Gomes, S. T. M., Souza, N. E. D., Visentainer, J. V., & Matsushita, M. (2016). Nutritional and bioactive compounds of adzuki beans cultivars using chemometric approach. *Ciência e Agrotecnologia*, *40*(1), 104–113.

Gohara, A. K., Souza, A. H., Rodrigues, Â. C., Stroher, G. L., Gomes, S., Souza, N. E., Visentainer, J.V. & Matsushita, M. (2013). Chemometric methods applied to the mineral content increase in chocolate cakes containing chia and azuki. *Journal of the Brazilian Chemical Society*, *24*(5), 771–776.

Gopalan, C. B., Ramasastri, V., & Balasubramanian, S. C. (1978). Nutritive value of Indian foods, Hyderabad, India. *National Institute for Nutrition*, 204.

Gopalan, C, Rama Sastri B.V. and Balasubramanian, S.C., 2004, Nutritive Value of Indian Foods, National Institute of Nutrition, ICMR, Hyderabad.

Gross, R. (1983). Composition and protein quality of winged bean (*Psophocarpus tetragonolobus*). Qual Plant Plant Foods Hum Nutr 32:117–124. doi:https://doi.org/10.1007/BF01091332.

Guerra-García, A., Suárez-Atilano, M., Mastretta-Yanes, A., Delgado-Salinas, A., & Piñero, D. (2017). Domestication genomics of the open-pollinated scarlet runner bean (*Phaseolus coccineus* L.). *Frontiers in plant science*, *8*, 1891.

Gupta, N., Shrivastava, N., Singh, P. K., & Bhagyawant, S. S. (2016). Phytochemical evaluation of moth bean (*Vigna aconitifolia* L.) seeds and their divergence. *Biochemistry research international, 2016*.

Hajika, M. (2016). An overview of legume cultivation in Japan. In *JIRCAS International Symposium Proceedings* (pp. 47–57).

Hall, J. B., Thomlinson, H. F., Oni, P. I., Buchy, M., & Aebischer, D. P. (1997). A monograph of *Parkia biglobosa*. School of Agricultural and Forest Sciences Publication N. 9.

Hamunyela, M. H., Nepolo, E., & Emmambux, M. N. (2020). Proximate and starch composition of marama (*Tylosema esculentum*) storage roots during an annual growth period. *South African Journal of Science, 116*(5–6), 1–6.

Hang, A. N., McClary, D. C., Gilliland, G. C., & Lumpkin, T. A. (1993). Plant configuration and population effects on yield of azuki bean in Washington State. *New crops. Wiley, New York*, 588–590.

Harder, D.K. (1992). Chromosome counts in Psophocarpus. Kew Bull 47:529. https://doi.org/10.2307/4110581.

Harder, D.K., & Smartt, J. (1992). Further evidence on the origin of the cultivated winged bean, *Psophocarpus tetragonolobus* (L.) DC. (Fabaceae): chromosome numbers and the presence of a host-specific fungus. Econ Bot 46:187–191. https://doi.org/10.1007/BF02930637.

Hardman, L. L., Oplinger, E. S., Doll, J. D., & Combs, S. M. (1989). Alternative field crops manual: adzuki bean. *University of Wisconsin-Cooperative Extension, Madison*.

Hardwick, R. C. (1972). The emergence and early growth of French and runner beans (*Phaseolus vulgaris* L. and *Phaseolus coccineus* h.) sown on different dates. *Journal of Horticultural Science, 47*(3), 395–410.

HDC. (2014). HDC planting recommendations for azuki beans. [Online]. Hensall District Co-operative, Hensall, ON. Available from http://www.hdc.on.ca/for-growers/26-forgrowers/322-hdc-azuki-bean-planting-recommedation-2014.html.

HDC. (2017). Food products. [Online]. HDC. Available from http://www.hdc.on.ca/foodproducts.html.

Herath, H. M. W. and Ormrod, D. P. (1979). Effects of temperature and photoperiod on winged beans [*Psophocarpus tetragonolobus* (L.) DC]. *Annals of Botany, 43*(6), 729–736.

Holm, J. M. (1940). The Bambara groundnut or Njugo bean. *Farming in South Africa, 15*, 195 200.

Holubowicz, R., & Dickson, M. H. (1989). Cold tolerance in beans (*Phaseolus spp*.) as analyzed by their exotherms. *Euphytica, 41*(1), 31–37.

Hołubowicz, R., & Khan, A. A. (1989). Ethylene production as a biochemical marker for bean cold tolerance. Folia Hortic. Ann. I, 1, 5–19.

Honda, Y., Saito, Y., Mishima, T., Katsumi, N., Matsumoto, K., Enomoto, T., & Miwa, S. (2020). Characterization of physicochemical and digestive properties of starches from various "Dainagon" adzuki beans (*Vigna angularis*) cultivated in Japan. International journal of biological macromolecules, 148, 1021–1028.

Hopkins, H. C. (1983). The taxonomy, reproductive biology and economic potential of Parkia (Leguminosae: Mimosoideae) in Africa and Madagascar. *Botanical Journal of the Linnean Society, 87*(2), 135–167.

Hoshikawa, K. (1994). Apios-the forgotten tuber. *Plants World, 45*, 270.

Hou, C. J., Chen, Z. C., Ferng, S. H., & Liao, H. F. (2018). Effects of organic cultivation with and without pest management of Adzuki bean (*Vigna angularis*) on yield characteristics, seed chemical composition, and antioxidant capacity. Net Journal of Agricultural Science, 6(2), 20–28.

Huang, Y. F. (2008). The characteristics of French sword bean and high yield techniques of French sword bean cultivation. *China Fruit and Vegetable, 1*, 18.

Ikpa, S. C. (2017). *Proximate and phytochemical composition of African yam bean (Sphenostylis stenocarpa) and its effect on the lipid profile of hypercholesterolemic rats* (Doctoral dissertation).

Indriani, N. P., Mustafa, H. K., Ayuningsih, B., & Rochana, A. (2019). Production and nitrogen, phosphorus and calcium absorption of sword bean leaf (*Canavalia gladiata*) in application of rock phosphate and VAM inoculation. Legume Research-An International Journal, 42(2), 238–242.

Itoh, T., Kobayashi, M., Horio, F., & Furuichi, Y. (2009). Hypoglycemic effect of hot-water extract of adzuki (*Vigna angularis*) in spontaneously diabetic KK-Ay mice. *Nutrition*, *25*(2), 134–141.

Jackson, J. C. (2017). Technology and Nutrition Opportunities for Healthful Foods from Morama Beans, an Emerging Crop in Botswana. In *Global Food Security and Wellness* (pp. 125–140). Springer, New York, NY.

Jackson, J. C., Duodu, K. G., Holse, M., de Faria, M. D. L., Jordaan, D., Chingwaru, W., Hansen, A., Cencic, A., Kandawa-Schultz, M., Mpotokwane, S.M. and Chimwamurombe, P. de Kock, H.L. & Minnaar, A. (2010). The Morama bean (*Tylosema esculentum*): a potential crop for southern Africa. *Advances in Food and Nutrition Research*, *61*, 187–246.

Jansen, P. C. M. (1989). *Macrotyloma uniflorum* (Lam.) Verdc. In *Plant resources of South-East Asia: I. Pulses* (pp. 53–54). Pudoc.

Jayaraj, A. P., Tovey, F. I., Lewin, M. R., & Clark, C. G. (2000). Duodenal ulcer prevalence: experimental evidence for the possible role of dietary lipids. *Journal of gastroenterology and hepatology*, *15*(6), 610–616.

Johnson, D. J. (1968). The Bambara groundnut. A review Rhodesia Agric. *Journal*, *65*, 149–155.

Kadam, S. S., Salunkhe, D. K., & Maga, J. A. (1985). Nutritional composition, processing, and utilization of horse gram and moth bean. *Critical Reviews in Food Science & Nutrition*, *22*(1), 1–26.

Kadam, S., & Smithard, R. (1987). Effects of heat treatments on trypsin inhibitor and hemagglutinating activities in winged bean. Plant Food Hum Nutr 37, 151–159. https://doi.org/10.1007/BF01092051.

Kaga, A., Isemura, T., Tomooka, N., & Vaughan, D. A. (2008). The genetics of domestication of the azuki bean (*Vigna angularis*). *Genetics*, *178*(2), 1013–1036.

Kains, M. G. (2020). *Propagation Of Plants-A Complete Guide For Professional And Amateur Growers Of Plants By Seeds, Layers, Grafting And Budding, With Chapters On Nursery And Greenhouse Management*. Read Books Ltd.

Kalberer, S., Belamkar, V., Singh, J., & Cannon, S. (2020). *Apios americana*: natural history and ethnobotany. *LEGUME*, 29–32.

Kaleem, A., Ahmed, S., & Hassan, M. M. (2020). *Vigna aconitifolia* (Jacq.) Marechal. (Papilionaceae): A review of medicinal uses, Phytochemistry and Pharmacology. *Journal of Pharmacognosy and Phytochemistry*, *9*(1), 1153–1155.

Kaloo, G. S. (1995). Runner bean–*Phaseolus coccineus* L. In “Genetic Improvement of Vegetable Crops”.

Kardono, L. B. S., Tsauri, S., Padmawinata, K., Pezzuto, J. M., & Kinghorn, A. D. (1990). Cytotoxic constituents of the seeds of Pachyrrhizus erosus. *Planta Medica*, *56*(06), 673–674.

Karikari, S.K. (1972). Pollination requirements of winged beans (*Psophocarpus spp.* Neck) in Ghana. Ghana J. Agric. Sci. 5, 235–239.

Karikari, S. K. (1978). Characters in the selection of winged beans (Psophocarpus tetragonolobus (L) DC) tolerant to drought. [Conference paper]. *Acta Horticulturae (Netherlands). no. 84*.

Kay, D. E. (1979). *Crop & Product Digest No. 3 Food Legumes* (No. 641.3565/K23).

Kendall, D. A., & Smith, B. D. (1976). The Pollinating Efficiency of Honeybee and Bumblebee Visits to Flowers of the Runner bean *Phaseolus coccineus* L. *Journal of Applied Ecology*, 749–752.

Khan, T. (1982). *Winged Bean Production in the Tropics*. Food and Agriculture Organization of the United Nations. p. 1.

Khan, T.N. (1976). Papua New Guinea: a centre of genetic diversity in winged bean (*Psophocarpus tetragonolobus* (L.)Dc.). Euphytica 25:693–705. https://doi.org/10.1007/BF00041608.

Kikuta, C., Sugimoto, Y., Konishi, Y., Ono, Y., Tanaka, M., Iwaki, K. Fujita, S. & Kawanishi Asaoka, M. (2011). Physicochemical and structural properties of starch isolated from *Apios americana* Medikus. *Journal of Applied Glycoscience*, *59*(1), 21–30.

Kimani, P. M., Njau, S., Mulanya, M., & Narla, R. D. (2019). Breeding runner bean for short-day adaptation, grain yield, and disease resistance in eastern Africa. *Food and Energy Security*, *8*(3), e171.

Kisambira, A., Muyonga, J. H., Byaruhanga, Y. B., Tukamuhabwa, P., Tumwegamire, S., & Gruneberg, W. J. (2015). Composition and functional properties of yam bean (Pachyrhizus spp.) seed flour.

Klu, G.Y.P. and Kumaga, F.K. (1999). Testing of induced mutants of winged bean (*Psophocarpus tetragonolobus* (L.) DC.) for nodulation and phenotypic performance. Ghana J Sci 39:55–62.

Knoblauch, M., Peters, W. S., Ehlers, K., & van Bel, A. J. (2001). Reversible calcium-regulated stopcocks in legume sieve tubes. *The Plant Cell*, *13*(5), 1221–1230.

Kochhar, S. L., 2016. Mat or Moth bean [*Phaseolus aconitifolius*]. In: Kochhar (Ed.), Economic botany: a comprehensive study, 5th edition, Cambridge Univ. Press, 430p.

Koley, T. K., Maurya, A., Tripathi, A., Singh, B. K., Singh, M., Bhutia, T. L., Tripathi, P.C. & Singh, B. (2019). Antioxidant potential of commonly consumed underutilized leguminous vegetables. *International Journal of Vegetable Science*, *25*(4), 362–372.

Kondo, N., & Tomooka, N. (2012). New Sources of Resistance to *Cadophora gregata* f. sp. adzukicola and *Fusarium oxysporum* f. sp. adzukicola in *Vigna spp*. Plant disease, 96(4), 562–568.

Koshy, E.P., Alex, B.K. and John, P. (2013). Pollen Viability Studies in *Psophocarpus tetragonolobus* (L.) DC. Int J Appl Nat Sci 2:65–68.

Kramer, C., Soltani, N., Robinson, D. E., Swanton, C. J., & Sikkema, P. H. (2012). Control of volunteer adzuki bean in soybean. Agric. Sci. 3: 501–509. https://doi.org/10.4236/as.2012.34059.

Krishnan, H. B. (1998). Identification of genistein, an anticarcinogenic compound, in the edible tubers of the American groundnut (*Apios americana* Medikus). *Crop science*, *38*(4), 1052–1056.

Kumalasari, I. D., Nishi, K., Harmayani, E., Raharjo, S., & Sugahara, T. (2014). Immunomodulatory activity of Bengkoang (*Pachyrhizus erosus*) fiber extract in vitro and in vivo. *Cytotechnology*, *66*(1), 75–85.

Kumar, R., Mittal, R. K., & Pandey, D. P. (2012). Genetic variability for yield and growth attributes in adzuki bean. *Research on Crops*, *13*(2), 562–565.

Kushwah, A., Rajawat, P., & Kushwah, H. S. (2002). Nutritional evaluation of extruded faba bean (*Vicia faba* L.) as a protein supplement in cereals based diet in rats.

Łabuda, H. (2010). Runner bean (*Phaseolus coccineus* L.)-biology and use. *Acta Scientiarum Polonorum-Hortorum Cultus*, *9*(3), 117–132.

Lam Sanchez, A., Durigan, J. F., de Campos, S. L., Silvestre, S. R., Pedroso, P. A. C., & Banzatto, D. A. (1990). Effects of planting dates on the chemical composition and physico-chemical properties of seeds of Phaseolus vulgaris, Vigna angularis and Vigna unguiculata. *Alimentos e Nutricao (Brazil)*, (2), 35–44.

Lawal, B., Shittu, O. K., Kabiru, A. Y., Jigam, A. A., Umar, M. B., Berinyuy, E. B., & Alozieuwa, B. U. (2015). Potential antimalarials from African natural products: A review. *Journal of intercultural ethnopharmacology*, *4*(4), 318.

Lawal, B., Shittu, O. K., Oibiokpa, F. I., Berinyuy, E. B., & Mohammed, H. (2017). African natural products with potential antioxidants and hepatoprotectives properties: a review. *Clinical Phytoscience*, *2*(1), 1–66.

Lee, A., Kim, G. N., Kim, H. O., Song, W., & Roh, S. S. (2017). Antioxidant activity and melanin inhibitory effects of yam bean (*Pachyrhizus erosus*) extract. *The Korea Journal of Herbology*, *32*(2), 57–64.

Lee, E., Lee, S., Park, J., Kim, E., Hong, Y., Lee, S., & You, Y. (2020). Correlation between a soil respiration and environmental factors, air temperature and precipitation in *Pinus densiflora* community in Namsan and meaning on an urban forest management. *Journal of Korean Society of Forest Science*, *109*(2), 136–144.

Lee, J.H., Ham, H., Kim, M.Y., Ko, J.Y., Sim, E.Y., Kim, H.J., Lee, C.K., Jeon, Y.H., Jeong, H.S., & Woo, K.S. (2018). Phenolic compounds and antioxidant activity of adzuki bean cultivars. *Legume Research*, *41*(5), 2018: 681–688. https://doi.org/10.18805/LR-381.

Lepcha, P., Egan, A.N., Doyle, J.J., & Sathyanarayana, N. (2017). A review on current status and future prospects of winged bean (*Psophocarpus tetragonolobus*) in tropical agriculture. Plant Foods Hum Nutr *72*:225–235. https://doi.org/10.1007/s11130-017-0627-0.

Li, J., Jiang, J., Stel, H. V., Homkes, A., Corajod, J., Brown, K., & Chen, Z. (2014). Phylogenetics and biogeography of Apios (Fabaceae) inferred from sequences of nuclear and plastid genes. *International Journal of Plant Sciences, 175*(7), 764–780.

Li, N., Li, X., Feng, Z. G., & Masayuki, Y. (2007). Chemical constituents from *Canavalia gladiata*. J. Shenyang Pharm. Univ, 24, 676–678.

Li, L., Yang, T., Liu, R., Redden, B., Maalouf, F., & Zong, X. (2017). Food legume production in China. *The Crop Journal*, 5(2), 115–126.

Lim, T. K. (2012). *Edible medicinal and non-medicinal plants* (Vol. 1, pp. 285–292). Dordrecht, The Netherlands: Springer.

Liu, R., Zheng, Y., Cai, Z., & Xu, B. (2017). Saponins and flavonoids from adzuki bean (*Vigna angularis* L.) ameliorate high-fat diet-induced obesity in ICR mice. *Frontiers in pharmacology, 8*, 687.

Luo, S.F., Hu, L.H., Chen, Y.Y., & Li, P.X. (2015). The effects of modified atmosphere packaging on the quality of sword bean storage and antioxidant enzyme activities (in Chinese). Food Science *17*, 1–8.

Maceno, E. (2019). The effects of jack bean [*Canavalia ensiformis* (L.) DC.] and sunn hemp (*Crotalaria juncea* L.) as legume cover crops on soil physical, biological and chemical properties in an agricultural field in Puerto Rico. *Master's thesis. University of Puerto Rico, Río Piedras Campus. College of Natural Sciences Department of Environmental Science. Río Piedras, Puerto Rico. 108 pp.*

Makri, E. A., & Doxastakis, G. I. (2006). Emulsifying and foaming properties of *Phaseolus vulgaris* and *coccineus* proteins. *Food chemistry, 98*(3), 558–568.

Martín-Cabrejas, M. Á. (Ed.). (2019). *Legumes: Nutritional Quality, Processing and Potential Health Benefits* (Vol. 8). Royal Society of Chemistry.

Minka, S. R., & Bruneteau, M. (2000). Partial chemical composition of Bambara pea [*Vigna subterranea* (L.) Verde]. *Food Chemistry, 68*(3), 273–276.

Minka, S. R., Mbofung, C. M., Gandon, C., & Bruneteau, M. (1999). The effect of cooking with kanwa alkaline salt on the chemical composition of black beans (*Phaseolus vulgaris*). *Food Chemistry, 64*(2), 145–148.

Mishra, S., Singh, C. P., Singh, K. P., Singh, N. K., & Sinha, U. P. (1993). Response of potassium application on yield, quality and cracking of yam bean tuber. *J. Root Crops, 19*(2), 118–121.

Miyazaki, H., Okamoto, Y., Motoi, A., Watanabe, T., Katayama, S., Kawahara, S., Makabe, H., Fujii, H., & Yonekura, S. (2019). Adzuki bean (*Vigna angularis*) extract reduces amyloid-β aggregation and delays cognitive impairment in Drosophila models of Alzheimer's disease. Nutrition Research and Practice, *13*(1), 64–69.

Mohan, V. R., & Janardhanan, K. (1994). The biochemical composition and nutrient assessment of less known pulses of the genus Canavalia. *International journal of food sciences and nutrition, 45*(4), 255–262.

Mohanty, C.S., Verma, S., & Singh, V. (2013) Characterization of winged bean (*Psophocarpus tetragonolobus* (L.) DC.) based on molecular, chemical and physiological parameters. Am. J. Mol. Biol. *3*, 187–197. https://doi.org/10.4236/ajmb.2013.34025.

Morris, J. B. (1999). Legume genetic resources with novel value added industrial and pharmaceutical use. *Perspectives on new crops and new uses*, 196–201.

Moteetee, A. N. (2016). Canavalia (Phaseoleae, Fabaceae) species in South Africa: naturalized and indigenous. *South African Journal of Botany, 103*, 6–16.

Muhammed, D., Adeyemi, A. G., & Adenike, A. R. (2019). Seed dormancy breakage, enhance germination and growth performances of *Parkia biglobosa* seeds using concentrated H2SO4. *GSC Biological and Pharmaceutical Sciences, 8*(3).

Mukai, Y., & Sato, S. (2009). Polyphenol-containing azuki bean (Vigna angularis) extract attenuates blood pressure elevation and modulates nitric oxide synthase and caveolin-1 expressions in rats with hypertension. *Nutrition, Metabolism and Cardiovascular Diseases, 19*(7), 491–497.

Mukai, Y., & Sato, S. (2011). Polyphenol-containing azuki bean (Vigna angularis) seed coats attenuate vascular oxidative stress and inflammation in spontaneously hypertensive rats. *The Journal of nutritional biochemistry, 22*(1), 16–21.

Mulanya, M. M., Kimani, P. M., Narla, R. D., & Ojwang, P. O. (2019). Genetic inheritance of photoperiod sensitivity in Runner bean (*'Phaseolus coccineous'* L.). *Australian Journal of Crop Science, 13*(9), 1511–1519.

Munro, D. B., & Small, E. (1997). *Vegetables of Canada*. NRC Research Press.

Munro, D. B., & Small, E. (1998). *Les légumes du Canada*. NRC Research Press.

Mupangwa, W., Thierfelder, C., & Ngwira, A. (2017). Fertilization strategies in conservation agriculture systems with maize-legume cover crop rotations in Southern Africa. *Experimental Agriculture, 53*(2), 288. https://doi.org/10.1017/S0014479716000387.

Musa, H. L. (1991). Ginger and locust bean tree: History, growth, use and potentials. In Tuk-Ham'Symposium, Kurmin-Musa (pp. 1–16).

Museler, D. L., & Schonfeldt, H. C. (2006). The nutrient content of the Marama bean (*Tylosema esculentum*), an under utilised legume from Southern Africa. *Agricola, 16*, 7–13

Myers, C. (Ed.). (1998). *Specialty and minor crops handbook* (Vol. 3346). UCANR Publications. Oakland, CA.

NARO, 2020. NARO Genebank Databases. Genebank Project, NARO, Japan.

NAS. (1975). The winged bean: a high-protein crop for the humid tropics. National Academy of Sciences, Washington, D.C.

National Research Council. (1979). Tropical legumes: resources for the future.

Neacsu, M., Vaughan, N. J., Perri, V., Duncan, G. J., Walker, R., Coleman, M., & Russell, W. R. (2021). Nutritional and chemical profiling of UK-grown potato bean (Apios americana Medik) reveal its potential for diet biodiversification and revalorisation. *Journal of Food Composition and Analysis, 98*, 103821.

Ng, N. Q., & Marechal, R. (1985). Cowpea taxonomy, origin and germplasm. *Cowpea research, production and utilization*, 11–21.

Niazi, N. K., Bibi, I., Fatimah, A., Shahid, M., Javed, M. T., Wang, H., Ok, Y.S., Bashir, S., Murtaza, B., Saqib, Z.A. & Shakoor, M. B. (2017). Phosphate-assisted phytoremediation of arsenic by *Brassica napus* and *Brassica juncea*: morphological and physiological response. *International journal of phytoremediation, 19*(7), 670–678.

Nurrochmad, A., Leviana, F., Wulancarsari, C. G., & Lukitaningsih, E. (2010). Phytoestrogens of *Pachyrhizus erosus* prevent bone loss in an ovariectomized rat model of osteoporosis. *International Journal of Phytomedicine, 2*(4).

Obasi, M. O., & Agbatse, A. (2003). Evaluation of nutritive value and some functional properties of Kerstings groundnut seeds for optimum utilisation as a food and feed resource. *East African Agricultural and Forestry Journal, 68*(4), 173–181.

Ojewumi, M. E., Ayomide, A. A., Obanla, O. M., & Ojewumi, E. O. (2014). Pozzolanic properties of Waste Agricultural Biomass-African Locust Bean Pod Waste. *World Journal of Environmental Biosciences, 6*(3), 1–7.

Ojewumi, M. E., James Omoleye, and A. A. Ajayi. "Optimum fermentation temperature for the protein yield of *Parkia biglobosa* seeds (Iyere)." (2016): 584–587.

Odunfa, S. A. (1982). Legume Based Fermented Foods. CRC Press Inc Boca Raton pg 173–189.

Okonkwo, J. C., & Udedibie, A. B. I. (1991). Preliminary observations on the yield performance of jack bean (*Canavalia ensiformis*) and sword bean (*Canavalia gladiata*) in the Guinea Savanna of Nigeria. Paper presented at the 27th Annual Conference of Agric. Soc. Of Nigeria, Minna, Nigeria. 1–4 September, 1991.

Ologhobo, A. D. (1992). Nutritive values of some tropical (West African) legumes for poultry. *Journal of Applied Animal Research, 2*(2), 93–104.

Olowokudejo, J. D., Kadiri, A. B., & Travih, V. A. (2008). An ethnobotanical survey of herbal markets and medicinal plants in Lagos State of Nigeria.

Oluwalana, T., Okeleke, S.O., & Akinbosoye, T.B.S. (2019). Comparative Effect of Poultry and Cow dung on the Early Growth of *Parkia biglobosa* (Jacq) Benth, Seedings. Direct Research Journal of Public Health and Environmental Technology, *4* (5), 42–46.

OMAFRA. (2016). Guide to weed control, 2016–2017. Publication 75. Ontario Ministry of Agriculture, Food and Rural Affairs, Toronto, ON.

Omotayo, A. O., & Aremu, A. O. (2021). Marama bean [*Tylosema esculentum* (Burch.) A. Schreib.]: an indigenous plant with potential for food, nutrition, and economic sustainability. *Food & Function*, *12*(6), 2389–2403.

Ong, H. C., & Nordiana, M. (1999). Malay ethno-medico botany in Machang, Kelantan, Malaysia. *Fitoterapia*, *70*(5), 502–513.

Oni, K. C. (1990). Shelling machine related properties for African Locust Bean fruit. *Transactions of the ASAE*, *33*(2), 572–0576.

Osuigwe, D. I., Obiekezie, A. I., & Ogunji, J. O. (2002). Preliminary Evaluation of Jack bean (*Canavalia ensiformes* L. DC) seed meal as a substitute for fishmeal in the diets of *Clarias gariepinus* (Burchell, 1822); Deutscher Tropentag 2002; 9–11 October 2002, University of Kassel-Witzenhausen. *Deutscher Tropentag, October*, 9–11.

Osuigwe, D. I., Obiekezie, A. I., & Onuoha, G. C. (2006). Effects of jack bean seed meal on the intestinal mucosa of juvenile *Heterobranchus longifilis*. *African Journal of Biotechnology*, *5*(13).

Otubo, S. T., Ramalho, M. A. P., Abreu, Â. D. F. B., & dos Santos, J. B. (1996). Genetic control of low temperature tolerance in germination of the common bean (*Phaseolus vulgaris* L.). *Euphytica*, *89*(3), 313–317.

Palombini, S. V., Maruyama, S. A., Claus, T., Montanher, P. F., Souza, N. E. D., Visentainer, J. V., Gomes, S.T.M. & Matsushita, M. (2013). Antioxidant activity of Brazilian bean cultivars. *Journal of the Brazilian Chemical Society*, *24*(5), 765–770.

Park, C. J., & Han, J. S. (2015). Hypoglycemic effect of jicama (*Pachyrhizus erosus*) extract on streptozotocin-induced diabetic mice. *Preventive nutrition and food science*, *20*(2), 88.

Parker, M. A. (1999). Relationships of bradyrhizobia from the legumes *Apios americana* and *Desmodium glutinosum*. *Applied and Environmental Microbiology*, *65*(11), 4914–4920.

Parthipan, S., & Kulasooriya, S.A. (1989). Effect of nitrogen- and potassium-based fertilizers on nitrogen fixation in the winged bean (*Psophocarpus tetragonolobus*). World J Microbiol Biotechnol 5, 335–341. https://doi.org/10.1007/BF01741764.

Patel, D. P., Dabas, B. S., Sapra, R. S., & Mandal, S. (1995). Evaluation of horse gram (*Macrotyloma uniflorum*)(Lam.) germplasm. *National Bureau of Plant Genetic Resources Publication, New Delhi, India*.

Patel, R., Singh, R.K.R., Tyagi, V., Mallesha, & Raju, P.S. (2016). Nutritional evaluation of *Canavalia ensiformis* (Jack bean) cultivated in North East region of India. *International Journal of Botany Studies 1*(6), 18–21.

Paull, R. E., & Jung Chen, N. (1988). Compositional changes in yam bean during storage. *HortScience*, *23*(1), 194–196.

Pereira, B. F. F., Abreu, C. A. D., Herpin, U., Abreu, M. F. D., & Berton, R. S. (2010). Phytoremediation of lead by jack beans on a Rhodic Hapludox amended with EDTA. *Scientia Agricola*, *67*(3), 308–318.

Peterson, G., & Barnes, S. (1993). Genistein and biochanin A inhibit the growth of human prostate cancer cells but not epidermal growth factor receptor tyrosine autophosphorylation. *The Prostate*, *22*(4), 335–345.

Phukan, P., Bawari, M., & Sengupta, M. (2015). Promising neuroprotective plants from north east India. *Int J Pharm Pharm Sci*, *7*(3), 28–39.

Phuthego, B. L. (2014). *Physico-Functional Properties of Wheat-Morama Bean Composite Flour and its Performance in Food Systems* (Doctoral dissertation, University of Ghana).

Pickersgill, B. (1980). Cytology of two species of winged bean, *Psophocarpus tetragonolobus* (L.) DC and P. scandens. (Endl.MLeguminoseae). Bot. J. Linn Soc. *80*, 279–291.

Potter, D., & Doyle, J. J. (1992). Origins of the African yam bean (*Sphenostylis stenocarpa*, Leguminosae): evidence from morphology, isozymes, chloroplast DNA, and linguistics. *Economic Botany*, *46*(3), 276–292.

Pugalenthi, M., & Vadivel, V. (2005). Nutritional evaluation and the effect of processing methods on antinutritional factors of sword bean (*Canavalia gladiata* (Jacq.) DC). *Journal of Food Science and Technology-Mysore*, *42*(6), 510–516.

Purseglove, J. W. (1968). *Canavalia gladiata* (Jacq.) DC. *Tropical Crops: Dicotyledons, 245*.

Purseglove, J. W. (1974). *Dolichos uniflorus. Tropical crops: Dicotyledons*. (pp. 263–264). London and New York: Longman.

Putnam, D. H., Heichel, G. H., & Field, L. A. (1991). Response of *Apios americana* to nitrogen and inoculation. *HortScience*, *26*(7), 853–855.

Quagliotti, L., & Marletto, F. (1987). Research on the pollination of Runner bean (*Phaseolus coccineus* L.) for dry grain production. *Advances in Horticultural Science*, 43–49.

Raai, M. N., Zain, N. A. M., Osman, N., Rejab, N. A., Sahruzaini, N. A., & Cheng, A. (2020). Effects of shading on the growth, development and yield of winged bean (*Psophocarpus tetragonolobus*). *Ciência Rural*, *50*(2).

Ramaswamy, N., Muthukrishnan, C. R., & Shanmugavelu, K. G. (1980, November). Varietal performance of Mishrikand [*Pachyrhizus erosus* (L.) Urban]. In *National Seminar on Tuber Crops Production Technology* (Vol. 21, p. 22).

Rassel, A. (1960). Voandzou, Voandzeia subterranea Thouars, and its cultivation in Kwango. *Bulletin agricole du Congo belge*, *51*, 1–26.

Rathore, D. K., Kumar, R., Singh, M., Meena, V. K., Kumar, U., Gupta, P. S., Yadav, T. & Makarana, G. (2015). Phosphorus and zinc fertilization in fodder cowpea-A review. *Agricultural Reviews*, *36*(4), 333–338.

Reddy, C. K., Luan, F., and Xu, B. (2017). Morphology, crystallinity, pasting, thermal and quality characteristics of starches adzuki bean (*Vigna angularis* L.) and edible kudzu (*Pueraria thomsonii* Benth). *International Journal of Biological Macromolecules*, 105:354–362.

Reed, M. J. (1985). Observation of the potential of *Apios americana* as a food crop. *Hort Sci.*, *20*, 557–558.

Revilla, P., Butrón, A., Cartea, M. E., Malvar, R. A., & Ordás, A. (2005). Breeding for cold tolerance. M. Ashraf and PJC Harris (ed.) Abiotic stresses: Plant resistance through breeding and molecular approaches. Haworth Press, New York. *Breeding for cold tolerance. In M. Ashraf and PJC Harris (ed.) Abiotic stresses: Plant resistance through breeding and molecular approaches. Haworth Press, New York.*

Reynolds, B. D., Blackmon, W. J., Wickremesinhe, E., Wells, M. H., & Constantin, R. J. (1990). Domestication of *Apios americana*. In *Advances in new crops. Proceedings of the first national symposium 'New crops: research, development, economics', Indianapolis, Indiana, USA, 23–26 October 1988.* (pp. 436–442). Timber Press.

Rodino, A. P., Lema, M., Pérez-Barbeito, M., Santalla, M., & De Ron, A. M. (2007). Assessment of runner bean (*Phaseolus coccineus* L.) germplasm for tolerance to low temperature during early seedling growth. *Euphytica*, *155*(1), 63–70.

Roy, F., Boye, J. I., & Simpson, B. K. (2010). Bioactive proteins and peptides in pulse crops: Pea, chickpea and lentil. *Food research international*, *43*(2), 432–442.

Rugare, J. T., Pieterse, P. J., & Mabasa, S. (2020). Evaluation of the potential of jack bean [*Canavalia ensiformis* (L.) DC.] and velvet bean [*Mucuna pruriens* (L.) DC.] aqueous extracts as post-emergence bio-herbicides for weed control in maize (Zea mays L.). *Asian Journal of Agriculture and Rural Development*, *10*(1), 420–439.

Sarijan, A., Surahman, M., Setiawan, A., & Giyanto, G. (2020). Pengaturan Arsitektur Tanaman untuk Menyeimbangkan Sink dan Source serta Meningkatkan Pertumbuhan dan Hasil Kacang Koro Pedang. *Jurnal Ilmu Pertanian Indonesia*, *25*(1), 43–51.

Sathe, S. K., & Venkatachalam, M. (2007). Fractionation and biochemical characterization of moth bean (*Vigna aconitifolia* L.) proteins. *LWT-Food Science and Technology*, *40*(4), 600–610.

Schwember, A. R., Carrasco, B., & Gepts, P. (2017). Unraveling agronomic and genetic aspects of runner bean (*Phaseolus coccineus* L.). *Field Crops Research, 206*, 86–94.

Seabrook, J. A., & Dionne, L. A. (1976). Studies on the genus Apios. I. Chromosome number and distribution of *Apios americana* and *A. priceana*. *Canadian Journal of Botany, 54*(22), 2567–2572.

Sen, H., & Mukhopadhyay, S. K. (1989). *Effect of nitrogen and potassium on tuber yield and quality of Yam Bean (Pachyrrhizus erosus)* (No. AV 635.1 no. 34).

Senayake, Y.D.A., & Sumanasinghe, V.A.D. (1978). Stigma receptivity in winged bean (*Psophocarpus tetragonolobus* (L.) DC.). SABRAO J Breed Genet *10*:116–119.

Shahrajabian, M. H., Soleymani, A., & Khoshkharam, M. (2017). Influence of green manuring from different cover crops and farm yard manures on quantitative and qualitative characteristics of forage corn in low input farming. *Research on Crop Ecophysiology, 12*(2), 62–68.

Shahrajabian, M.H., Sun, W., Khoshkharam, M., Zandi, P., & Cheng, Q. (2019). Adzuki beans (*Vigna angularis*), a Traditional Chinese Legume for Sustainable Agriculture and Food Production. *J. Biol. Environ. Sci., 13*(38), 79–84.

Shariff, A. F., Sajjan, A. S., Babalad, H. B., Nagaraj, L. B., & Palankar, S. G. (2017). Effect of organics on seed yield and quality of green gram (*Vigna radiata* L.). *Legume Research, 40*(2), 388–392. https://doi.org/10.18805/lr.v0i0f.11297.

Sherasia, P. L., Garg, M. R., & Bhanderi, B. M. (2017). *Pulses and their by-products as animal feed.* Food and Agriculture Organization of the United Nations (FAO).

Siddhuraju, P. (2006). The antioxidant activity and free radical-scavenging capacity of phenolics of raw and dry heated moth bean (*Vigna aconitifolia*)(Jacq.) Marechal seed extracts. *Food Chemistry, 99*(1), 149–157.

Siddhuraju, P., & Becker, K. (2001). Species/variety differences in biochemical composition and nutritional value of Indian tribal legumes of the genus Canavalia. *Food/Nahrung, 45*(4), 224–233.

Siddhuraju, P., Vijayakumari, K., & Janardhanan, K. (1996). Chemical composition and protein quality of the little-known legume, velvet bean (Mucuna pruriens (L.) DC.). *Journal of Agricultural and Food Chemistry, 44*(9), 2636–2641.

Singh, S., & Ansari, I. (2018). A pharmacognostic and pharmacological review on *Vigna aconitifolia* (Moth bean). The Pharma Innovation Journal, 7(10): 491–495.

Smartt, J. (1976). Tropical pulses, pp. 56–58. London: Longman.

Smartt, J. (1985). Evolution of grain legumes. II. Old and new world pulses of lesser economic importance. *Experimental Agriculture, 21*(1), 1–18.

Sodani, S. N., Paliwal, R. V., & Jain, L. K. (2006). Phenotypic stability for seed yield in rainfed horse gram [*Macrotyloma uniflorum* (Lam.) Verdc.]. *Arid Legumes for Sustainable Agriculture and Trade (Vol. 2), 4*(53.336), 340.

Sørensen, M. (1996). *Yam Bean: Pachyrhizus DC.-Promoting the conservation and use of underutilized and neglected crops.* 2 (Vol. 2). Bioversity International.

Stanton, W. R., Doughty, J., Orraca-Tetteh, R., & Steele, W. (1966). Grain legumes in Africa. *Grain legumes in Africa.*

Stanton, W.R. (1966). Grain legumes in Africa, pp. 93–95. Rome, Italy: FAO.

Suma, A., Latha, M., John, J. K., Aswathi, P. V., Pandey, C. D., & Ajinkya, A. (2021). Yard-long bean. In *The Beans and the Peas* (pp. 153–172). Woodhead Publishing.

Suttie, J. M. (1969). The butter bean (*Phaseolus coccineus* L.) in Kenya. *East African Agricultural and Forestry Journal, 35*, 211–212.

Takahashi, S., Hori, K., Shinbo, M., Hiwatashi, K., Gotoh, T., & Yamada, S. (2008). Isolation of human renin inhibitor from soybean: soyasaponin I is the novel human renin inhibitor in soybean. Biosci. Biotechnol. Biochem. *72*, 3232–3236. https://doi.org/10.1271/bbb.80495.

Tanzi, A. S., Eagleton, G. E., Ho, W. K., Wong, Q. N., Mayes, S., & Massawe, F. (2019). Winged bean (*Psophocarpus tetragonolobus* (L.) DC.) for food and nutritional security: synthesis of past research and future direction. *Planta, 250*(3), 911–931.

Tateishi, Y. (1983). Leguminosae collected in the Arun valley, East Nepal. *Structure and dynamics of vegetation in Eastern Nepal.*

Tateishi, Y. (1984). Contribution to the genus Vigna (Leguminosae) in Taiwan. *Sci. Rep. Tohoku Univ. 4th Ser.(Biology), 38*, 335–350.

Tazawa, H., Kano, H., Matsumoto, N., & Murakami, M. (2018). Aroma compounds contributing to the flavor of Adzuki bean jam used for Dorayaki sweets. *Journal of the Integrated Study of Dietary Habits, 29*(1):45–52.

Termote, C., Odongo, N. O., Dreyer, B. S., Guissou, B., Parkouda, C., & Vinceti, B. (2020). Nutrient composition of *Parkia biglobosa* pulp, raw and fermented seeds: a systematic review. *Critical Reviews in Food Science and Nutrition*, 1–26.

Thaptimthong, T., Kasemsuk, T., Sibmooh, N., & Unchern, S. (2016). Platelet inhibitory effects of juices from *Pachyrhizus erosus* L. root and *Psidium guajava* L. fruit: a randomized controlled trial in healthy volunteers. *BMC complementary and alternative medicine, 16*(1), 1–12.

Thayer, S. (Summer/Fall 2002) Hopniss: North America's best wild tuber? Also known as Indian Potato or Groundnut. The Forager's Harvest Newsletter 2(3).

Tiamiyu, L. O., Okomoda, V. T., & Akpa, P. O. (2016). Nutritional profile of toasted *Canavalia ensiformis* seed and its potential as partially replacement for soybean in the diet of *Clarias gariepinus*. Brazilian Journal of Aquatic Science and Technology, 20(2), 12–17. https://doi.org/10.14210/bjast.2016vn202.

Tomooka, N. (2009). The origin of rice bean (*Vigna umbellata*) and azuki bean (*V. angularis*): \ the evolution of two lesser-known Asian beans. *An illustrated eco-history of the Mekong River Basin. Bangkok: White Lotus Co.*

U.S. Department of Agriculture, Agricultural Research Service. (2010). USDA national nutrient database for standard reference, release 23. Nutrient data laboratory home page. http://www.ars.usda.gov/ba/bhnrc/ndl.

Uchehara, C. P., Onyeonagu, C. C., & Asiegbu, J. E. (2013). Effect of plant population and N fertilizer on the growth and yield of Bambara groundnut (*Vigna subterranean* (L.) Verdc.). *Agro-Science, 12*(3), 25–34.

USDA. USDA national nutrient database for standard reference, release 18. [Internet] U.S. Department of Agriculture, Agricultural Research Service, Nutrient Data Laboratory, Beltsville, Maryland, United States, 2005, http://www.nal.usda.gov/fnic/foodcomp. September.

Uzabakiriho, J. D. (2016). *Diversity and distribution of Tylosema esculentum (Marama bean) endophytic bacteria communities in Omitara, Harnas and Otjinene, eastern Namibia* (Doctoral dissertation, University of Namibia).

Vadivel, V. & Janardhanan, K. (1998). Genetic resources of some south Indian tribal pulses. IPGRI Newsletter for Asia, the Pacific and Oceania, 26, 21–22.

Vadivel, V., & Janardhanan, K. (2005). Nutritional and antinutritional characteristics of seven South Indian wild legumes. *Plant Foods for Human Nutrition, 60*(2), 69–75.

Vadivel, V., Doss, A., & Pugalenthi, M. (2010). Evaluation of nutritional value and protein quality of raw and differentially processed sword bean [*Canavalia gladiata* (Jacq.) DC.] seeds. *African Journal of Food, Agriculture, Nutrition and Development, 10*(7).

Vadivel, V., Janardhanan, K., & Vijayakumari, K. (1998). Diversity in sword bean (*Canavalia gladiata* (Jacq.) DC.) collected from Tamil Nadu, India. *Genetic Resources and Crop Evolution, 45*(1), 63–68.

Vavilov, N.I. (1951). The origin, variation, immunity and breeding of cultivated plants. In: Chronica Botanica 13. Ronald Press, New York.

Woods, M. (2005). A revision of the North American species of Apios (Fabaceae). *Castanea, 70*(2), 85–100.

Xianzong, X. I. A., Ruoxi, Y. I. N., Wei, H. E., Hołubowicz, R., & Górna, B. (2017). Seed Yield and Quality of Sword Bean (*Canavalia gladiata* (Jacq.) DC.) Produced in Poland. *Notulae Botanicae Horti Agrobotanici Cluj-Napoca, 45*(2), 561–568.

Yadav, M., & Patel, J. J. (2019). Influence of dates of sowing and irrigation scheduling (IW: CPE ratio) on yield attributes and yield of summer moth bean. *Journal of Pharmacognosy and Phytochemistry*, *8*(6), 1175–1178.

Yamaguchi, H. (1992). Wild and weed azuki beans in Japan. *Economic Botany*, *46*(4), 384–394.

Yang, D., Lei, Y.G., Liu, M., & Zhang, S.X. (2014). Therapeutic effect of 5% amino oligosaccharide water on the chilling injury of sword bean (in Chinese). Modern Agricultural Technology, *7*, 145–149.

Yang, S., Grall, A., & Chapman, M.A. (2018). Origin and diversification of winged bean (*Psophocarpus tetragonolobus* (L.) DC.), a multipurpose underutilized legume. Am J Bot *105*, 888–897. https://doi.org/10.1002/ajb2.1093.

Yang, X.H., & Ma, S.B. (2010). The control techniques of sword bean root rot (in Chinese). Vegetables, *4*, 19.

Yao, Y., Cheng, X., Wang, S., Wang, L., & Ren, G. (2012). Influence of altitudinal variation on the antioxidant and antidiabetic potential of azuki bean (*Vigna angularis*). *International journal of food sciences and nutrition*, *63*(1), 117–124.

Yin, Z. C., Guo, W. Y., Liang, J., Xiao, H. Y., Hao, X. Y., Hou, A. F., Zong, X.X., Leng, T.R., Wang, Y.J., Wang, Q.Y., & Yin, F. X. (2019) Effects of multiple N, P, and K fertilizer combinations on adzuki bean (*Vigna angularis*) yield in a semi-arid region of northeastern China. *Scientific reports*, 9(1), 1–13.

Yong, A., Hu, Y., Shahrajabian, M. H., Ren, C., Guo, L., Wang, C., & Zeng, Z. (2018). Changes in dry matter, protein percentage and organic matter of soybean-oat and groundnut-oat intercropping in different growth stages in Jilin province, China. *Acta Agriculturae Slovenica*, *111*(1), 1–7.

Yong, Y., and Shahrajabian, M. H. (2017). Evaluation of the benefits of oat-soybean and oat groundnut intercropping in Jilin province, China. Research on Crop Ecophysiology. *12*(1), 98–106.

Yong, Y., Hu, Y., Shahrajabian, M. H., Ren, C., Guo, L., Wang, C., & Zeng, Z. (2017). Organic matter, protein percentage, yield, competition and economics of oat-soybean and oat groundnut intercropping systems in Northern China.

Yoshida, K., Sato, H., & Sato, M. (1995). The extent and its source of variation for characteristics related to seed quality of adzuki beans [*Vigna angularis*], 3: The water uptake of seeds and hard seededness. *Japanese Journal of Crop Science (Japan)*. https://doi.org/10.1626/jcs.64.7.

Zha, L. Y., Mao, L. M., Lu, X. C., Deng, H., Ye, J. F., Chu, X. W., Sun, S.X. & Luo, H. J. (2011). Anti-inflammatory effect of soyasaponins through suppressing nitric oxide production in LPS-stimulated RAW 264.7 cells by attenuation of NF-κB-mediated nitric oxide synthase expression. *Bioorganic & medicinal chemistry letters*, *21*(8), 2415–2418. https://doi.org/10.1016/j.bmcl.2011.02.071.

Zhang, H.Y., & Yang, S.S. (2007). Characteristics of sword bean seeds germination (in Chinese). Crops *6*, 38–39.

Zhang, W., & Popovich, D. G. (2010). Group B oleanane triterpenoid extract containing soyasaponins I and III from soy flour induces apoptosis in HepG2 cells. *J. Agric. Food Chem*, 58, 5315–5319. https://doi.org/10.1021/jf9037979.

Zhu, J.B., & Wang, H.X. (2002). The comprehensive development and utilization of *Canavalia ensiformis* (in Chinese). Chinese Wild Plants Research *3*:17–19.

Zohary, D. (1970). Centers of diversity and centers of origin. In O. H. Frankel & F. Bennett (Eds.), *Genetic resources in plants* (pp. 33–42). Oxford: Blackwell Scientific Publications.

https://www.google.com/imgres?imgurl=https%3A%2F%2Fwww.ayurtimes.com%2Fwp-content%2Fuploads%2F2015%2F01%2FVigna-Subterranea.jpg&imgrefurl=https%3A%2F%2Fwww.ayurtimes.com%2Fbambara-groundnut%2F&tbnid=BGd9eJbv-_YVBM&vet=12ahUKEwjdm9b20sr3AhXAxqACHYsAAh0QMygFegUIARDnAQ..i&docid=Zs8lDLfLPcXIzM&w=765&h=510&q=vigna%20subterranea&hl=en&ved=2ahUKEwjdm9b20sr3AhXAxqACHYsAAh0QMygFegUIARDnAQ

https://www.google.com/imgres?imgurl=https%3A%2F%2Fupload.wikimedia.org%2Fwikipedia%2Fcommons%2Fthumb%2Fd%2Fd2%2FBambara_nut_unearthed..JPG%2F220px-Bambara_nut_unearthed..JPG&imgrefurl=https%3A%2F%2Fen.wikipedia.org%2Fwiki%2FVigna_subterranea&tbnid=Mn34ZrH0hlv8tM&vet=12ahUKEwjdm9b20sr3AhXAxqACHYsAAh0QMygAegUIARDdAQ..i&docid=fbBnjKITrR8UMM&w=220&h=147&q=vigna%20subterranea&hl=en&ved=2ahUKEwjdm9b20sr3AhXAxqACHYsAAh0QMygAegUIARDdAQ

Heuzé V., Thiollet H., Tran G., Edouard N., Lebas F., 2019. *African locust bean (Parkia biglobosa & Parkia filicoidea)*. Feedipedia, a programme by INRAE, CIRAD, AFZ and FAO. https://www.feedipedia.org/node/268 *Last updated on March 21, 2019, 10:22*

https://www.google.com/imgres?imgurl=http%3A%2F%2Fwww.atdforum.org%2Fjournal%2Fhtml%2F2009-34%2F9%2Fplate-1.jpg&imgrefurl=http%3A%2F%2Fwww.atd-forum.org%2Fjournal%2Fhtml%2F2009-34%2F9%2F&tbnid=axpfS6PNohM0jM&vet=12ahUKEwiFx8GO0Mr3AhVI_jgGHdJoA4oQMygKegUIARDRAQ..i&docid=UiX-nfgz5gmzgM&w=400&h=316&q=African%20yam%20bean%20flower&hl=en&ved=2ahUKEwiFx8GO0Mr3AhVI_jgGHdJoA4oQMygKegUIARDRAQ

https://www.google.com/imgres?imgurl=https%3A%2F%2Fpbs.twimg.com%2Fmedia%2FEyaeXh_U4AMcp3-.jpg%3Alarge&imgrefurl=https%3A%2F%2Ftwitter.com%2Ftravisparker91%2Fstatus%2F1379965091181850624&tbnid=1id5t0DdiwNlTM&vet=12ahUKEwiFx8GO0Mr3AhVI_jgGHdJoA4oQMygUegUIARDlAQ..i&docid=DiveWRXcQyGHjM&w=2048&h=1536&itg=1&q=African%20yam%20bean%20flower&hl=en&ved=2ahUKEwiFx8GO0Mr3AhVI_jgGHdJoA4oQMygUegUIARDlAQ

https://www.google.com/imgres?imgurl=https%3A%2F%2Fwww.slcu.cam.ac.uk%2Ffiles%2Fmedia%2Fyam_bean_plant_in_flower_tubers_and_bean_pods-nadia_radzman.png&imgrefurl=https%3A%2F%2Fwww.slcu.cam.ac.uk%2Fnews%2Fpeas-n-chips-creating-food-security-african-yam-bean&tbnid=hxK8ukYiYBgckM&vet=12ahUKEwif_92H0Mr3AhV1_TgGHf2LACoQMygGegUIARDbAQ..i&docid=R52CmmnkVCf5xM&w=800&h=400&q=African%20yam%20bean&hl=en&ved=2ahUKEwif_92H0Mr3AhV1_TgGHf2LACoQMygGegUIARDbAQ

https://www.agrifarming.in/yard-long-beans-farming-cultivation

https://www.healthbenefitstimes.com/yardlong-bean/

https://www.neseed.com/shop/vegetable-seeds/beans/pole-snap/yardlong/

https://en.wikipedia.org/wiki/Macrotyloma_uniflorum

https://nl.pinterest.com/pin/99219998010316968/

https://www.google.com/imgres?imgurl=https%3A%2F%2Fwww.hobbyseeds.com%2Fimage%2Fcache%2Fdata%2Fproducts%2Ftwin-flower-dolichos-dolichos-biflorus_0-250x250.jpg&imgrefurl=https%3A%2F%2Fwww.hobbyseeds.com%2Fdolichos-biflorus-horse-gram-100.html&tbnid=RjxFatVVCYveiM&vet=12ahUKEwiF4ZO7z8r3AhU4_TgGHe0YAhUQMygKegQIARAy..i&docid=LG3pYQUvTJiNuM&w=250&h=250&q=Dolichus%20uniflorus%20Lam.%20(Dolichus%20biflorus%20Auct.%20L.)%20pods&hl=en&ved=2ahUKEwiF4ZO7z8r3AhU4_TgGHe0YAhUQMygKegQIARAy

Heuzé V., Tran G., Hassoun P., Renaudeau D., Bastianelli D., 2015. *Velvet bean (Mucuna pruriens)*. Feedipedia, a programme by INRAE, CIRAD, AFZ and FAO. https://www.feedipedia.org/node/270 *Last updated on October 13, 2015, 13:42*

https://en.wikipedia.org/wiki/Phaseolus_polystachios

https://www.google.com/imgres?imgurl=https%3A%2F%2F4.bp.blogspot.com%2F-Ltil7-5REwY%2FWnbPPpCdFYI%2FAAAAAAAAHeQ%2FXM1rJ1cCg5soqoxMAYcgvt3NnlT8XpwjACLcBGAs%2Fs1600%2FP1151912.JPG&imgrefurl=http%3A%2F%2Funusualediblesandtheirwildrelatives.blogspot.com%2F2018%2F02%2Fwild-perennial-hardy-bean-phaseolus.html&tbnid=qgfV1-SFAjSYYM&vet=12ahUKEwj9o6nezcr3AhXH_jgGHSA0Ap0QMygAegUIARCoAQ..i&docid=hTZGyUdOt0Ay8M&w=1600&h=1066&q=Thicket%20bean%20seeds&hl=en&ved=2ahUKEwj9o6nezcr3AhXH_jgGHSA0Ap0QMygAegUIARCoAQ

http://www.southeasternflora.com/view_flora.php?plantid=82

https://truelyf.com/product/agathi-vegetable-hummingbird-powder/

https://www.shevalonvarmakalai.com/post/sesbania-grandiflora

https://www.google.com/imgres?imgurl=https%3A%2F%2Fmedhyaherbals.com%2Fwp-content%2Fuploads%2F2019%2F11%2FSesbania-Grandiflora-white-1.jpg&imgrefurl=https%3A%2F%2Fmedhyaherbals.com%2Fsesbania-grandiflora%2F&tbnid=7elgshtDu70spM&vet=12ahUKEwjY_MK1zMr3AhVo_jgGHZmUD-cQMygWegUIARCXAg..i&docid=nH_gGkGu_ibC7M&w=1200&h=900&q=agathi&hl=en&ved=2ahUKEwjY_MK1zMr3AhVo_jgGHZmUD-cQMygWegUIARCXAg

https://climbers.lsa.umich.edu/?p=416

https://www.fireflyforest.com/flowers/3492/phaseolus-acutifolius-var-tenuifolius-tepary-bean/

https://www.plant-ark.com/vegetables/tepary-bean-phaseolus-acutifolius.html

https://www.google.com/imgres?imgurl=https%3A%2F%2Fi0.wp.com%2Fhappyacres.blog%2Fwp-content%2Fuploads%2F2015%2F09%2Fteparybeanpod.jpg%3Fssl%3D1&imgrefurl=https%3A%2F%2Fhappyacres.blog%2F2015%2F09%2F14%2Ftepary-bean-update%2F&tbnid=8ncDzQaCJtrQ4M&vet=12ahUKEwjL7NSay8r3AhWTgGMGHVWlBLwQMygBegUIARC7AQ..i&docid=8imh1vT91DByMM&w=600&h=461&q=tepary%20bean%20pods&hl=en&ved=2ahUKEwjL7NSay8r3AhWTgGMGHVWlBLwQMygBegUIARC7AQ

https://awaytogarden.com/apios-americana-the-potato-bean-or-groundnut/

https://www.google.com/imgres?imgurl=https%3A%2F%2F1.bp.blogspot.com%2F-W2Cqxuptf1o%2FVLs49kJthyI%2FAAAAAAAALMM%2Fhxp4X4IEgsg%2Fw1200-h630-p-k-no-nu%2FPachyrhizus-erosus-Jicama.jpg&imgrefurl=http%3A%2F%2Fmedplants.blogspot.com%2F2015%2F01%2Fpachyrhizus-erosus-jicama-sankalu.html&tbnid=MqTkkWqD205m-M&vet=12ahUKEwie8djzysr3AhXN_jgGHbwjCLMQMygGegUIARDIAQ..i&docid=NjLqeQeNPL_meM&w=1200&h=630&q=Pachyrhizus%20erosus%20pods&hl=en&ved=2ahUKEwie8djzysr3AhXN_jgGHbwjCLMQMygGegUIARDIAQ

http://chengailimfruittrees.blogspot.com/2014/09/yam-bean.html

https://www.caribbeangardenseed.com/collections/yam/products/jicama-seeds-climbing-yam-bean-aka-mexican-potato-asian-vegetables

https://www.dreamstime.com/photos-images/yam-bean.html

https://www.eattheweeds.com/scarlet-runner-bean/

https://hort.extension.wisc.edu/articles/scarlet-runner-bean-phaseolus-coccineus/

https://nl.pinterest.com/pin/233272455672872966/

https://hort.extension.wisc.edu/articles/scarlet-runner-bean-phaseolus-coccineus/

Heuzé V., Tran G., Lebas F., 2020. *Moth bean (Vigna aconitifolia)*. Feedipedia, a programme by INRAE, CIRAD, AFZ and FAO. https://www.feedipedia.org/node/237 *Last updated on October 27, 2020, 14:38*

Heuzé V., Tran G., 2015. *Jack bean (Canavalia ensiformis)*. Feedipedia, a programme by INRAE, CIRAD, AFZ and FAO. https://www.feedipedia.org/node/327 *Last updated on September 7, 2015, 13:09*

https://alchetron.com/Canavalia-ensiformis

http://www.flowersofindia.net/catalog/slides/Horse%20Bean.html

https://www.checkyourfood.com/ingredients/ingredient/1194/winged-bean-leaves

https://www.exotic-seeds.store/ro/home/adzuki-bean-finest-seeds-vigna-angularis.html

https://balconygardenweb.com/how-to-grow-adzuki-beans-growing-adzuki-beans/

https://davesgarden.com/guides/pf/showimage/243397/

Chapter 4
Production Technology of Underutilized Crops of Cucurbitaceae Family

Khushboo Kathayat and Monisha Rawat

Introduction

Those crops/species whose potential is unexplored for different parameters like food security, health benefits, general uses, and income generation are known as underutilized crops (Jaenicke and Hoeschle-Zeledon 2006). In dicots, *Cucurbits* genus is the biggest genera belong to the family of Cucurbitaceae, having 118 genera and 825 species (Jeffery 1980). India is a primary and secondary center for many cucurbits like gourd and melon (Choudhury 1996). Some cucurbits like snapmelon, oriental pickling melon, chow-chow, and gherkin are underutilized. These underutilized crops are more famous in northeastern states of India; however, their popularity can be increased by promotions of their health benefits. Their medicinal and nutritional uses are not known to people despite that these vegetables are rich in nutritional source (Parvathi and Kumar 2002). These vegetables are less costly, easily available, and contain different minerals and vitamins. These crops are hardy, can be grown in simple soils, and with minimum inputs (Hughes and Ebert 2013). This chapter deals with nutritional uses, medicinal benefits, production technology, and plant protection of different underutilized *Cucurbits* species.

Origin and Distribution

Knowing about the origin, history, and process of domestication increases the chances of better cultivar development by exploiting the genetic diversity.

K. Kathayat (✉) · M. Rawat
Department of Horticulture, School of Agriculture, Lovely Professional University, Phagwara, Punjab, India
e-mail: khushboo.22214@lpu.co.in

Savita et al. (eds.), *Production Technology of Underutilized Vegetable Crops*,
https://doi.org/10.1007/978-3-031-15385-3_4

Snapmelon

India is considered as origin of snapmelon (*Cucumis melo* var. *momordica*, $2n = 2x = 24$) (Duthie 1905; Vishnu-Mittre 1974). It is cultivated in Southeast Asia, i.e., Myanmar and Vietnam. It is widely cultivated in various Indian states such as Rajasthan, Gujarat, Punjab, Haryana, Uttar Pradesh (UP), West Bengal (WB), and some other northeastern states. It is commonly called as "phut," which means "to split," whereas other names are phootkakari, pottuvellari, or kakadia. It is a hardy and short-duration indigenous crop with much resistance to biotic and abiotic stress (Maurya et al. 2006).

Oriental Pickling Melon

Oriental pickling melon (*Cucumis melo* var. *conomon* Thunberg, $2n = 2x = 24$) (Munshi and Alvarez 2005) was reported to be originated from wild melon (var. *agrestis*) (Walters 1989) and is considered an ancient melon which is grown in China (Walters and Wehner 1994) from 1000 and 500 years BCE. It is cultivated in different parts of the world (Japan, Korea, and Southeast Asia). In India, it is particularly grown in Southern parts of India. It is also known by other names like Japanese pickling melon, golden melon, culinary melon, Indian yellow cucumber, yellow cucumber, and lemon cucumber.

Chow-Chow

Chow-chow or chayote (*Sechium edule*, $2n = 2x = 28$) is a single-seeded, perennial, viviparous vegetable (Aung et al. 1990) of which leaves, stem, and immature part are edible. The origin place is the mountainous cool region of Central America. It was first grown in the Aztecs Empire (Newstrom 1991) from where it was introduced in India by Western Missionaries (Singh et al. 2012). It is the most popular vegetable in the Northeastern states as well as in different parts of Kerala, Tamilnadu, and Karnataka (Sanwal et al. 2008). High moisture content (89.3–94.2%) and soft flesh are important features of this crop (Mishra and Das 2015).

Gherkin

Gherkin/burr gherkin/West Indian gherkin/pickling cucumber (*Cucumis anguria*, $2n = 2x = 14$) is a trailing warm-season vine originated from Southern Africa. In India, it was introduced for export purposes in the late 1980s (Anonymous 1975).

Fruits are served raw, cooked, or pickled. Ninety percent of gherkin preserve is exported from India by Karnataka state, which grows this crop mainly in contract farming (Singh 2000).

Nutritional Importance and Uses

Many underutilized cucurbits vegetables are an important part of the human diet in cooked form or raw form. They are also nutritionally rich vegetables. Many health benefits are mentioned in different literature.

Snapmelon

One hundred grams of fruit contains 79.93 g of moisture, 15.60 g carbohydrates, 0.369 g crude proteins, 1.12 g fats, 1.34 g crude fibers, 1.64 g total ash, 0.76 mg Ca, 0.088 mg P, 0.843 mg Fe, and 0.2023 mg Zn (Samadia 2003). Ripe fruits are consumed as a dessert, raita, or salad, whereas non-sweet types are consumed as a vegetable. It is also grown as an ornamental plant. Its seed kernel is also used for the preparation of sweets and bakery products. A popular light traditional drink, i.e., *Thandai*, is also prepared from its ripe fruit (Pareek et al. 1999). It is used for the moisturization of skin and as a healing agent for burns and abrasions. It is also used to cure stomach pain and helps in relieving constipation and vomiting. It is cultivated as a mixed crop along with maize, sorghum, and pearl millet in the rainy season or as a sole crop in the summer season (Seshadri and More 2001). Khelra, a dry preserved product, is prepared from mature fruits after peeling off the skin (Pareek and Samadia 2002).

Oriental Pickling Melon

It is a good source of vitamins (vitamin A and vitamin C), minerals (calcium, phosphorus, and iron), and also contains different types of antioxidants such as beta-carotene. It does not contain so much sugar, hence it is good for pickling purposes.

It is also used in the preparation of chutney, vegetables, curry, and pickles. Fruits are also used for the preparation of skin moisturizers as well as used to resolve digestion-related problems (George 2008).

Chow-Chow

It is a rich source of vitamins (B_6 and K), minerals (potassium, manganese, zinc, and copper), several amino acids, and rich in antioxidants (Ordonez et al. 2006). It is good for pregnant women and control diabetes (Maity et al. 2013), reduces hypertension (Lombardo-Earl et al. 2014), control ulcers (Sateesh et al. 2012), and has antimicrobial properties (Ordoñez et al. 2003) (Table 4.1).

All plant parts like a fruit, tender shoots, leaves, and tuberize roots (like a potato) of chow-chow are edible. It is consumed either alone in the form of vegetables, mixed with meat, or in soups (Sharma et al. 1995; Singh et al. 2012). Its older leaves are used as feed for pigs. It is also an important part of baby food, juices, sauces, and pastes.

Gherkin

It is a good source of water, vitamins A, B, C, K, manganese, potassium, magnesium, calcium, silica, and fiber. It is also very low in fat content, so it is good for weight loss. Pickles, vegetables, and curry are prepared from unripe fruits (Purseglove 1969). It is also used for the treatment of different diseases like jaundice, stomach pain, and kidney stone (Baird and Thieret 1988).

Table 4.1 Chemical composition (per 100 g) of chow-chow

Component	Fruit	Shoot and leaf	Tuberize root
Water (%)	94	90	80
Calorie (kcal)	28	60	79
Starch (mg)	200	700	13,600
Protein (mg)	1000	4000	2000
Fat (mg)	200	400	200
Carbohydrate (mg)	5600	4700	17,800
Fiber (mg)	700	1200	400
Ashes (mg)	500	1200	1000
Calcium (mg)	15.5	58.0	7
Phosphorus (mg)	17.0	108.0	34
Iron (mg)	0.40	2.50	0.8
Vitamin A (g)	0.005	0.62	–
Thiamin (mg)	0.03	0.08	0.05
Riboflavin (mg)	0.04	0.18	0.03
Niacin (mg)	0.50	1.10	0.9
Vitamin C (g)	0.016	0.016	19.0

Source: Aung et al. (1990) and Verma et al. (2014)

Botany and Taxonomy

Snapmelon

It is an annual climber growing to a height of 1.5 m and has a shallow taproot system. Leaves are simple or somewhat palmately lobed, arranged alternately. It contains monoecious flowers – staminate flowers in cluster and the pistillate flowers are in solitary. Five lobed calyx and corolla are found. Flower and seed ripening are observed in July–September and August–October, respectively. Fruit type is pepo (average weight 700 g) having flesh color cream and yellow to orange having different shapes of fruits (Dhillon et al. 2007). It is an entomophilous crop and pollination is generally done by the honeybee.

Oriental Pickling Melon

It is a climbing dioecious vine and the length of the vine ranges from 100 to 200 cm. The vines bear dark, spiny leaves and sericeous ovary. The flower color is yellow and blooms in August. Fruits are white, green and has an oblong shape. The length of the fruit is found to be 20–30 cm.

Chow-Chow

It is also called chayote, Cho-Cho, vegetable pear, and custard marrow. It is a perennial vine that climbs by clinging with tendrils and has tuberous roots. The leaves are triangulated about 5–8 in length, with shallow lobes along with 10–15 cm long sulcate petioles and 3–5 divided tendrils. The flowers are pentamerous and unisexual, with small whitish flowers are observed. Racemose inflorescence (10–30 cm long) has male flowers in groups, while female flowers are in solitary. The fruits are pear-shaped wrinkled fruits (a mixture between potato and cucumber) in which large-sized seed is present, which is consumed along with the flesh. Different shapes, color, sizes of fruits are available (Rai et al. 2006). The fruit color is pale apple-green and the size varies between 10 and 15 cm in length.

Gherkin

Palmately lobed leaves have a length of 2.5 m. Small flowers and fruits are approximately 5 cm long.

Agronomic Practices

Snapmelon

The optimum temperature for seed germination and growth is 18–25 °C and 15–32 °C, respectively. Frost or low temperature (leads to femaleness) and very high temperature (leads to maleness) is not desirable for its growth. Well-drained organic matter rich sandy loam or loamy soil is recommended for cultivation. It requires a pH of 6.5–7.5. It is sensitive to acidic soil.

In Goa, Konan areas of Maharashtra, and Karnataka, snapmelon are planted in months of June–July, whereas in Rajasthan and West Bengal seed sowing is done in months of January–March. February–March is the optimum time for seed sowing in Kerala and Tamilnadu. Hill sowing (2–3 seeds/hill) of seed (2–3 kg/seed) is the best method. Row-to-row distance and hills in-row spacing are 1.5–2.0 m and 75–100 cm.

Well decomposed farmyard manure (FYM) should be incorporated @ 20–25 t/ha at the time of field preparation. In addition to this, nitrogen @ 40 kg/ha in two splits (half dose of nitrogen is applied at the point of field preparation whereas the rest half at flowering time) is applied. It is a moderately deep-rooted crop and requires irrigation at an interval of 5–7 days interval if it is grown as a summer season crop. However, in the time of rainy season, irrigation is not applied unless a long dry spell is observed. Light hoeing and weeding are essential operation in the early stages of growth. Preplant application of fluchloralin @ 1.2 kg/ha is done if weeds are not controlled by weeding. Mulching is also applied in the field to reduce water evaporation, soil temperature as well as reduce weed growth. Plant growth regulators are also applied to increase fruit set, fruit quality, and yield. Spraying of GA3 @ 20 ppm at the 2–4 leaf stage increases yield.

Oriental Pickling Melon

It requires an optimum temperature of 24–27 °C (min. 18 °C). Extreme low temperature is not feasible for growth. Sandy loam soil (rich in organic matter) having pH of 6.0–6.7 (plants can prefer pH up to 8) is feasible for its production. Sowing can be done from June or January to April. Furrow method (30–45 cm distance) of seed sowing is done. Pit of 45 cm^3 size, filled with FYM and soil are used for sowing of seeds (5–6 seeds per pit). A total of 0.75 kg seed is required for pit sowing spaced at 2.00 m × 1.50 m. Only three healthy plants are allowed to grow, and the remaining is removed. Apply basal dose of 40 t/ha FYM along with 150:75:75 kg NPK/ha. Different biofertilizers like 2 kg/ha each of *Azospirillum* and *Phosphobacteria* along with 100 kg neem cake is applied. At an interval of 3–4 days, water is applied in furrows in the initial period of growth, whereas alternate days are in the flowering and fruiting stage. At the time of fertilizer application, weeding by hand/hoe should be done and when needed. Earthing up is an important operation, which is performed during the rainy season.

Chow-Chow

Warm season (30 °C optimum temperature) is suitable for its cultivation. Fruit can tolerate full sunlight in 12 h of the day (yellow color fruit) to shade conditions (dark green fruits, used as a vegetable). It does not tolerate dry as well as frost conditions. It is grown at an elevation of 1200–1500 MSL (Engels and Jeffrey 1993; Saade 1996). Well-drained soil having 5–6.5 pH is most suitable for its production. The best time of sowing is April–May or the rainy season. Sowing is done in 50 cm^3 size pits (2–3 sprouted fruit/pit) filled with soil and FYM at a spacing of 6 feet × 9 feet. For sowing in 1 ha of land, 1500–1600 sprouted vegetables are required. A total of 50–100 kg N (half nitrogen as basal and half nitrogen in flowering stage), 40–60 kg P (full), and 30–60 kg K (full) are applied into the field. Water is applied at an interval of 3–4 days in summer and a weekly interval in winter. Drip irrigation is a popular method of irrigation in the greenhouse. Weeding is done as per requirement. Trellis's method of training is followed. Trellis is prepared from bamboo, and the height of the structure is such that any one person can walk easily below that. A Bower system (Criss Crosswire network at height of 5 feet) is also used. It is done at the end of each fruiting season (2 seasons per year). Only 1.5 m of the stem is left in each season.

Gherkin

A minimum temperature of approximately 12 °C is required for germination (frost is not desirable), whereas 15 °C is required for flowering. Well-drained soil having a 6.0–6.8 pH is desired. The optimum time of sowing of gherkin is February–March and June–August. Hill sowing (2 seeds per hill) of seeds @ 800 g/ha (treated with 4 g *Trichoderma viride*, 10 g *Pseudomonas*, or 2 g per kg carbendazim seed) at 30 cm spacing is done. Light irrigation is given immediately after sowing. Seeds germinate within a week after sowing. Gap filling is done after 10 days of germination. A total of 25 t/ha FYM in basal along with 150:75:100 kg NPK/ha (three doses, i.e., basal, 3 and 5 weeks after sowing) is required. After 25 days of sowing, earthing up is done and it also requires support after 15–20 days of sowing (Table 4.2).

Table 4.2 Propagation material and popular varieties of major underutilized cucurbits species

Cucurbits sp.	Propagation	Varieties
Snapmelon	Seeds	Pusa Shandar, ASH 10, ASH 82, Shantiniketan SM-1, Konkan Madhur
Oriental pickling melon	Seeds	Sambhagya, Mudichode, Arunima
Chow-chow	Seeds (whole fruit/ vegetable is planted as seed)	Broad green, pointed green, oval greens are Florida green and Monticello white
Gherkin	Seeds	Mejorado de Bourbonne, no de Meaux Pequeño Verde de Paris, Verde de Massy, Verde Grueso, Vorgebirge, Winsconsin, along with the hybrid varieties Epros mix, Fanto, Hyclos mix, Levo, Parafin mix, Parigyno, Pioneer, Uwy, Vorifin, Wisco, Witlo

Harvesting, Yield, and Postharvest Management

Snapmelon

Fruits are usually harvested when the fruit color of fruit skin changes from green to cream or dull yellow and before crack starts on skin. It means that fruit is harvested at the half-ripe stage because at full ripe stage fruit bursts and it is not fit for marketing purposes. If fruits start cracking, individual fruits are to be tied by keeping arecanut sheath or like materials on one side or both sides of fruits. Fruits should be harvested in the morning hours before fruits get warm. Fruit storage quality is very poor. The yield of snapmelon fruit is 150–200 q/ha. If the snapmelon is grown as an intercrop, it gives a yield of about 60–80 q/ha as an intercrop. Seed yield is 1–3 q/ha. At room temperature, storage life of perishable snapmelon fruits are 2–4 days. However, at 2–4 °C temperature and 85–90% relative humidity, storage life of the fruits is 2–3 weeks.

Oriental Pickling Melon

The plant starts fruiting 45 days after sowing. Fruits are harvested after attaining full size and when the color changes from green to golden yellow. The yield of pickling melon fruit is 250 q/ha, whereas seed yield is 175–250 kg/ha.

Chow-Chow

Fruit maturity started 10–12 days after anthesis. It is characterized by no pubescence existing on the skin, cylindrically shaped fruits, and soft seeds, which means that the fruit is mature. Fruit harvesting starts around 50–55 DAS (days after sowing) and is done usually at an interval of 3–4 days. It is done by sharp knife by keeping a small part of fruit stalk along with fruit. Avoid injury of fruits and vine while harvesting. The yield of OP (open pollinated) varieties is 200–250 g/ha, whereas hybrid gives 400–500 g/ha. Cool and moist conditions for 3–5 days are required for storage.

Gherkin

The maturity period of fruit is 30–35 days from the date of transplanting. Harvesting should be done in the evening time or early morning at a frequent interval of 1 day. A little stalk should remain on the plant. Throughout the period, 10–12 t/ha of fruits were obtained. Immediately after harvesting, fruits are shifted in shade on the same day. No water is applied to the harvested product because if any traces of water remain on the surface of the fruit it would lead to different fungus problems. A jute bag (avoid plastic bag) is the best material for the collection of harvested products. Gherkin is mainly exported to countries such as the United States, Belgium, Spain, France, and Russia in the form of ready-to-eat form packed in HDPE (High density polyethylene) drums and smaller packs of jars and cans.

Plant Protection

Important diseases and insect pest are powdery mildew, downy mildew, red pumpkin beetle, aphids, and mites, which is explained in Table 4.3.

Table 4.3 Important disease and insect pest of Cucurbitaceae family diseases

Name	Causal organism	Symptoms	Control
Powdery mildew	*Erysiphe cichoracearum*	The disease appeared as white-colored powdery spots which are formed on leaf surfaces, shoots, flowers, and fruit. Later, these spots enlarged and cover a large portion of affected parts	Good sunlight and air circulation is needed as well as to avoid extra use of fertilizer. For control, chemically Karathane @ 0.1% is applied at an interval of 14 days

(continued)

Table 4.3 (continued)

<table>
<tr><th>Name</th><th>Causal organism</th><th>Symptoms</th><th>Control</th></tr>
<tr><td>Downy mildew</td><td>Pseudoperonospora cubensis</td><td>On the lower side of leaves, black grayish downy fungal growth is observed. In severe stage, rolling of leaves occurs</td><td>Wider spacing, proper air circulation, and a spray of 0.2% metalaxyl are done</td></tr>
<tr><td colspan="4">Insect pests</td></tr>
<tr><td>Red pumpkin beetle (Aulacophora foveicollis)</td><td>Grubs (root feeder), beetle (cotyledons and foliage), and adults (seedling stage) attack the crop. In later stages, the whole destruction of the crop is observed</td><td colspan="2">Repeat ploughing of the field in summer seasons is done. Application of Rogor 0.1% twice, 0.01% NSK extract, and 0.4% neem oil is recommended</td></tr>
<tr><td>Mites</td><td>The sap of leaves is sucked by nymphs and adults. Because of that curling of leaves occurs. The color of leaves changes from green to dark gray and its petiole increases their size. Therefore, a reduction in flowering and yield is observed. Hard white strips are also observed in later stages</td><td colspan="2">Application of Phosalone (3 ml/l), wettable sulfur (3 g/l of water), Dicofol 5 ml/l of water, phorate 10% G @ 10 kg/ha, and Dimethoate 30% EC 1.0 ml/l is done. Amblyseius ovalis predator is used to control</td></tr>
<tr><td>Aphid</td><td>The sap is sucked by the aphid; therefore, a sweety substance releases, which is a good agent for ants and ultimately it results in sooty mold development. After that fruit color changes to black</td><td colspan="2">Spray alternatively with 0.1% Dimethoate or Methyl demeton (2 ml in 1 l of water) or 1.5 ml or Acephate (1 g in 1 l of water) at 10 days intervals</td></tr>
</table>

References

Anonymous. (1975). Gherkin (*Cucumis sativus* L.). *Farm Digest*, 1–3:6–8.

Aung, L. H., Ball, A. and Kushad, M. (1990). Developmental and nutritional aspects of chayote (*Sechium edule*, Cucurbitaceae). *Economic botany*, 44(2), 157–164.

Baird, J. R., & Thieret, J. W. (1988). The bur gherkin (*Cucumis anguria* var. *anguria*, Cucurbitaceae). *Economic Botany*, 447–451.

Choudhury, B. (1996). Vegetables. National Book Trust, New Delhi, India.

Dhillon, N. P. S., Ranjana, R., Singh, K., Eduardo, I., Monforte, A. J., Pitrat, M., Dhillon, N.K. & Singh, P. P. (2007). Diversity among landraces of Indian snapmelon (*Cucumis melo* var. *momordica*). *Genetic Resources and Crop Evolution*, 54(6), 1267–1283.

Duthie, J. F. (1905). Flora of the upper Gangetic plain and the adjacent Sivalik and sub Himalyan tracts, vol. I Ranunculaceae to Companulaceae. Office of the Superintendent of Government of Printing, Calcutta, 41–43.

Engels, J.M.M., & Jeffrey, C. (1993). *Sechium edule* (Jacq.) Swartz. In: Siemonsma, J.S. and Piluek, K. (eds) Plant resources of south-east Asia Vol.8-Vegetables. Pudoc Scientific Publishers, Wageningen, pp 246–248.

George, T.E. (2008). Kanivellari Visual treat, Kerala Calling Monthly, 3:36–37.

Hughes, J.D.A., & Ebert, A.W. (2013). Research and Development of Underutilized Plant Species: The Role of Vegetables in Assuring Food and Nutritional Security. In Proceedings of the 2nd International Symposium on Underutilized Plant Species: Crops for the Future—Beyond Food Security; Massawe, F., Mayes, S. and Alderson, P., Eds.; International Society for Horticultural Sciences (ISHS): Korbeek-Lo, Belgium, 2:79–91.

Jaenicke, H., & Hoeschle-Zeledon, I. (2006). Strategic Framework for Underutilized Plant Species Research and Development, with Special Reference to Asia and the Pacific, and to Sub-Saharan Africa. ICUC, Colombo and GFU, Rome 33p.

Jeffrey, C. (1980). A review of the Cucurbitaceae. *Botanical Journal of the Linnean Society*, 81(3), 233–247.

Lombardo-Earl, G., Roman-Ramos, R., Zamilpa, A., Herrera-Ruiz, M., Rosas-Salgado, G., Tortoriello, J., & Jiménez-Ferrer, E. (2014). Extracts and fractions from edible roots of *Sechium edule* (Jacq.) Sw. with antihypertensive activity. Evidence-Based Complementary and Alternative Medicine, 2014.

Maity, S., Firdous, S. M., & Debnath, R. (2013). Evaluation of antidiabetic activity of ethanolic extract of *Sechium edule* fruits in alloxan-induced diabetic rats. *World J Pharm Sci*, 2(5), 3612–3621.

Maurya, I. B., Arvindakshan, K., Sharma, S. K., & Jalwania, R. (2006, December). Status of indigenous vegetables in southern part of Rajasthan. In *I International Conference on Indigenous Vegetables and Legumes. Prospectus for Fighting Poverty, Hunger and Malnutrition 752* (pp. 193–196).

Mishra, L. K., & Das, P. (2015). Nutritional evaluation of squash (*Sechium Edule*) germplasms collected from Garo Hills of Meghalaya-Northeast India. *International Journal of Agriculture, Environment and Biotechnology*, 8(4), 971.

Munshi, A.D., & Alvarez, J.M. (2005). Hybrid melon development. *Journal of new seeds*, 6(4), 321–360.

Newstrom, L. E. (1991). Evidence for the origin of chayote, *Sechium edule* (Cucurbitaceae). *Economic Botany*, 45(3), 410–428.

Ordonez, A. A. L., Gomez, J. D., & Vattuone, M. A. (2006). Antioxidant activities of Sechium edule (Jacq.) Swartz extracts. Food chemistry, 97(3), 452–458.

Ordoñez, A. A. L., Gómez, J. D., Cudmani, N. M., Vattuone, M. A., & Isla, M. I. (2003). Antimicrobial activity of nine extracts of *Sechium edule* (Jacq.) Swartz. Microbial ecology in health and disease, 15(1), 33–39.

Pareek, O. P., Vashistha, B. B., & Samadia, D. K. (1999). Genetic diversity in drought hardy cucurbits from hot arid zone of India. *IPGRI Newslett. Asia Pacific Oceania*, 28, 22–23.

Pareek, O.P., & Samadia, O.K. (2002). For arid Zone farmers Promising indigenous cucurbit varieties. *Indian Journal of Horticulture,* 47(2), 15–18.

Parvathi, S., & Kumar, V. J. F. (2002). Studies on chemical composition and utilization of the wild edible vegetable athalakkai (*Momordica tuberosa*). *Plant Foods for Human Nutrition*, 57(3), 215–222.

Purseglove, J.W. (1969). Tropical Crops Dicotyledons-1. Longmans Green and Co. Ltd,, 109–110.

Rai, N., Sanwal, S.K, Yadav, R.K., & Phukan, R.M. (2006). Diversity in Chow-chow in north eastern region. *Indian Hort*, 51(2), 11–12.

Saade, R.L. (1996). Chayote. *Sechium edule* (Jacq.) Sw.: Promoting the conservation and use of underutilized and neglected crops. International Plant Genetic Resources Institute, Rome, Italy.

Samadia, D. K. (2003). Snap melon Production Technology for arid and semi-arid regions. *Int. J. Agri*, 23–29.

Sanwal, S. K., Yadav, R. K., Singh, P. K., & Rai, N. (2008). Variability and genetic diversity studies in indigenous chow-chow genotypes of northeast India. *Indian Journal of Horticulture*, 65(2), 167–170.

Sateesh, G., Hussaini, S. F., Kumar, G. S., & Rao, B. S. S. (2012). Anti-ulcer activity of *Sechium edule* ethanolic fruit extract. *The Pharma Innovation,* 1(5, Part A), 77.

Seshadri, V. S., & More, T. A. (2001, September). Indian land races in *Cucumis melo*. In *II International Symposium on Cucurbits 588* (pp. 187–193).

Sharma, M.D., Newstrom-Lloyd L, & Neupane, K.R. (1995). Nepal's new chayote gene bank offers great potential for food production in marginal lands. Diversity. 11, 7–8.

Singh, B. K., Pathak, K. A., & Ngachan, S. V. (2012). Exploring underutilized chow–chow in Mizoram. *Indian Hortic*, 57(5), 3–5.

Singh, S. (2000). Contract farming for agricultural diversification in the Indian Punjab: A study of performance and problems. *Indian Journal of Agricultural Economics*, 55(3), 283–294.

Verma, V. K., Jha, A. K., & Singh, B. K. (2014). Nutritional properties of different fruit parts of popular chow–chow genotype grown in NEH Region of India. *Veg Newslett*, 1(1), 7.

Vishnu-Mittre. (1974). Palaeobotanical evidence in India. In Evolutionary Studies in World Crops, Diversity and Changes in Indian Subcontinent, J. Hutchinson, ed. (Cambridge, UK: Cambridge Univ. Press), p.3–30.

Walters, S. A., & Wehner, T. C. (1994). Evaluation of the US cucumber germplasm collection for root size using a subjective rating technique. *Euphytica*, 79(1), 39–43.

Walters, T. W. (1989). Historical overview on domesticated plants in China with special emphasis on the Cucurbitaceae. *Economic Botany*, 43(3), 297–313.

Chapter 5
Production Technology of Underutilized Vegetables of Dioscoreaceae Family

Sandeep Kaur and Monisha Rawat

Introduction

Potato yam or true yam is an underutilized tuber crop as it is not usually grown. The scientific name of potato yam is *Dioscorea bulbifera* L., which belongs to the family Dioscoreaceae. It is commonly known as air potato, cheeky yam, wild yam, bitter yam, aerial yam, and bulbil yam. It can be quickly grown anywhere because of its broader adaptation under all types of climatic conditions. It can be easily seen in the south-eastern United States and West Indies. It is originated in Florida and is used to make medicines approximately 115 years ago (USDA.gov). It has medicinal properties that can cure sores, diabetes, wound, ulcers, etc. (Shriram et al. 2008). This plant is a source of ayurvedic medicines used by Europeans and Unani to prevent various diseases. Owing to its elevated demand for Ayurvedic medicines and fewer side effects, it spreads worldwide (Kumar et al. 2016; Dhiman and Bhattacharya 2020). Air potato accounts for many health benefits but is still not consumed by the people (Afiukwa and Igwe 2015). The bulbils are an excellent source of nutrients like K, Ca, Cl, and Na and can be used for food processing because these minerals are suitable for human health (Gutierrez 2021; Kumar 2011).

The fruits of its species are spicy and bitter, but the fruits of cultivated species have a sweeter taste after boiling. Humans do not prefer it, so market demand is significantly less (http://issg.org/) (ifas.ufl.edu). It is a herbaceous vine (Sonibare and Adeniran 2014) that spreads rapidly under warm conditions. It is a militant weed of Florida that favors forestland areas. It cannot tolerate saline and marine conditions (Hammer 1998) (Table 5.1).

S. Kaur · M. Rawat (✉)
Department of Horticulture, School of Agriculture, Lovely Professional University, Phagwara, Punjab, India
e-mail: monisha.22692@lpu.co.in

Savita et al. (eds.), *Production Technology of Underutilized Vegetable Crops*,
https://doi.org/10.1007/978-3-031-15385-3_5

Table 5.1 The international and local common name of potato yam

Countries	Common name
Spanish	Papa voladora
Chinese	Huang du
German	Knollen yam
India	Hoei-oepas
Malaysia	Kaachil
USA	Hoi
Brazil	Cara-de-espinho
Portuguese	Batata-de-rama

Source: https://www.cabi.org

Origin and Distribution

Potato yam is commonly found in approximately 67 countries of Florida, from northern to western regions and significantly in southern areas (Jameson 2001). It was introduced from West Africa basically for its edible bulbils. After that, it was domesticated and brought under cultivation for human consumption in both Africa and Asia (Hammer 1998; Florida Exotic Pest Plant Committee Council 2009). It occurs as a wild species and is considered a minor vegetable crop. It is included in the oldest group of angiosperms. Potato yam was introduced as an ornamental plant in North America and Florida by USDA scientists. Now, it is widely cultivated in subtropical and tropical areas of the West Indies, Pacific Islands, and America (ISSG 2012; USDA-ARS 2012). Dioscoreaceae family has 600 species but only ten are cultivated as a vegetable and considered edible. Odisha is the main center of wild species of *Dioscorea*, commonly known as banal, gotcha all, pita Kanda, kukuralu, and pita all. Sri Lanka, India, Nepal, and Bhutan are the countries where potato yam is grown. In Australia, it is found growing in the Northern region, Queensland, and Western region (Martin and FKS 1974). It has many species like white yam (*Dioscorea alata*), identified by its pink-colored winged stem in Florida. It appears like an air potato (Hammer 1998). In other countries like China, it is found commonly in southern regions such as Henan, Gansu, Anhui, Guizhou, etc. (EBCMM 1999; Guan et al. 2017) (Table 5.2) (Fig. 5.1).

Nutritional Importance and Uses

Potato yam is considered a nutritious food. Cooked or roasted bulbils of *Dioscorea* are consumed as a vegetable and are used to cure diabetes, dysentery, ulcers, cough, etc. (Dutta 2015). It contains 20.38 mg/g allantoin (Lebot et al. 2019). The fermented and unfermented bulbils of potato yam contain 2.50 mg/g, 1.16 mg/g, and

Table 5.2 Different species with common names of potato yam

Various species of *Dioscorea*	Common name
Dioscorea batatas	Chinese yam
Dioscorea deltoidea	Yam
Dioscorea japonica	Japanese yam, glutinous yam
Dioscorea tokoro	–
Dioscorea villosa	Wild yam
Dioscorea alata	Greater yam, white yam, purple yam, water yam
Dioscorea cayennansis	Yellow guinea yam, yellow yam
Dioscorea esculenta	Wild yam, lesser yam, potato yam, Chinese yam
Dioscorea trifida	Sweet yam, cush-cush yam
Dioscorea communis	Black bryony

Source: https://pfaf.org

2.57 mg/g of tannins, phenols, and saponins, respectively. A large group of secondary metabolites is present in the plants. Apart from this, many terpenoids occur in the form of glycosides, which are used by food processing companies based on spices, flavors, and natural perfumes (Guo et al. 2018). A study revealed that 0.70–10.81 mg/g of nonessential and essential amino acids are present in potato yam. For example, potato yam fermented with *Pleurotus ostreatus* contains a high amount of alkaloids and flavonoids (3.05 mg/g and 2.77 mg/g, respectively), and unfermented potato yam contains 2.57 mg/g saponins, 1.16 mg/g phenols, and 2.50 mg/g tannins (Bolaniran et al. 2019). The flour of potato yam contains 4.63% minerals and 7.38% fats. Drying potato yam flour at three different temperatures viz. 50 °C, 60 °C, and 70 °C. It was observed that with increased temperature, the carbohydrate percentage was also enhanced by 69.81%, 73.97%, and 74.34%, respectively (Rudito et al. 2018) (Table 5.3).

Potato yam is used to make flour by drying, and flour can be used as an ingredient of various value-added products like making cakes, noodles, bread, etc. (Rudito et al. 2018). Its bulbils are used to make French fries and chips (Table 5.4).

Botany and Taxonomy

The Dioscoreaceae family includes *Tacca*, *Dioscorea*, *Trichopus*, and *Stenomeris* (Caddick et al. 2002). In the family, *Dioscorea* is the largest genus with 350–400 species (Caddick et al. 2002) distributed in the tropical regions of Australia, Asia, Africa, and America (Ayensu and Coursey 1972). It is further subdivided into 60 sections (Knuth 1924). There are four foreign species viz. *Dioscorea sansibarensis*, *Dioscorea alata*, *Dioscorea polystachya*, and *Dioscorea bulbifera* (Al-Shehbaz and Schubert 1989; USDA, NRCS 2002; Wunderlin and Hansen 2011) (Tables 5.5 and 5.6).

Fig. 5.1 Different species of *Dioscorea*

Table 5.3 Nutritional component of air potato bulbils from Nueva Vizcaya and Quirino

Nutrient content	Amount
Total lipid (g/100 g)	0.1593–0.2032
Moisture (%)	17.52–26.84
Total protein (mg/ml)	1.0
Total ash content (g/100 g)	0.737–7.28
Crude fiber (g/100 g)	0.29–2.7942
Total carbohydrates (g/100 g)	0.8189–0.8721

Source: Gutierrez (2021)

Potato yam is an evergreen vine with more fibrous roots that can grow from 10 to 20 m in length (Langeland and Meisenburg 2008; Lim 2016). The vine climbs on other trees and sometimes suppresses the whole tree because of its faster growth rate, i.e., up to 8 inches a day, and can reach up to 65 feet (Langeland and

Table 5.4 Uses of *Dioscorea bulbifera*

Plant parts	Uses	References
Tuber	Used to cure skin infections	Tiwari and Pande (2006)
	Dried bulbils powder is used as a contraceptive	Swarnkar and Katewa (2008), Kamble et al. (2010), Das et al. (2014)
	Crushed and roasted bulbils mixed with salt can cure cough	Singh et al. (2009)
	Bulbils can be used to heal throat pain	Mbiantcha et al. (2011)
	Bulbils are cut into pieces and dried pieces can cure piles Boiled bulbils are used as a remedy to relieve abdominal pain	Abhyankar and Upadhyay (2011)
	In Uganda, boiled bulbils are used to cure HIV patients	Nabatanzi and Nakalembe (2016)
	Raw bulbils are used to increase appetite	Mishra et al. (2008)
	Bulbils boiled a couple of times with ash are used to treat tuberculosis	Sharma and Bastakoti (2009)
Leaves	Paste of leaves used to treat skin infections	Girach et al. (1999)
Stem	Crushed tender shoots and twigs used on the scalp for dandruff problems	Dutta (2015)

Source: Kundu et al. (2020)

Table 5.5 Classification

Kingdom	Plantae
Phylum	Spermatophyta
Subphylum	Angiospermae
Class	Monocotyledonae
Order	Dioscoreales
Family	Dioscoreaceae
Genus	*Dioscorea*
Species	*Dioscorea bulbifera*

Source: www.cabi.org

Burks 1998). During summer, the vine produces rhizomes (vegetative propagules) called "bulbils" or "air potato." Each vine can produce approximately 100 bulbils, and the bulbils that drop from the vine again sprout in the following year giving rise to numerous bulbils (Kerr and Poffenberger 2016). Leaves of potato yam are arranged alternately on the stem (Kerr and Poffenberger 2016). Leaves are heart-shaped, 10–20 cm long, 5–15 cm wide with narrow, prolonged tips, and shiny (Coursey 1967).

The stem is nonwoody, smooth, and round. Stem twines from right to left and takes the support of other plants. The stem and petiole are often purple-colored

Table 5.6 Some *Dioscorea* species of North America

Species	Bulbils present or not	Characters	Origin
Dioscorea alata (winged yam)	Pear-shaped	Square-shaped stem twines left to right	Asia
Dioscorea polystachya (Chinese yam)	Bulbils are present	Short projections present at the tip of leaves having three lobed margins	India
Dioscorea floridana (Florida yam)	Bulbils not present	Less plant height, yellowish rhizomes, and articulate nodes	USA
Dioscorea bulbifera (air potato)	Round shaped bulbils present	Round-shaped stem twines right to left	Asia, Africa
Dioscorea sansibarensis (Zanzibar yam)	Bulbils are present	3–5 lobed leaf margins, leaf apex expanded	Africa
Dioscorea villosa (four leaf yam, wild yam)	Bulbil not present	Nonarticulate nodes, brown-colored rhizomes	USA

Source: Overholt et al. (2014)

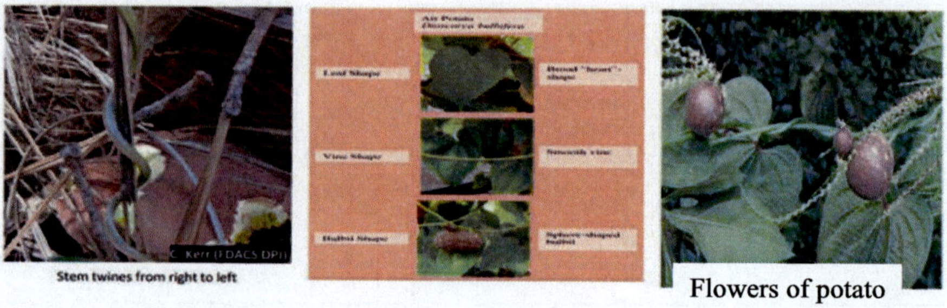

Fig. 5.2 Plant parts of *Dioscorea bulbifera*

(Miller 2003). In the growing season, bulbils produce thick roots, and the primary function of these rigid roots is to spread on the ground, firmly hold the soil, and develop quick shoots. Bulbils give rise to branched and fibrous roots (Coursey 1967).

Bulbils of potato yam vary from species to species. Some species produce very small bulbils like pea, but some species produce large-sized bulbils. Bulbils are smooth, tan-colored, but some species have dark brown colored bulbils with a rough surface. Apart from this, some bulbils are round-shaped, while some are pear-shaped (Kerr and Poffenberger 2016).

Flowers are fragrant, dioecious, and greenish-white in color. *Dioscorea* species contain more male flowers than females. Pistillate flowers are small, 2–4 mm in diameter, and 5–7 mm in length (Coursey 1967). Tepals have three sepals and three petals (Raz 2003) (Fig. 5.2).

Varieties

TDA 291, TDA 297 (resistant to anthracnose).

Climate and Soil

Potato yam mostly prefers tropical climate and moist weather conditions. The optimum temperature required is 20–30 °C, but it can tolerate temperatures ranging from 12 to 38 °C. Growth is suppressed at 9 °C or low temperature. The needed average rainfall is 1200–2600 mm per annum, but it can tolerate 900–4000 mm.

It favors well-drained, deep, sandy loam soil for best production, free from waterlogging conditions with a pH of 6–6.7. For rapid growth, a day length of more than 12 h is required. Initiation of bulbils formation occurs 5–6 months after planting (Tindall 1983).

Agronomic Practices

Propagation

Corms or aerial bulbils can easily propagate potato yam, although vegetatively propagated plants have significantly rapid growth than the seedlings. Large-sized bulbils with a diameter of more than 4 cm sprout 2 days earlier than smaller bulbils. November to December is the ideal month for collecting bulbils for propagation. Bulbils are stored in the soil and used for propagation in the next season.

Direct planting of the bulbils or corms in the field is more effective than transplanting seedlings from the nursery. In the mid-hills, the suitable time for planting bulbils and corms is April–May. Seeds are separated from mature fruits in November for nursery raising and are sown during early summer or spring. It takes about 15–30 days for germination.

For planting a one-hectare area, about 80–100 kg of well-developed corms or bulbils is required. The field should be appropriately tilled, well-pulverized, and free from weeds. Before planting, 15–20 t/ha of farmyard manure (FYM) should be applied. For optimum growth and good yield, 50 cm × 50 cm spacing is suitable. This spacing can accommodate 40,000 plants per hectare. Potato yam does not prefer intercropping, but it needs staking for proper sunlight. The plants may be supported by strings or trellis or bamboo pillars, or wooden stakes. During dry season, light irrigation is preferred through sprinklers to maintain humidity. Flood irrigation should be avoided as it may result in waterlogging. Weed growth is at its peak during July–September. Therefore, regular weeding is required (Table 5.7).

Harvesting and Yield

After about 180–200 days of sprouting, the bulbils become ready to harvest. After 2–3 years, the mother corm may be harvested. During winter, the crop undergoes dormancy. Hence, the ideal months to harvest bulbils are October–November. In tropical and subtropical areas, the most suitable time for harvesting is mid-September to mid-October. Fully matured bulbils are harvested when the leaves start yellowing. The average yield is about 3–5 tons of tubers/hectare (Tindall 1983).

Plant Protection

Insect Pests

For good growth and yield of the plant, effective management of insect pests is necessary. Some of the most common pests affecting this crop include nematodes, yam beetle, mealybugs, and termites.

Nematodes (*Scutellonema bradys* and *Meloidogyne* spp.)

Nematodes are serious insect pests of potato yam that attack the emerging seedlings and cause rotting of the sets in the soil. It is an endoparasite of bulbils and roots and exists in the vicinity of the host plants.

Table 5.7 Weed plant similar to potato yam vine (Kerr and Poffenberger 2016)

Weeds	Characters	Images
Smilax sp. (Green briar)	Bear thorns	
Ipomea sp. (Morning glory)	Wedge-shaped, purple, white, and blue colored flowers	
Parthenocissus quinquefolia (Virginia creeper)	5 leaflets joined to the stalk at each node	
Pueraria montana (Kudzu)	3 leaflets joined to leaf stalk at each node	
Vitis rotundifolia (Muscadine grape)	Tendrils for clamp	

Symptoms Nematodes attack the shoots through roots and wounds in the bulbil's skin (Bridge 1972). These feed extracellularly on the yam tissues, causing cell wall burst and forming cavities (Adesiyan et al. 1975; Goodey 1935; Bridge 1973). Creamish and light-yellow colored scratches are found under the external skin. Lesions can transmit up to 2 cm.

Control/Management Before planting, mixing cow dung or wood ash with bulbils can reduce the nematode population (Adesiyan and Adeniji 1976). Seed treatment with hot water (50–55 °C) for 40 min can help minimize the nematode infection in the field and after harvest during storage. Following crop rotation and intercropping with *Crotalaria* spp. be effective (Fig. 5.3).

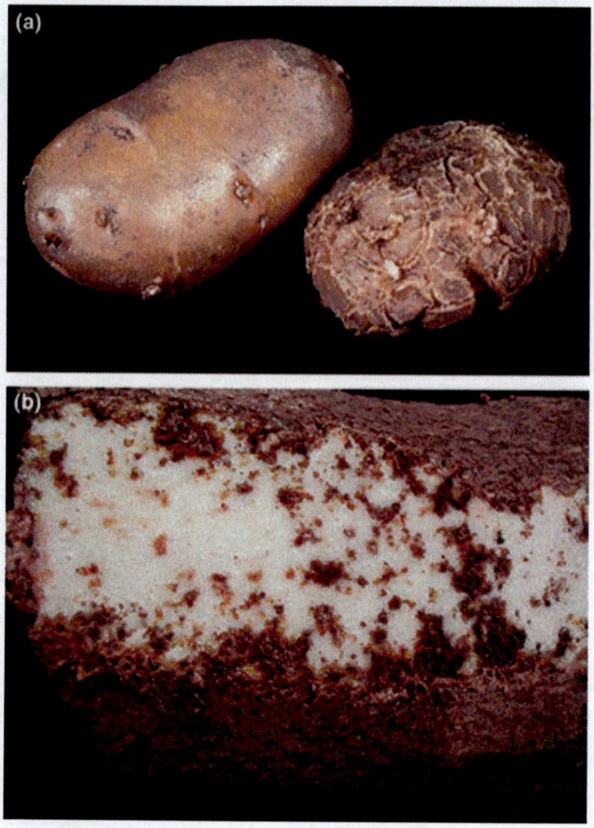

Fig. 5.3 Nematode infested tuber

Yam Beetle (*Heteroligus meles*)

Beetle is the most destructive insect pest of *Dioscorea*. During the months of May–July, the adults attack the bulbils by piercing (Onwueme and Sinhad 1991).

Symptoms The adult beetles feed on the planting sets, cause wilting, and kill the whole plant. These make holes in the leaves and tubers, reducing their market value (Fig. 5.4).

Control/Management Fresh lemongrass (*Cymbopogon citratus*) and African basil (*Ocimum viride*) as mulch can control yam beetle. Spraying neem oil or neem-based botanicals have also been found to be effective.

Fig. 5.4 Yam beetle

Mealybugs (*Planococcus dioscorea*)

These are tiny, flattened, smooth bodies of insects enfolded with a distinct segmentation, and the body is covered with a fleecy snow secretion. It is a sucking insect which causes a reduction in the weight of tubers.

Symptoms These draw the sap from fresh leaves, stalks, and bulbils. The injured leaves appear yellow and get dried, leading to flowers, leaves, and bulbils. These exude honeydew, and the bulky coating of honeydew causes the blackening of leaves, bulbils, and branches, thus reducing the photosynthetic rate.

Control/Management Remove the infected plant parts from the field before flowering. Use of natural enemies (*Gyranusoidea tebygi*) and a spray of soap products, neem-based products, and mineral oil @ 1–2% can be done.

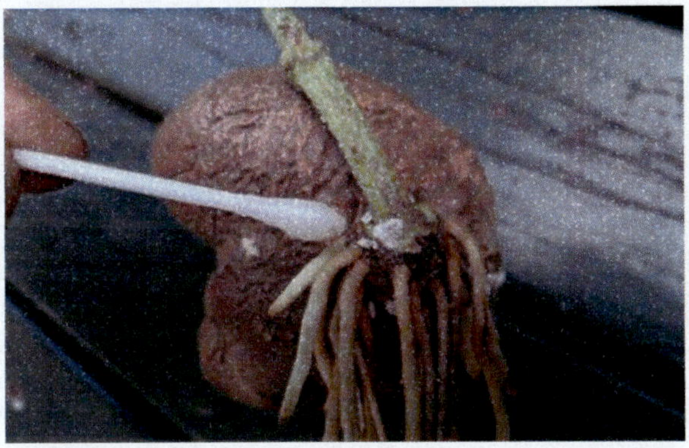

Mealybug

Termites

Termites mostly attack during October but during December, the attack is severe. It attacks the plant at all the growth stages from seedling to maturity of the plants.

Symptoms These are minor insects that damage roots, shown first in the upper parts of the plant as wilting. They make the entire bulbils hollow within a few weeks of infestation. Some plants lodge during strong wind and they are often covered with soil, under which termites may be found.

Control/Management Neem extract is generally used by the farmers of India to reduce termite attacks (Kumari et al. 2013).

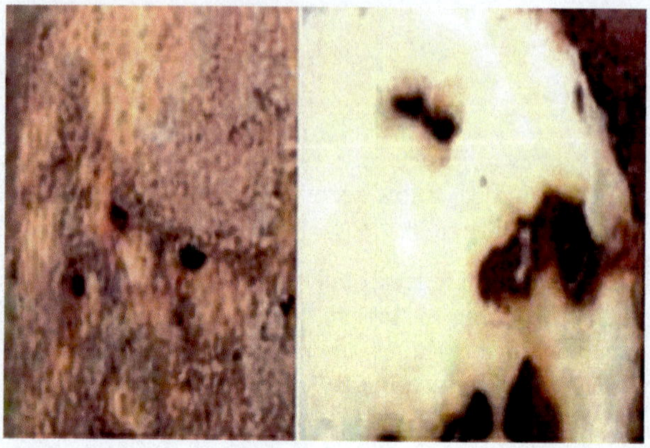

Termites

Diseases

Yam Mosaic Virus (YMV)

This virus is mainly transmitted by aphids and causes a 40% reduction in yield.

Symptoms The affected leaves turn yellow causing stunting of the plant.

Control/Management Use vigorous and virus-free bulbils. Use weed-free fields for planting. Immediately uproot the infected plants, either bury them deep into the soil or burn them.

Anthracnose (*Colletotrichum gloeosporioides*)

Symptoms Black-colored blotches appear in the middle of veins. It spreads mainly in humid and wet conditions.

Control/Management Follow crop rotation and grow resistant varieties like TDA 297 and TDA 291.

Postharvest Handling and Storage

The bulbils or corms can be stored in paper bags or gunny bags without causing any adverse effect on the sprouting. Bulbils can be stored for an extended period in dried and properly ventilated stores at a temperature ranging from 29 to 32 °C with 90–95% relative humidity.

Conclusion

Potato yam being an underutilized vegetable crop has several nutritional and medicinal properties, which are good for human health. Its proper utilization can not only benefit a lot of farmers in the rural areas but it will also help in contributing to the national economy. Therefore, there is a need to give more attention to this crop and also to develop different value-added products from its tubers.

References

Abhyankar, R. K., & Upadhyay, R. (2011). Ethnomedicinal studies of tubers of Hoshangabad. *MP Bulletin of Environment, Pharmacology, and Life Sciences, 1*(1), 57–59.

Adesiyan, S. O., & Adeniji, M. O. (1976). Studies on some aspects of yam nematode (Scutellonema bradys). *Ghana Journal of agricultural science, 9*(2), 131–136.

Adesiyan, S. O., Odihirin, R. A., & Adeniji, M. O. (1975). Economic losses caused by the yam nematode, Scutellonema bradys, in Nigeria. *Plant Disease Reporter, 59*(6), 477–480.

Afiukwa, C., & Igwe, D. (2015). Comparative nutritional and phytochemical evaluation of the aerial and underground tubers of air potato (*Dioscorea bulbifera*). British Journal of Applied Science & Technology, *11*(4), 1–7

Al-Shehbaz, I. A., & Schubert, B. G. (1989). The Dioscoreaceae in the southeastern United States. *Journal of the Arnold Arboretum, 70*(1), 57–95.

Ayensu, E. S., & Coursey, D. G. (1972). Guinea yams: the botany, ethnobotany, use and possible future of yams in West Africa. *Economic botany, 26*(4), 301–318.

Bolaniran, T., Ogidi, C. O., & Akinyele, B. J. (2019). Nutritional value and safety of air potato Dioscorea bulbifera L. fermented with Pleurotus ostreatus and Calocybe indica. *Brazilian Journal of Biological Sciences, 6*(13), 467–482.

Bridge, J. (1972). Nematode problems with yams (Dioscorea spp.) in Nigeria. *PANS Pest Articles & News Summaries, 18*(1), 89–91.

Bridge, J. (1973). Nematodes as pests of yams in Nigeria. *Mededelingen Fakulteit Landbouwwetenschappen, 38*(3), 841–852.

Caddick, L. R., Wilkin, P., Rudall, P. J., Hedderson, T. A., & Chase, M. W. (2002). Yams reclassified: a recircumscription of Dioscoreaceae and Dioscoreales. *Taxon, 51*(1), 103–114.

Coursey, D. G. (1967). Yams. An account of the nature, origins, cultivation and utilisation of the useful members of the Dioscoreaceae. *Yams. An account of the nature, origins, cultivation and utilisation of the useful members of the Dioscoreaceae.*

Das, B. A. N. A. N. I., Talukdar, A. D., & Choudhury, M. D. (2014). A few traditional medicinal plants used as antifertility agents by ethnic people of Tripura, India. *International Journal of Pharmacy and Pharmaceutical Sciences, 6*(3), 47–53.

Dhiman, N., & Bhattacharya, A. (2020). Nardostachys jatamansi (D. Don) DC.-Challenges and opportunities of harnessing the untapped medicinal plant from the Himalayas. *Journal of ethnopharmacology*, *246*, 112211.

Dutta, B. (2015). Food and medicinal values of certain species of Dioscorea with special reference to Assam. *Journal of Pharmacognosy and Phytochemistry*, *3*(5).

Editorial board of Chinese Materia Medica, (1999). Huangyaozi. Zhong Hua Ben Cao (Chinese Materia Medica). Shanghai Science and Technology Press, Shanghai, pp.7278–7280. Vol 8.

Florida Exotic Pest Plant Committee Council 2009, 'Florida Exotic Pest Plant Council's 2009 list of invasive species', *Wildland Weeds*, 12, 13 16.

Girach, R. D., Shaik, A. A., Singh, S. S., & Ahmad, M. (1999). The medicinal flora of Simlipal forests, Orissa state, India. *Journal of ethnopharmacology*, *65*(2), 165–172.

Goodey, T. (1935). Observations on a Nematode Disease of Yams. *Journal of Helminthology*, *13*(3), 173–190.

Guan, X. R., Zhu, L., Xiao, Z. G., Zhang, Y. L., Chen, H. B., & Yi, T. (2017). Bioactivity, toxicity and detoxification assessment of *Dioscorea bulbifera* L.: a comprehensive review. *Phytochemistry Reviews*, *16*(3), 573–601.

Guo, K.W., Zeng, J., Yun, P.S., WenFang, J., Hiu-wen, L., Yang, Y., Zhong-yu, Z., Jin-song, L., Jin-song, L. (2018). New norclerodane diterpenoids from *Dioscorea bulbifera.* Phytochemistry Letters. *27*, 59–62.

Gutierrez, R. L. V. (2021). Nutritional, phytochemical and cytotoxicity analyses of air potato *Dioscorea bulbifera* L. bulbils. *Plant Science Today*, *8*(2), 357–364.

Hammer, R. L. (1998). Diagnosis: Dioscorea. *Wildland Weeds*, *2*(1), 8–10.

ISSG, 2012. Global Invasive Species Database (GISD). Global Invasive Species Database (GISD). Auckland, New Zealand: University of Auckland. http://www.issg.org/database

Jameson, A. (2001). Control of pest species: 29—Freezing controls the exotic invasive vine, air potato (Florida). *Ecol. Restor*, *19*, 51–52.

Kamble, S. Y., Patil, S. R., Sawant, P. S., Sawant, S., Pawar, S. G., & Singh, E. A. (2010). Studies on plants used in traditional medicine by Bhilla tribe of Maharashtra.

Kerr, C., & Poffenberger, R. (2016). Quick identification guide: Air Potato, *Dioscorea bulbifera.*

Knuth, R. (1924). Dioscoreaceae In: Engler, HGA (Ed.) Das Planzenreich 87 (IV. 43). *HR Engelmann (J. Cramer), Leipzig*.

Kumar, A. (2011). Moisture content: A stability problem in pharmaceutical products. PHARMATUTOR-ART-1187.

Kumar, V., Moyo, M., & Van Staden, J. (2016). Somatic embryogenesis in Hypoxis hemerocallidea: an important African medicinal plant. *South African Journal of Botany*, *108*, 331–336.

Kumari, K., Patil, K,, & Sharma, S. (2013). Farmer friendly ways to control termites. Popular Kheti. 1:2529.

Kundu, B. B., Vanni, K., Farheen, A., Jha, P., Pandey, D. K., & Kumar, V. (2020). *Dioscorea bulbifera* L.(Dioscoreaceae): A review of its ethnobotany, pharmacology and conservation needs. *South African Journal of Botany*.

Langeland, K. A., & Burks, K. C. (1998). Identification & biology of non-native plants in Florida's natural areas.

Langeland, K. A., & Meisenburg, M. J. (2008). *Natural area weeds: air potato (Dioscorea bulbifera)*. University of Florida Cooperative Extension Service, Institute of Food and Agriculture Sciences, EDIS.

Lebot, V., Faloye, B., Okon, E., & Gueye, B. (2019). Simultaneous quantification of allantoin and steroidal saponins in yam (*Dioscorea spp.*) powders. *Journal of Applied Research on Medicinal and Aromatic Plants*, *13*, 100200.

Lim, T. K. (2016). Dioscorea bulbifera. In *Edible Medicinal and Non-Medicinal Plants* (pp. 235–252). Springer, Dordrecht.

Martin, F. W., & FKS, K. (1974). Stimulation of yam (Dioscorea) tuber growth by gamma irradiation. J. Amer. Soc. Hortic. Sci.; USA, *99*(3), 282–284.

Mbiantcha, M., Kamanyi, A., Teponno, R. B., Tapondjou, A. L., Watcho, P., & Nguelefack, T. B. (2011). Analgesic and anti-inflammatory properties of extracts from the bulbils of Dioscorea bulbifera L. var sativa (Dioscoreaceae) in mice and rats. *Evidence-based complementary and alternative medicine, 2011*.

Miller, J. H. (2003). Nonnative invasive plants of southern forests: a field guide for identification and control. *Gen. Tech. Rep. SRS-62. Asheville, NC: US Department of Agriculture, Forest Service, Southern Research Station. 93p., 62*.

Mishra, R. K., Upadhyay, V. P., & Mohanty, R. C. (2008). Vegetation ecology of the Similipal biosphere reserve, Orissa, India. *Applied Ecology and Environmental Research, 6*(2), 89–99.

Nabatanzi, A., & Nakalembe, I. (2016). Wild food plants used by people living with HIV/AIDS in Nakisunga sub-county, Uganda. *African Journal of Food, Agriculture, Nutrition and Development, 16*(4), 11310–11330.

Onwueme, I.C., & Sinhad, T.D. (1991). Field Crop Production in Tropical Africa. CTA, Ede-Wageningen, The Netherlands, pp: 337–342.

Overholt, W.A., Markle, L., Meisenberg, M., Raz, L., Wheeler, G., Pemberton, R., Taylor, J., King, M., Schmitz, D., Parks, G.R., & Rayamajhi, M. (2014). Air potato management plan for Florida. Florida Exotic Pest Plant Council. www.fleppc.org

Raz, L. (2003). Dioscoreaceae: R. Brown: Yam Family. In: Flora of North America Editorial Committee (eds.). Flora of North America. Oxford University Press, New York, pp. 479–485.

Rudito, R., Naibaho, N. M., Suwarto, S., Jayus, J., Witono, Y., Saragih, B., & Arung, E. T. (2018). Physical and Chemical Characteristics of Fermented 'Dayak' Wild Yam (Dioscorea hispida Dennst), Purple Yam (*Dioscorea alata* var. purpurea) and Air Potato (*Dioscorea bulbifera* L.) Flour as Food Ingredient.

Sharma, L. N., & Bastakoti, R. (2009). Ethnobotany of Dioscorea L. with emphasis on food value in Chepang communities in Dhading district, central Nepal. *Botanica Orientalis: Journal of Plant Science, 6*, 12–17.

Shriram, V., Jahagirdar, S., Latha, C., Kumar, V., Puranik, V., Rojatkar, S., Dhakephalkar, P.K. & Shitole, M. G. (2008). A potential plasmid-curing agent, 8-epidiosbulbin E acetate, from Dioscorea bulbifera L. against multidrug-resistant bacteria. *International Journal of Antimicrobial Agents, 32*(5), 405–410.

Singh, N. E. H. A., Pangtey, Y. P. S., Khatoon, S., & Rawat, A. K. S. (2009). Some ethnomedicinal plants of Ranikhet region, Uttaranchal. *J. Econ. Taxon. Bot, 33*, 198–204.

Sonibare, M. A., & Adeniran, A.A. (2014). Comparative micromorphological study of wild and micropropagated Dioscorea bulbifera Linn. Asian Pacific journal of tropical biomedicine, 4(3), 176–183.

Swarnkar, S., & Katewa, S. S. (2008). Ethnobotanical observation on tuberous plants from tribal area of Rajasthan (India). *Ethnobotanical leaflets, 2008* (1), 87.

Tindall, H. D. (1983). *Vegetables in the Tropics*. Macmillan International Higher Education.

Tiwari, L., & Pande, P. C. (2006). Indigenous veterinary practices of Darma valley of Pithoragarh district, Uttaranchal.

USDA, NRCS. (2002). The PLANTS Database, Version 3.5, http://plants.usda.gov., National Plant Data Center. Baton Rouge, LA.

USDA-ARS, 2012. Germplasm Resources Information Network (GRIN). Online Database. Beltsville, Maryland, USA: National Germplasm Resources Laboratory.

Wunderlin, R. P., & Hansen, B. F. (2011). *Guide to the vascular plants of Florida*. University Press of Florida.

Chapter 6
Production Technology of Underutilized Vegetables of Basellaceae Family

Monisha Rawat

Introduction

Malabar spinach, scientifically known as *Basella alba* L. or *Basella rubra* Roxb., belongs to the Basellaceae family (Deshmukh and Gaikwad 2014) and has chromosome number $2n = 24$. It is synonymously known as *Basella lucida* L. and *Basella cordifolia* Lam. There are two different forms of *Basella*, one has red-colored petioles and stems and another with green-colored leaves, petioles, and stems. Both these are consumed, whereas the green types are commercially cultivated and belong to the variety *alba*. It is reported that the coloring agent present in the red variety was used as a dye in China. A glycoprotein containing 20% carbohydrate and 80% protein has been identified from the leaves of *Basella alba* which inhibits tobacco mosaic virus (Ushasri et al. 1982). It may be planted in any corner, and given support to climb upon, or it may be planted against trees or in the hedges surrounding the garden. The crop becomes ready to harvest in 3 months.

It is known by different common names, including Indian spinach, Malabar nightshade, Malabar climbing spinach (Sen et al. 2010), Ceylon spinach, Poi sag, buffalo spinach, Gambian spinach, vine spinach (Roy et al. 2010), East-Indian spinach, Chinese spinach (Bamidele et al. 2010), cyclone spinach (Nirmala et al. 2011), country spinach, Indian saag, red vine spinach, slippery vegetable, and Suriname spinach.

M. Rawat (✉)
Department of Horticulture, School of Agriculture, Lovely Professional University, Phagwara, Punjab, India
e-mail: monisha.22692@lpu.co.in

Savita et al. (eds.), *Production Technology of Underutilized Vegetable Crops*,
https://doi.org/10.1007/978-3-031-15385-3_6

Origin and Distribution

India and Indonesia are believed to be its center of origin and it is also found growing naturally in the areas of tropical Africa and Asia (Vanaliya et al. 2012). It is grown throughout India exclusively in North-East and South India. It is cultivated as annual in temperate regions, while perennial crop in the tropical and subtropical areas.

Nutritional Importance and Uses

It is widely cultivated for its fresh tender leaves and soft succulent stem, which are generally consumed raw as salad, boiled, stewed, or stir-fried or in soups, curries, or cooked and eaten as saag. It is a succulent vine, having climbing growth habit, leaves are circular to ovate in shape. While cooking, the leaves and young tender shoots exude a gelatinous or mucilaginous substance (Palada and Crossman 1999). This plant has numerous health benefits because of the high content of minerals, proteins, oil, carbohydrates, fibers, carotenoids, organic acids, vitamins, and basella saponins (Deshmukh and Gaikwad 2014; Grubben and Denton 2004; Palada and Crossman 1999; Duke and Ayensu 1985; Murakami et al. 2001; Lyimo et al. 2003; Khare 2007). In East African regions, the plant is fed to livestock, which helps in increasing the milk production. Red-colored forms are commonly grown as ornamental plants.

One hundred grams of fresh Malabar spinach contains water (93 g), energy (19 kcal), protein (1.8 g), fat (0.3 g), Ca (109 mg), P (52 mg), Fe (1.2 g), Mg (65 mg), K (510 mg), Na (24 mg), Zn (0.43 mg), vitamin A (8000 IU), thiamine (0.05 mg), riboflavin (0.16 mg), niacin (0.50 mg), ascorbic acid (102 mg) (Yang et al. 2008; USDA 2018; Lyimo et al. 2003), ash (15.9 g), carbohydrate (42.1 g), and fiber (11.3 g) (Maisuthisakul et al. 2008). Furthermore, it was also reported that inclusion of Malabar spinach in the daily diet positively affects the storage of vitamin A in the body (Haskell et al. 2004). The mucilage is a good source of soluble fibers, which helps to remove mucus and toxins out of the body.

This plant has been tremendously explored in the Chinese as well as Indian traditional medical practices due to its medicinal properties and has been used as an anti-inflammatory and also for treating constipation (Murakami et al. 2001). The nutritional potential as well as nutraceutical properties of its leaves have been confirmed through recent studies (Kumar et al. 2015a). The extract of its fruits have antioxidant activity and cytotoxicity, which can protect from cervical cancer (Kumar et al. 2015b). Along with this, betalains have been extracted from its fruits and are used in different food preparations (Kumar et al. 2015c, d). Owing to its beneficial effects on human health, betalains are regarded as a substitute to synthetic coloring agents in different food industries (Khan et al. 2012; Lin et al. 2010) and thus have found their use in the pharmaceuticals and cosmetics (Khan 2016).

Malabar spinach is rich in kaempferol, a flavonoid which protects the body against cancers and cardiovascular diseases (Yang et al. 2008). It also has antioxidants and phenolics (Olajire and Azeez 2011; Reshmi et al. 2012). Its shoots, leaves, and flowers have a natural color pigment, i.e., anthocyanin (Glässgen et al. 1993). It also possesses wound-healing effects (Haneefa et al. 2012), antibacterial activities (Oyewole and Kalejaiye 2012), and antiviral properties, particularly in the forms having red-colored stem (Dong et al. 2011), anti-inflammatory effects (Kumar et al. 2011), and anti-ulcer effects (Venkatalakshmi and Senthamaraiselvi 2012). Its leaves have several triterpene oligoglycosides, which include basella saponins, beta vulgaroside 1, spinacoside C, and momordins. Two peptides having antifungal properties and two ribosome-inactivating proteins having antiviral activities have also been isolated from its seeds.

Fruits are used to prepare dye that is used as rouge, ink for official seals, food-coloring agent, and in cosmetics (Hanelt 2017; Useful Tropical Plants 2017), in addition to the use of its leaves as food (Encyclopedia of Life 2017; PROTA 2017). It is suggested that its seeds have antifungal properties (Wang and Ng 2001) and thus can be used in pharmaceuticals, and also as an alternate source of nonconventional oils that can have varied uses in cosmetics and food industries (Diemeleou et al. 2014). Its leaves are brewed and used as a substitute of tea. Due to the presence of mucilaginous substances, it is used as a thickener in soups and stews (Hanelt 2017; PROTA 2017).

It has various therapeutic properties, i.e., anti-ulcer, antioxidant, cytotoxic, antibacterial, anti-inflammatory, nephron-protective properties. Apart from this it also heals wounds and functions as a depressant of central nervous system as reported by Kumar et al. (2013). The extract of its fruits have the potential for treating cancers (Kumar et al. 2015b).

Practically, all the plant parts of Malabar spinach can be used to prepare traditional medicines, which can be effectively used as laxative, demulcent, rubefacient, febrifuge, diuretic, astringent, and also helps in treating conjunctivitis, dysentery, catarrh, diarrhea, stomach-related problems like improper digestion, constipation, boils, and sores. It is also used as a powerful antidote to poison (Useful Tropical Plants 2017; National Parks Board 2020). In India, people chew its leaves to cure mouth ulcers (Hebbar et al. 2004). A popular nourishing and restorative tonic for postpartum *wellness* is prepared in Thailand for mothers and babies (Panyaphu et al. 2011).

Botany and Taxonomy

There are four genera and 20 species reported in the Basellaceae family. *Basella rubra* and *Basella alba* were mentioned as two separate species based on its leaf characteristics and the stem color by Linnaeus in *Species Plantarum*. Though, afterwards, no such evidence was reported indicating *Basella rubra* a different species; therefore, it is considered as an alternate of *B. alba* (Deshmukh and Gaikwad 2014).

Fig. 6.1 Malabar spinach, *Basella alba*: (**a**) plant, (**b**) flowers, and (**c**) fruits

It has fibrous roots with lateral growth. Its stem is thin, succulent, twining, and smooth with bright green or purple color, which can be 8–10 m in length.

The leaves are simple, fleshy, dark green, or purplish; stipules are absent; petiole is sessile or short stout 4–9 cm in length, alternately/spirally arranged; margin is entire; leaf blade is often broad; the length of leaves is more than the width, ovate to heart-shaped; and the juvenile leaves are often large, round at the apex and cordate at the base. The presence of betalains causes red-violet color at the under surface of the leaves (Cyunel 1989), stalks, petioles, and fruits (Palada and Crossman 1999).

Flowers are hermaphrodite, pentamerous, sepals white or pink or red or purple in color, style with three linear stigmas; ovary is superior and one celled. Fruits are small, dark purple, or purplish black in color, pseudo berry/baccate drupe, 4–10 mm in diameter, having violet-colored juice and one seed. Seed is globose, having a diameter of 3 mm, dark brown to black in color; testa is thick with bright rough surface. Under suitable conditions, the seeds can be stored for 4 years (Deshmukh and Gaikwad 2014; Almeida 1996; Grubben and Denton 2004; Mahr 2014) (Fig. 6.1).

Varieties

Three different types of Malabar spinach have been identified:

1. Plants with dark green colored leaves, which are ovate to almost round in shape and are most commonly grown.
2. Plants having red-colored stems and leaves, which are ovate to almost round in shape, mostly grown as ornamental.
3. Plants with dark green colored leaves, which are heart-shaped (Elzebroek and Wind 2008).

Climate and Soil

Malabar spinach grows without any difficulty under suitable soil and climatic conditions (Palada and Crossman 1999). It does well in the tropical lowland areas at an altitude of 500 m, but can also survive at an elevation of 2600 m and in temperate regions. It cannot tolerate extremes of temperature.

The plant grows well at an optimum temperature of 32 °C, but growth and development are arrested if the temperature drops to 26 °C. The growth of plant is affected by low temperature and fluctuation in day and night temperature (PIER 2017). It performs better if grown under partial shade. The optimum temperature requirement for its cultivation is 10–35 °C, whereas 18–23 °C is good for seed germination. Germination is epigeal. It favors an annual rainfall ranging from 2000–2500 mm, but can also tolerate 700–4200 mm of rainfall. It cannot tolerate salinity and water logging conditions (PROTA 2017).

Well-drained sandy loam to clay soil, which is friable and rich in humus, is ideal for its growth. It prefers to grow in soils with 5.5–7 pH but can also be grown with pH range of 4.3–7.5. It can even grow in poor soils with small periods of drought (Useful Tropical Plants 2017) but water stress induces the plant to flower early. Higher relative humidity plays a significant role in its cultivation as it is crucial for inhibiting flowering, which can lead to bitterness in leaves.

Agronomic Practices

Malabar spinach can be propagated using the seeds, stem, and root cuttings. Seeds can be sown in the month of June, while in rainy season, stem cuttings are planted in the plains of India. The crop can be cultivated either by direct sowing or by transplanting method. Seed rate required for one hectare area is 12.5–16 kg. Soaking of seeds a day prior to sowing is recommended and the seeds germinate within 10–21 days under suitable conditions.

The plants can either be staked on bamboo machans or trained on trellises. For trailing on the ground, a spacing of 60 cm × 30–60 cm is adopted. Root cuttings can also be used for planting. Mulching with straw is useful especially during the early stages of growth and development, which also helps in moisture conservation particularly during the drier periods. To decrease overcrowding, its shoots are tied in bunches. In general, the plant requires 5–6 irrigations.

Nutrient Management

Malabar spinach responds well to application of chemical fertilizers and organic manures. Soil rich in nutrients is essential for a good crop; therefore, 20–30 tons of farm yard manure along with 60:60:40 kg/ha NPK can be applied during sowing or transplanting (Thamburaj 2000). It does not require much care except for regularly providing support to the stem as it has a twining growth habit. The roots are also very weak; therefore, the collar region of roots should be filled well.

Harvesting and Yield

The young shoots, branches, fresh tender leaves, and new flower buds can be continuously picked throughout the growing season, and the yield of approximately 1.5 kg per plant can be obtained.

Postharvest Handling and Storage

Malabar spinach can be stored for a day at 20–30 °C temperature. For enhancing the shelf life, it must be stored at a lower temperature. For seed production purpose, dried fruits are selected and it can yield 1000–2000 kg seeds/ha (Grubben and Denton 2004).

Conclusion

Malabar spinach is an important leafy vegetable and due to its adaptation to warmer climatic conditions, it can be used as a substitute for spinach. Owing to its higher nutrient content and various medicinal properties, it has the potential to cure various diseases and thus can help in improving human health. It is therefore important to cultivate this minor crop on a wider scale to get maximum health benefits.

References

Almeida, M. (1996). Flora of Maharashtra, Blatter Herbarium, *St. Xavier's College, Mumbai.*

Bamidele, O., Akinnuga, A. M., Olorunfemi, J. O., Odetola, O. A., Oparaji, C. K., & Ezeigbo, N. (2010). Effects of aqueous extract of Basella alba leaves on haematological and biochemical parameters in albino rats. *African Journal of Biotechnology*, 9(41), 6952–6955.

Cyunel, E. (1989). *Basella alba* L.: in vitro culture and the production of betalains. In *Medicinal and aromatic plants II* (pp. 47–68). Springer, Berlin, Heidelberg.

Deshmukh, S. A., & Gaikwad, D. K. (2014). A review of the taxonomy, ethnobotany, phytochemistry and pharmacology of Basella alba (Basellaceae). *Journal of Applied Pharmaceutical Science*, *4*(01), 153–165.

Diemeleou, C. A., Zoue, L. T., & Niamke, S. L. (2014). Basella alba seeds as a novel source of non-conventional oil with beneficial qualities. *Romanian Biotechnological Letters*, *19*(1), 8966.

Dong, C. X., Hayashi, K., Mizukoshi, Y., Lee, J. B., & Hayashi, T. (2011). Structures of acidic polysaccharides from Basella rubra L. and their antiviral effects. *Carbohydrate polymers*, *84*(3), 1084–1092.

Duke, J. A., & Ayensu, E. S. (1985). *Medicinal plants of China*. Reference Publications.

Elzebroek, T., & Wind, K. (2008). Oil crops. *Guide to cultivated plants*, 182–225.

EOL. (2017). Encyclopedia of Life on-line database, http://www.eol.org. (Accessed 14 October 2020).

Glässgen, W. E., Metzger, J. W., Heuer, S., & Strack, D. (1993). Betacyanins from fruits of Basella rubra. *Phytochemistry*, *33*(6), 1525–1527.

Grubben, G. J. H., & Denton, O. A. (2004). Plant resources of tropical Africa 2. Vegetables. *Plant resources of tropical Africa 2. Vegetables*. PROTA Foundation, Backhuys, CTA pp: 103–111.

Haneefa, M. K., Abraham, A., Saraswathi, R., Mohanta, G. P., & Nayar, C. (2012). Formulation and evaluation of herbal gel of Basella alba for wound healing activity. *Journal of Pharmaceutical Sciences and Research*, *4*(1), 1642–1648.

Hanelt, P. (2017). Mansfeld's World Database of Agricultural and Horticultural Crops. *Mansfeld's World Database of Agricultural and Horticultural Crops*.

Haskell, M. J., Jamil, K. M., Hassan, F., Peerson, J. M., Hossain, M. I., Fuchs, G. J., & Brown, K. H. (2004). Daily consumption of Indian spinach (Basella alba) or sweet potatoes has a positive effect on total-body vitamin A stores in Bangladeshi men. *The American journal of clinical nutrition*, *80*(3), 705–714.

Hebbar, S. S., Harsha, V. H., Shripathi, V., & Hegde, G. R. (2004). Ethnomedicine of Dharwad district in Karnataka, India—plants used in oral health care. *Journal of ethnopharmacology*, *94*(2–3), 261–266.

Khan, M. I. (2016). Plant betalains: Safety, antioxidant activity, clinical efficacy, and bioavailability. *Comprehensive Reviews in Food Science and Food Safety*, *15*(2), 316–330.

Khan, M. I., Harsha, P. S., Giridhar, P. S. C. P., & Ravishankar, G. A. (2012). Pigment identification, nutritional composition, bioactivity, and in vitro cancer cell cytotoxicity of Rivina humilis L. berries, potential source of betalains. *LWT*, *47*(2), 315–323.

Khare, C. P. (2007). Indian medicinal plants: an illustrated dictionary. Springer-Verlag Berlin/ Heidelberg, pp. 436.

Kumar, S. S., Manoj, P., & Giridhar, P. (2015a). Nutrition facts and functional attributes of foliage of Basella spp. *LWT-Food Science and Technology*, *64*(1), 468–474.

Kumar, S. S., Manoj, P., & Giridhar, P. (2015c). A method for red-violet pigments extraction from fruits of Malabar spinach (Basella rubra) with enhanced antioxidant potential under fermentation. *Journal of food science and technology*, *52*(5), 3037–3043.

Kumar, S. S., Manoj, P., Giridhar, P., Shrivastava, R., & Bharadwaj, M. (2015b). Fruit extracts of Basella rubra that are rich in bioactives and betalains exhibit antioxidant activity and cytotoxicity against human cervical carcinoma cells. *Journal of Functional Foods*, *15*, 509–515.

Kumar, S. S., Manoj, P., Shetty, N. P., Prakash, M., & Giridhar, P. (2015d). Characterization of major betalain pigments-gomphrenin, betanin and isobetanin from Basella rubra L. fruit and evaluation of efficacy as a natural colourant in product (ice cream) development. *Journal of food science and technology*, *52*(8), 4994–5002.

Kumar, S., Prasad, A. K., Iyer, S. V., & Vaidya, S. K. (2013). Systematic pharmacognostical, phytochemical and pharmacological review on an ethnomedicinal plant, Basella alba L. *Journal of Pharmacognosy and Phytotherapy*, *5*(4), 53–58.

Kumar, V., Bhat, Z. A., Kumar, D., Bohra, P., & Sheela, S. (2011). In-vitro anti-inflammatory activity of leaf extracts of Basella alba Linn. var. alba. *Int J Drug Dev Res*, *3*(2), 176–179.

Lin, S. M., Lin, B. H., Hsieh, W. M., Ko, H. J., Liu, C. D., Chen, L. G., & Chiou, R. Y. Y. (2010). Structural identification and bioactivities of red-violet pigments present in Basella alba fruits. *Journal of agricultural and food chemistry*, *58*(19), 10364–10372.

Lyimo, M., Temu, R. P. C., & Mugula, J. K. (2003). Identification and nutrient composition of indigenous vegetables of Tanzania. *Plant Foods for Human Nutrition*, *58*(1), 85–92.
Mahr, S. (2014). Malabar spinach. *Basella alba*.
Maisuthisakul, P., Pasuk, S., & Ritthiruangdej, P. (2008). Relationship between antioxidant properties and chemical composition of some Thai plants. *Journal of Food Composition and Analysis*, *21*(3), 229–240.
Murakami, T., Hirano, K., & Yoshikawa, M. (2001). Medicinal Foodstuffs. XXIII. 1) Structures of New Oleanane-Type Triterpene Oligoglycosides, Basellasaponins A, B, C, and D, from the Fresh Aerial Parts of Basella rubra L. *Chemical and pharmaceutical bulletin*, *49*(6), 776–779.
National Parks Board. (2020). Flora and Fauna Web. In: Flora and fauna web, Singapore: National Parks Board. http://florafaunaweb.nparks.gov.sg/Home.aspx (Accessed 23 December 2020).
Nirmala, A., Saroja, S., & Gayathri, S. (2011). Phytochemical screening and antihyperglycemic activity of Basella rubra. *Recent Research in Science and Technology*, 3(11), 80–83.
Olajire, A. A., & Azeez, L. (2011). Total antioxidant activity, phenolic, flavonoid and ascorbic acid contents of Nigerian vegetables. *African Journal of Food Science and Technology*, *2*(2), 22–29.
Oyewole, O. A., & Kalejaiye, O. A. (2012). *The antimicrobial activities of ethanolic extracts of Basella alba on selected microorganisms* (No. RESEARCH). Scientific J Microbiol 1: 113–118.
Palada, M. C., & Crossman, S. M. (1999). Evaluation of tropical leaf vegetables in the Virgin Islands. *Perspectives on new crops and new uses*. ASHS Press, *Alexandria, VA*, 388–393.
Panyaphu, K., Van On, T., Sirisa-Ard, P., Srisa-Nga, P., ChansaKaow, S., & Nathakarnkitkul, S. (2011). Medicinal plants of the Mien (Yao) in Northern Thailand and their potential value in the primary healthcare of postpartum women. *Journal of Ethnopharmacology*, *135*(2), 226–237.
PIER. (2017). Pacific Islands Ecosystems at Risk. In: Pacific Islands Ecosystems at Risk Honolulu, Hawaii, USA: HEAR, University of Hawaii. http://www.hear.org/pier/index.html (Accessed 23 December 2020)
Plants, U. T. (2017). Useful tropical plants database. *Useful tropical plants database: K Fern*. http://tropical.theferns.info.
PROTA. (2017). PROTA4U web database. In: PROTA4U web database Wageningen and Nairobi, Netherlands\Kenya: Plant Resources of Tropical Africa. https://www.prota4u.org/database/ (Accessed 23 December 2020)
Reshmi, S. K., Aravinthan, K. M., & Devi, P. S. (2012). *Antioxidant analysis of betacyanin extracted from Basella alba fruit* (No. RESEARCH). Int J Pharm Tech Res 4: 900–913.
Roy, S. K., Gangopadhyay, G., & Mukherjee, K. K. (2010). Is stem twining form of Basella alba L. a naturally occurring variant?. *Current Science*, 1370–1375.
Thamburaj, S., & Singh, N. (2001). *Textbook of vegetables, tubercrops and spices*. Indian Council of Agricultural Research.
USDA, U. (2018). National nutrient database for standard reference, Legacy Release.
Ushasri, V., Nagarajan, K., & Reddy, T. S. N. (1982). Isolation and characterisation of inhibitor to tobacco mosaic virus from Basella alba L. Tobacco research.
Vanaliya, S., Rao, P. S., Rao, S. K., & Sameja, K. (2012). Pharmacognostical study of Basella alba stem. *International Journal of Research in Pharmaceutical and Biomedical Sciences*, *3*(3), 1093–1094.
Venkatalakshmi, P., & Senthamaraiselvi, V. (2012). Anti-ulcer effect of Basella alba leaf extract in aspirin induced albino rats. *International Journal of Pharmaceutical Sciences and Research*, *3*(8), 2539–2542.
Wang, H., & Ng, T. B. (2001). Novel antifungal peptides from Ceylon spinach seeds. *Biochemical and biophysical research communications*, *288*(4), 765–770.
Yang, R. Y., Lin, S., & Kuo, G. (2008). Content and distribution of flavonoids among 91 edible plant species. *Asia Pacific journal of clinical nutrition*, *17*.

Chapter 7
Production Technology of Underutilized Vegetables of Lamiaceae Family

Monisha Rawat

Introduction

The Lamiaceae or formerly called Labiatae is a family of flowering plants generally known as the *Ocimum* or sage family, which is dispersed almost worldwide and most of the species are cultivated for their aromatic leaves and attractive flowers. These herb plants are valued for their flavor, fragrance, and medicinal properties. The plants of this family are usually cultivated by people, not only for the aroma but also for their easier cultivation, since these are readily propagated. The plants of this family have varied economic importance as the leaves of mint (*Mentha viridis*) are used as salad, tubers of Chinese artichoke (*Stachys affinis*) are consumed, and plants of other species such as *Mentha*, *Ocimum*, *Melissa officinalis*, and *Coleus forskohlii* are also used as condiments. In this chapter, various health benefits, nutritional importance, and cultivation technology of important members of the Lamiaceae family have been discussed.

Origin and Distribution

Coleus

Chinese potato/coleus or *koorka* or *ratala* in Hindi, *Plectranthus rotundifolius* (Poir.) Spreng Syn. *Solenostemon rotundifolius* (Poir.) Morton, *Coleus rotundifolius*, *Plectranthus tuberosus*, *Coleus parviflorus* Benth. belong to the family

M. Rawat (✉)
Department of Horticulture, School of Agriculture, Lovely Professional University, Phagwara, Punjab, India
e-mail: monisha.22692@lpu.co.in

Savita et al. (eds.), *Production Technology of Underutilized Vegetable Crops*,
https://doi.org/10.1007/978-3-031-15385-3_7

Lamiaceae. The chromosome number of coleus is $2n = 64$. East Africa is the native place of coleus, then distributed to tropical West Africa and Southeast Asian regions such as India, Indonesia, Malaysia, and Sri Lanka (Harlan et al. 1976), but is now well adapted to Southeast Asian regions, i.e., India and Sri Lanka. The word coleus has come from Koloes, in preference to the filaments being joined and forming a tube around the style. It has aromatic, thick, and succulent leaves. Small, round to oval-shaped tubers which are dark brown are found at the base of the stem. It is extensively grown in South Indian states (Karnataka, Kerala, and Tamil Nadu) and has also been found growing on a smaller scale in Odisha, Madhya Pradesh and north eastern regions, where it is generally eaten by the tribals. It has over 200 species having both medicinal and ornamental importance. The duration of the crop is about 5 months.

Chinese Artichoke

Stachys affinis Bunge (*S. sieboldii*), commonly known as crosne, Japanese artichoke, knotroot, artichoke betony, and Chorogi, is inherent to China, Japan, and Korea. It is a perennial plant in which the tubers are consumed. It has been cultivated in Europe since the nineteenth century and now it is grown throughout the world. The diploid chromosome number of Chinese artichoke as reported by different workers varies from 66 to 70 (Fedorov 1974).

Hoary Basil

It is commonly known as lemon basil and American basil, while in Hindi, it is called *ban-tulsi*, *vantulsi*, or *rantulsi*. Botanically, it is *Ocimum americanum* L. syn. *Ocimum canum* Sims. having a chromosome number of $2n = 24$. *Ocimum* is derived from the Greek word "ozo," which means fragrance or smell (Paxton 1868). It has a lemon aroma that is exceptionally pure and fresh (Morales and Simon 1997) with a slight fragrance of camphor, cinnamon, and lavender. Due to its varied use in the preparation of traditional medicines, perfumes, and pharmaceutical industries, it is also known as the "king of herbs" (Simpson and Conner 1986). The genus *Ocimum* is the largest genera of the Lamiaceae family including more than 160 species (Pushpangadan et al. 1995). Approximately 65 species are related to *Ocimum* and others are regarded as synonyms (Ashraf et al. 2021; Pushpangadan et al. 1995). *Ocimum* is cultivated all over the world, but there are three major centers of diversity in its distribution. The tropical regions of America and Asia, as well as the tropical and subtropical regions of Africa, have the most *Ocimum* species (Paton et al. 1999).

Nutritional Importance and Uses

Coleus

The tubers have a special aromatic flavor and delicious taste upon cooking. In India, because of its aromatic flavor and sweetness, it is used in many curry preparations and is also baked and made into chips like a potato. The tubers also contain 20–24% starch. The tubers are a good source of minerals (calcium and iron) and vitamins (ascorbic acid, niacin, riboflavin, and thiamine) (Jayapal et al. 2016). The tuber's scented flavor makes it a popular vegetable and apart from that it also possesses medicinal properties because of enzyme inhibitors (Prathibha et al. 1995) and flavonoids results in lowering the blood cholesterol level (Horvath et al. 2004; Abraham and Radhakrishnan 2005; Sandhya and Vijayalakshmi 2000). Tubers are also used for curing dysentery and certain eye disorders. It contains *beta-carotene* and *α-tocopherol* (vitamin E), which are known to have anticancer and antioxidant properties.

Chinese Artichoke

Chinese artichoke is valued for its rhizome, which is consumed without peeling and can be eaten raw, as salad, added to any cooked dish, curries, stir-fries, soups, casseroles, sweet and sour seafood, pickled and dried. It contains 63.6 calories, 14.3 g carbohydrates, 10.3 g fibers, 3.5 g proteins, 0.4 g total fats, 0.7 g iron, and 25.2 mg calcium per 100 g of edible portion. It is a tasty, culinary vegetable with a crisp, crunchy texture and an artichoke-like sweetness and nutty flavor. The tubers have a novel appearance, so these are used as a whole, rather than chopped. As the tubers do not have a proper shape, it is very difficult to clean the tubers. Tubers are pickled in China and Japan.

Vacuoles in the tuber are rich in stachyose, a tetrasaccharide, consisting of galactose, glucose, and fructose. Stachyose can be up to 80–90% in dry tubers.

Hoary Basil

It contains 87 g water, 3.3 g protein, 2.0 g fiber, 320 mg Ca, 4.5 mg Fe, 27 mg vitamin C, and 180 kJ energy per 100 g edible part. It contains aroma compounds like citral, camphor, and methyl cinnamate in varying proportions, which is responsible for its cinnamon, clove, and lemon-like fragrance. Lemon basil is used widely in soups, stews, curries, and stir-fried dishes. It is used mostly in Indonesian cuisine, where it is called *kemangi*, and is often used for seasoning curries, soups, stew, steamed, and grilled dishes. In Thailand, it is known as *maenglak* and is among the several basils used in Thai cuisine (Prakash and Gupta 2005). In patients with gastric, chronic fever, dysentery, hemorrhage, dyspepsia, and hepatic disorders, a juice of tulsi leaves has been recommended to help them recover faster. Its leaf juice along with *Triphala* is used in the preparation of *Ayurvedic* eye drop, which is suggested for treating various painful eye diseases like glaucoma, cataract, and chronic conjunctivitis. Its essential oil is extracted and used in soaps and cosmetics. It is also used to extract essential oils, which has both medicinal and industrial applications. The whole plant has hemolytic activity (Sutili et al. 2016), antifungal activity (Sethi et al. 2013), bactericidal activity (Vieira et al. 2014), antimicrobial activity (Thaweboon and Thaweboon 2009), antioxidant activity (Aluko et al. 2013a), and hepatoprotective activity (Aluko et al. 2013b), and the leaves possess analgesic and anti-inflammatory activity (Sripriya and Venkanna 2013), immunomodulatory activity (Sunitha and Begum 2013), and anesthetic activity (Silva et al. 2015).

Botany and Taxonomy

Coleus

It is a small annual bushy herb or under the shrub, can reach to 30–60 cm height, and forms dark brown colored round to oval-shaped small underground tubers with aromatic flavor in about 5 months. The tips of the adventitious roots are modified into small round tubers. It is also called a Hausa potato and has a prostrate or ascending stem. It has simple thick and aromatic leaves (exstipulate) and plants have a purple-colored mark on the central portion of the leaf lamina. Opoku-Agyeman et al. (2007) reported that the color of leaves is primarily green (>90%) with other deviations, i.e., green, light-green, and olive-green. Three landraces of *Plectranthus esculentus* are known, namely Bebot, Riyom, and Longat, whereas *Solenostemon rotundifolius* has two varieties, *S. rotundifolius* var. *nigra* and *S. rotundifolius* var. *alba* (Agyeno et al. 2014). The flowers are either hermaphrodite or male-sterile, pedicellate, and ebracteate. Small flowers are arranged in racemose cymes. The corolla is uniformly greenish-brown and violet colored, sympetalous, two-lipped, and rarely one-lipped. There are four stamens, violet-colored anthers, and stigma, which bear a cluster of dark brown, heteromorph tubes and superior four-lobed ovaries which yield four one-seeded nutlets. The fruit is carcerulus with persistent calyx and the seed is non-endospermic. The plants parts of Chinese potato are shown in Fig. 7.1.

At the base of the primary stem, the coleus produces small clusters of starchy, brown, or black aromatic tubers. *S. rotundifolius* produces small-sized tubers (Prematilake 2010), whereas in India and Sri Lanka, large-sized tubers are produced (Opoku-Agyeman et al. 2007). The skin color of tubers can be red or white or black (Dittoh et al. 1998). As per the skin color, *S. rotundifolius* has three varieties, i.e., var. *nigra* A. Chev. has a black color, var. *rubra* A. Chev. has a reddish-gray or

Fig. 7.1 Chinese potato: (a) plant, (b) flowers, and (c) tubers

Fig. 7.2 Chinese artichoke, *Stachys affinis*: (a) plant, (b) flowers, and (c) tubers

reddish-yellow color, and var. *alba* A. Chev has a white color (Tindall 1983). Despite the differences in skin color, the flesh color of tubers in all three varieties is white (Opoku-Agyeman et al. 2007).

Chinese Artichoke

Stachys affinis is a perennial herbaceous plant originating from north-western China. It is a perennial, erect, or inclined deciduous plant having a stem height of 30–120 cm. Leaves are ovate-cordate to ovate-oblong, 2.5–9.5 cm long and 1.5–3.5 cm wide (Fig. 7.2).

Flowers are ovoid, 1.5 cm in diameter, tuberculate, and red to purple. The tubers are small, convoluted, and indented. The tuber skin is thin, whitish-brown, or ivory-white. The tuber flesh is tender and of white color. Chinese poets compare the plant to jade beads since its structure is like beads.

Hoary Basil

It is valued for its young leaves and tips, which have a strong minty aroma, and it is used in various cuisines in Arab, Indonesia, Philippines, Lao, Malay, Persia, and Thailand. The plant is perennial, bushy, and vigorous which grows to about 15–60 cm in height and has subquadrangular striate branches (Fig. 7.3).

Leaves are elliptical lanceolate, entire or faintly toothed, glabrous, and gland-dotted. The flowers' color may be white or pink or purple arranged in elongated racemes. Fruits are small notelets, pitted, and mucilaginous when wet and generally found growing in open fields and wastelands.

Climate and Soil

Coleus

It can grow over a varied range of climate and soil conditions (Agyeno et al. 2014). Unlike other tuber crops, it is a transplanted vegetable. Coleus can grow well under both tropical and subtropical conditions but comes up well in a hot, humid climate where there is no incidence of frost. Cooler nights and hot humid days favor tuberization. Moreover, it requires a frost-free growth period. Like yam bean, it requires evenly distributed rainfall during its growth period and cannot withstand drought conditions. The ideal temperature for its growth and development ranges from 25 to 30 °C. It can grow in any type of soil, provided it is rich in organic matter and well-drained. However, for optimum yield, sandy loam to alluvial soils is the best. Waterlogged soils should be avoided as too much moisture reduces the tuber yields. For this reason, it is desirable to grow the crop on ridges than on flatbeds. It thrives best in soils with pH ranging from slightly acidic to neutral (5.5–7.0), which is suitable for its cultivation.

Fig. 7.3 Hoary basil, *Ocimum americanum*: (**a**) plant, (**b**) flowers, and (**c**) fruits

Chinese Artichoke

It prefers light, well-drained soil with good sunlight and can thrive in normal garden soil. It grows best in well-manured soil that does not dry out during the growing season. Plants can even withstand waterlogging conditions during winter.

Hoary Basil

The optimum temperature required for germination is 20–24 °C. It takes about 5–14 days for germination. Its growth occurs best in an area that receives full sunlight for at least 6 h. The plants are very hardy and can tolerate winter when grown in subtropical and tropical regions.

Agronomic Practices

Coleus

Propagation is mostly done by vine cuttings; therefore, for the production of vine cuttings, approximately 150–170 kg tubers are sufficient for 500 m^2 area which produce cuttings for one hectare area. Usually, tuber cuttings are taken from the previous season and before planting of tubers, 100–200 kg of well-decomposed farmyard manure is added in the nursery bed. The planting of tubers is done in April–May in the nursery bed and then transplanted the cuttings in the main field after 30–40 days. Usually, 1 kg of the tuber is sufficient to plant about 75–100 m^2 area. To minimize the requirement for seed tubers, the tubers are first planted in the nursery. The tubers start sprouting within 15 days and give rise to several sprouts. These sprouts grow to a height of 15–20 cm in about 3 weeks and can be readily transplanted in the field. Stem cuttings from these sprouts with a length of 10–20 cm can also be used as planting material, but planting suckers directly is desired for better establishment. Top suckers having 4–5 leaves are picked up and transplanted in the field. Within another 15 days, the initially planted tubers again form a cluster of sprouts ready to transplant in the field. This way, when the top suckers are picked at intervals from the nursery bed, 50 kg seed tubers can form enough suckers in about 2 months to plant a one-hectare area. However, in this case, staggered planting has to be followed.

The land for growing coleus should be thoroughly ploughed, leveled, and brought to fine tilth after adding 10–25 tons of well-decomposed farmyard manure or compost. The crop can be raised in both upland and lowland areas provided there is no waterlogging. The crop can be raised either on flat raised beds or small ridges. Seed tubers weighing about 7–10 g are planted at a spacing of 90 cm × 20 cm to achieve high yield (Bayorbor and Gumah 2007). In Kerala, however, the recommended spacing for coleus cultivation is 45 cm × 30 cm (Ravindran et al. 2013). While planting, a spacing of 60 cm × 30 cm is maintained, thus accommodating 55,000 plants per hectare. Fifteen to 20 cm long stem cuttings with 5 leaves are planted on 15–20 cm high ridges. The ridges are made at a distance of 30–50 cm. A plant spacing of 60 cm × 45 cm (Hrishi and Kumar 1976) or 60 cm × 15 cm (Geetha 1983) is also suggested for coleus.

Besides the ordinary ridge method, coiled planting and horizontal planting methods are also practiced. In the coiled planting method, 22 cm long stem cutting is used and about 12 cm of the more mature portion is coiled and planted in 7 cm wide and 5 cm deep holes, whereas in the horizontal planting method, 30 cm long stem cuttings are used and two similar cuttings are kept sidewise in opposite directions touching each other across the ridges. A two-thirds portion of these stem cuttings is placed in the soil, while the remaining portion projects outside the ridges. This way, planting the stem cuttings in pairs at a distance of 7–10 cm requires more planting material. Among these three methods, the coiled planting method gives maximum tuber yield. It has been observed that dense planting produces smaller-sized tubers but suppresses the weeds.

July–August or November–December is the ideal season for planting coleus in the main field. However, planting during the first week of October gives the highest tuber yield followed by planting in January and September (Singh and Mandal 1976). The later planting season can be followed in areas where irrigation facilities are available. In most parts of India, its cultivation is done during the rainy season. Two or three weeding can be done as per the requirement. Two months after planting, earthing up can be done.

Chinese Artichoke

Usually, it is propagated by tubers and very rarely by cuttings. It can also be propagated via seeds but seeds are scarce. It is vegetatively propagated using tubers in the spring season, i.e., from March to May. The tubers are planted into a pit of 30 cm × 30 cm size at a depth of 5–8 cm. Weeding is compulsory but care should be taken that the roots are not damaged. It is important to ensure that adequate water is supplied during the summer season. The plants are very tolerant to a higher temperature during the summer season. The tubers sprout when the temperature is more than 5 °C. Tuber development requires about 5–7 months. Flowering starts from July to August, while fruit set takes place in September.

Hoary Basil

It is usually propagated by seeds, which are sown in a seedbed and germinate after 1–2 weeks of sowing with epigeal germination. For indoor planting, March to April is the suitable time, while for outdoor conditions April to June is suitable. The seedlings are transplanted after 3–4 weeks at a distance of 20–30 cm. The young leaves should be picked before flowering, for promoting the growth of more leaves and branches. June to September is the flowering season. After 8–12 weeks of sowing, when the plants are about 25 cm tall, they begin to flower, and the plant continues to flower until it dies.

Nutrient Management

Coleus

The field is manured and fertilized before planting by adding 10 tons of farmyard manure per hectare. According to a field experiment conducted by Kerala Agriculture University, one hectare of cuttings can be fertilized with 1–2 doses of 40–60:60:100 kg/ha N:P:K. In addition to 10 tons of FYM, 30 kg of nitrogen and 50 kg of potash along with 60 kg of phosphorus per hectare can be added during earthing up. The application of fertilizer in split doses is very important. Thirty kilograms of nitrogen/ha along with 2 kg of *Azospirillum* can be applied 30 days after planting at the time of earthing up. If the soil gets removed from the basal portion of the plant, earthing up can be done again to encourage tuber production. Earthing up covers a basal portion of the stem with soil, which promotes tuberization and tuber development. The second earthing up can be done 1 month after the first earthing up. Danya and Potty (2007) reported siderophore production by *Pseudomonas fluorescens* from the rhizosphere of Chinese potato.

Varieties

Coleus

In different regions of India, generally, local cultivars are popular, which are of two types, small-sized tubers with good flavor and large-sized tubers having higher yield.

The cultivar Sree Dhara, which was released by Kerala Agriculture University (KAU), recorded a higher yield ranging from 25 to 29 tonnes/ha over 6 months.

Nidhi is a medicinal cultivar resistant to major pests; nematodes have been released by KAU for cultivation, which matures 15 days earlier than conventional types and takes approximately 135 days for maturity, and produces 27 tons/ha yield.

Harvesting and Yield

Coleus

When the vines get dried, about 4–5 months after planting, harvesting can be done by pulling out the plants, and the remaining tubers are removed from the soil. The yield may range from 15 to 20 tons/ha in 120 days.

Chinese Artichoke

The tubers are manually harvested after 6–8 months of planting. The smaller size and dispersed growth of tubers in the soil make the harvesting more time-consuming. Tubers can be harvested from October onwards. Though the above-ground portion of the plant can be damaged by frost, the tubers being very hardy can be left underground during the winter and are harvested as and when required. Under optimum growing conditions, a plant may give rise to about 40–220 tubers depending on the planting material, and the yield of 8–20 tons/ha can be obtained. The tubers dry out and get discolored if kept in the open air; therefore, tubers should be marketed immediately after harvesting.

Hoary Basil

Harvesting can be done after 2–3 months of planting. Harvesting of leaves can be done from July to September at regular intervals but the plant should never be completely defoliated. Usually, young shoots of 10 cm in length are harvested, but the whole plant can also be cut along with roots. Seeds mature after 14–20 weeks from sowing. The dried inflorescences are cut, sun-dried, and threshed by beating.

Plant Protection

Coleus

Root-knot nematode is a serious coleus problem that causes severe swelling or gall formation in the roots, thereby resulting in suppressed root growth, stunting, and wilting. The nematodes enter the seedling's roots when it is most vulnerable. Therefore, nematodes can be controlled by sowing tubers that are free from nematodes. Deep ploughing is done to expose the soil and kill the nematodes. Summer fallowing and soil solarization can be practiced. Sweet potato variety Sree Bhadra can be cultivated as a preceding crop during May–June, which helps in trapping root-knot nematodes in the soil.

Before planting, dip the vines in an insecticidal solution for about 10 min to control leaf folding caterpillars and vine borers. In case of severe damage in the field, insecticides can be sprayed at 1 ml/l.

Conclusion

Chinese potato, Chinese artichoke, and hoary basil are important minor vegetables of the Lamiaceae family having several uses. Being nutritionally important, there is good potential to bring these vegetables under cultivation in a suitable cropping pattern. There is a need to increase awareness about the cultivation and consumption of these valuable vegetables so that more people can include these in their balanced diet to get multiple health benefits. Furthermore, the different value-added products prepared by utilizing these vegetables can also be popularized among urban residents, simultaneously enhancing the income of resource-poor farmers.

References

Abraham, M., & Radhakrishnan, V. V. (2005). Assessment and induction of variability in coleus (*Solenostemon rotundifolius*). *Indian journal of agricultural science*, *75*(12), 834–836.

Agyeno, O. E., Jayeola, A. A., Ajala, B. A., & Mamman, B. J. (2014). Exo-morphology of vegetative parts support the combination of *Solenostemon rotundifolius* (Poir) JK Morton with *Plectranthus esculentus* NE Br. Natal (Lamiaceae) with insight into infra-specific variability. *Advances in Agriculture & Botanics*, *6*(1), 16–25.

Aluko, B. T., Oloyede, O. I., & Afolayan, A. J. (2013b). Hepatoprotective activity of *Ocimum americanum* L Leaves against paracetamol–induced liver damage in rats. *American Journal of Life Sciences*, *1*(2), 37–42.

Aluko, B. T., Oloyede, O. I., & Afolayan, A. J. (2013a). Polyphenolic contents and free radical scavenging potential of extracts from leaves of *Ocimum americanum* L. *Pakistan journal of biological sciences: PJBS*, *16*(1), 22–30.

Ashraf, K., Haque, M. R., Amir, M., Ahmad, N., Ahmad, W., Sultan, S., Shah, S.A.A., Alafeefy, A.M., Mujeeb, M. & Shafie, M. F. B. (2021). An overview of phytochemical and biological activities: *Ficus deltoidea* Jack and other *Ficus spp*. *Journal of pharmacy & bioallied sciences*, *13*(1), 11.

Bayorbor, T. B., & Gumah, A. Y. (2007). Effects of 'seed' tuber weight and spacing on the yield of Frafra potato (*Solenostemon rotundifolius*). *Ghana Journal of Horticulture*, *6*, 41–48.

Danya, M. K., & Potty, V. P. (2007). Sideriphore Production by *Pseudomonas fluorescens* isolated from rhizosphere *of Solestemon rotundifolius*. *Journal of Root Crop*, *33*, 138–140.

Dittoh, J. S., Bayorbor, T. B., Yidana, J. A., Abapol, R. R., & Otoo, J. A. (1998, November). The potential and constraints of persa (Frafra potato) as a food security crop in Northern Ghana. In *Proceedings of the 1st Biennial National Research Systems (NARS) Workshop* (pp. 2–15).

Fedorov, A. (1974). Chromosome numbers of flowering plants. Reprint by Otto Koeltz Science Publishers. West Germany. 926p.

Geetha, K. (1983). *Nutritional management in coleus (Coleus parviflorus Benth)* (Doctoral dissertation, Department of Agronomy, College of Agriculture, Vellayani).

Harlan, J. R., Dewet, J. M. J., & Stemler, A. B. L. (1976). *Origins of African plants domestication.* Monton, The Hague, The Netherlands.

Horvath, T., Linden, A., Yoshizaki, F., Eugster, C. H., & Rüedi, P. (2004). Abietanes and a Novel 20-Norabietanoid from *Plectranthus cyaneus* (Lamiaceae). *Helvetica Chimica Acta, 87*(9), 2346–2353.

Hrishi, N., & Kumar, C. R. M. (1976). Coleus for homestead gardens. *Indian farming.*

Jayapal, A., Swadija, K., & Anju, V. S. (2016). Effect of Organic Nutrition on Quality Characters of Chinese potato (*Plectranthus rotundifolius*). *Journal of Root Crops, 41*(1), 56–58.

Morales, M. R., & Simon, J. E. (1997). *'Sweet Dani': a new culinary and ornamental lemon basil* (No. RESEARCH).

Opoku-Agyeman, M. O., Bennett-Lartey, S. O., Vodouhe, R. S., Osei, C., Quarcoo, E., Boateng, S. K., & Osekere, E. A. (2007). Morphological characterization of Frafra potato (*Solenostemon rotundifolius*) germplasm from the savannah regions of Ghana. In *Plant Genetic Resources and Food Security in West and Central Africa. Regional Conference, 26-30 April 2004. Bioversity International, Rome, Italy* (Vol. 472, p. 116).

Paton, A., Harley, R. M., & Harley, M. M. (1999). Ocimum-an overview of relationships and classification. *Ocimum Aromatic Plants-Industrial Profiles. Amsterdam: Harwood Academic.*

Paxton, J. (1868). *Paxton's botanical dictionary: comprising the names, history, and culture of all plants known in Britain: with a full explanation of technical terms.* Bradbury, Evans, & Company.

Prakash, P. A. G. N., & Gupta, N. (2005). Therapeutic uses of *Ocimum sanctum* Linn (Tulsi) with a note on eugenol and its pharmacological actions: a short review. *Indian journal of physiology and pharmacology, 49*(2), 125.

Prathibha, S., Nambisan, B., & Leelamma, S. (1995). Enzyme inhibitors in tuber crops and their thermal stability. *Plant Foods for Human Nutrition, 48*(3), 247–257.

Prematilake, D. P. (2010). Inducing genetic variation of innala (*Solenostemon rotundifolius*) via in vitro callus culture. *Journal of the National Science Foundation of Sri Lanka, 33*(2).

Pushpangadan, P., Bradu, B. L., Chadha, K. L., & Gupta, R. (1995). Advances in Horticulture. Medicinal and Aromatic Plants.

Ravindran, C. S., Ramanathan, S. & Eswaran, M. (2013). Coleus. In: Agro techniques of tuber crops. Central Tuber Crops Research Institute. pp. 26–28.

Sandhya, C., & Vijayalakshmi, N. R. (2000). Antioxidant activity of flavonoids from Solenostemon rotundifolius in rats fed normal and high fat diets. *Journal of nutraceuticals, functional & medical foods, 3*(2), 55–66.

Sethi, S., Prakash, O., Chandra, M., Punetha, H., & Pant, A. K. (2013). Antifungal activity of essential oils of some *Ocimum species* collected from different locations of Uttarakhand.

Silva, L. D. L., Garlet, Q. I., Koakoski, G., Abreu, M. S. D., Mallmann, C. A., Baldisserotto, B., Barcellos, L.J.G. & Heinzmann, B. M. (2015). Anesthetic activity of the essential oil of *Ocimum americanum* in Rhamdia quelen (Quoy & Gaimard, 1824) and its effects on stress parameters. *Neotropical Ichthyology, 13*, 715–722.

Simpson, B. B., & Conner, O. M. (1986). Economic Botany-Plants in Our World McGraw-Hill Book Company. *Hamburg, 640.*

Singh, K. D., & Mandal, R. C. (1976). Performance of coleus and sweet potato in relation to seasonal variations, time of planting. *J. Root crops, 2*(2), 17–22.

Sripriya, D., & Venkanna, L. (2013). Estari Mamidala. *Analgesic and anti-inflammatory effects of Ocimum americanum (Linn) In Laboratory Animals. IJPSR, 201*(2), 8.

Sunitha, K., & Begum, N. (2013). Immunomodulatory activity of methanolic extract of *Ocimum americanum* seeds. *International Journal of Research in Pharmacy and Chemistry, 3*(1), 95–98.

Sutili, F. J., Velasquez, A., Pinheiro, C. G., Heinzmann, B. M., Gatlin III, D. M., & Baldisserotto, B. (2016). Evaluation of *Ocimum americanum* essential oil as an additive in red drum (*Sciaenops ocellatus*) diets. *Fish & shellfish immunology, 56*, 155–161.

Thaweboon, S., & Thaweboon, B. (2009). In vitro antimicrobial activity of *Ocimum americanum* L. essential oil against oral microorganisms. *The Southeast Asian Journal of Tropical Medicine and Public Health, 40*(5), 1025–1033.

Tindall, H. D. (1983). *Vegetables in the Tropics*. Macmillan International Higher Education.

Vieira, P. R., de Morais, S. M., Bezerra, F. H., Ferreira, P. A. T., Oliveira, Í. R., & Silva, M. G. V. (2014). Chemical composition and antifungal activity of essential oils from *Ocimum species*. *Industrial Crops and Products, 55*, 267–271.

Chapter 8
Production Technology of Underutilized Vegetables of Solanaceae Family

T. Chamroy

Introduction

India is bestowed with a varied range of agro-climatic conditions that make it possible to grow diverse kinds of vegetation. However, only a few common vegetables are cultivated or adopted by farmers and grown on a large scale due to their well-known improved agro-techniques. There are many more nutritious vegetables that are lesser known and unexplored neither commercially on large scale nor traded widely despite their varied uses and importance. The unavailability of quality planting materials, unawareness of nutritional and medicinal properties, and meager information on technical know-how on the production of these crops might be the reason for their underutilization. Among the solanaceous underutilized crops are, for instance, tree tomato/tamarilo, husk tomato/tomatillo, Thai brinjal, bitter brinjal, and bird's eye chilli, having immense potential in meeting the nutrient requirements being rich in various vitamins and minerals. According to Jaenicke and Höschle-Zeledon (2006), the underutilized crop being considered as food for the poor, their consumption is not socially acceptable by some communities. Therefore, to untapped the potentiality of these vegetables to meet the shortage of food per capita consumption, thereby ensuring food and nutritional security among the rural community in particular and the nation in general, more emphasis should be given to taking up programs on genetic resources exploration, management, utilization, and improvement.

T. Chamroy (✉)
School of Agriculture, Lovely Professional University, Phagwara, Punjab, India

Savita et al. (eds.), *Production Technology of Underutilized Vegetable Crops*,
https://doi.org/10.1007/978-3-031-15385-3_8

The lists of some underutilized solanaceous vegetables are as follows:

Crop	Botanical name
Tree tomato	*Solanum betaceum/Cyphomandra betacea/Cyphomandra hartwegi*
Husk tomato	*Physalis philadelphica* and *Physalis ixocarpa*
Bitter eggplant	*Solanum indicum*
Thai eggplant	*Solanum melongena*
Bird's eye chilli	*C. minimum* Syn. *C. fastigiatum*

The production technologies of some underutilized solanaceous vegetable crops are described in this chapter as follows.

Tree Tomato (*Solanum betaceum/Cyphomandra betacea/Cyphomandra hartwegi*)

Tree tomato, also commonly known as tamarillo, is a small perennial tree or shrub bearing an egg-shaped edible fruit. Its chromosome no. is $2n = 24$.

Origin and Distribution

The crop is native to South America, most likely in the Peruvian Andes. Other distinct locations of origin are Argentina, Chile, Bolivia, Colombia, Venezuela, and Ecuador. It was late in the 1800s, the crop was introduced in India by explorers, where it was grown relatively unknown throughout the hilly regions. The large-scale production of these crops was found in the 1930s, and it became considerably popular during World War II. It was in the year 1967, the commercial name of "tree tomato" was changed to "tamarillo" so as a way to distinguish it from the classic tomato. At present, countries such as Colombia and Ecuador produce tamarillo commercially. A handful of countries in Africa, Asia, Australia, and California are other growing regions of these crops.

In India, tamarillos are seen growing in the mid-hills of Assam, Uttarakhand, West Bengal (Darjeeling), Maharashtra, Sikkim, Nagaland, Manipur, Himachal Pradesh, and Nilgiri Hills in the South (Shaiyelee 2016).

Nutritional Importance and Uses

The fruits have a high nutritive value with low calories. The fruit is a rich source of various vitamins and minerals, acts as a protective food by providing good skin health, boosting immunity, bone health, ideal red blood cell levels, anticarcinogenic as the proteins in the fruit *reduced oxidative damage*, and prevents the formation of uric acid. Tree tomatoes also contain *lycopene*, which has important antioxidant properties that enhance the skin to withstand UV radiation. As per the analysis made in Ecuador, Guatemala, and India, 100 g of the edible portion of the fruit contains 82.7–87.8% moisture, 1.5 g protein, 10.3 g carbohydrates, 1.4–4.2 g fiber, 3.9–11.3 mg calcium, 52.5–65.5 mg phosphorus (with seeds), 0.66–0.94 mg iron, 0.371–0.653 mg carotene, and 23.3–33.9 mg ascorbic acid.

The unripe fruits are used for pickling, making chutneys and sauces for use in ice creams, and ripe fruits are eaten as such by scooping out the flesh. Chutneys are commercially prepared in Auckland, New Zealand. The fruit is rich in pectin content and can be utilized in the preparation of jelly.

Botany and Taxonomy

Tree tomato is a perennial, fast-growing tree, growing up to 5 m and forming a single vertical trunk with lateral branches on which flowers and fruits are found hanging down. The leaves are somewhat cordiform at the base and pointed at the apex, 4–12 cm broad and 10–35 cm long with conspicuous coarse veins, evergreen, arranged alternately, and has a strong pungent smell. The tree blooms in the early spring, producing fragrant flowers that are pink to white, and forming clusters of 10–50 flowers. Only about 12.3% of the flowers in an inflorescence set fruit on an average out of which only 3.3% develop to maturity. The earlier inflorescences borne on lower branches produce more fruits than those at successive positions along a branch (Lewis and Considine 1999).

The tree bears 1–6 egg-shaped fruits per cluster, which are about 4–10 cm long. The color of the fruit varies from yellow and orange-red to almost purple, sometimes having dark longitudinal stripes. Red-colored fruits are sourer than yellow and orange-colored fruits. The flesh of tree tomato is firmer in texture and the fruit contains more seeds that are larger compared to the common tomato. The tree has a very shallow roots system that is not very pronounced; therefore, it cannot tolerate drought stress and also the branches are easily breakable when fully loaded with fruits. Thus, a location with shelters should be selected or rows of windbreaks must be planted (Verhoeven 2016).

Genetic Resources

The NBPGR from Argentina collected one accession IC349909 from the local gene bank and introduced two accessions, i.e., EC25897 and EC39826 in 1962 and 1967, respectively. Other important cultivars are Red Beau, Gold Mine, Bold Gold, Inca Gold, Oratia Red, Rothman, Ruby Red, and Solid Gold (NBPGR, Phagli, Shimla).

Due to their attractive color, red fruit cultivars are more in demand for the fresh market, whereas yellow fruits are ideal for canning due to their better flavor (Morton 1987; Verhoeven 2016).

Climate and Soil

Tree tomatoes are a subtropical shrub, growing between elevations of 1000 and 7500 ft. The crop prefers a good amount of rainfall ranging from 600 to 4000 mm and an average temperature between 15 and 20 °C. It cannot withstand drought and frost (<–2 °C) stress. The night temperature may affect the fruit set. The plants require deep and fertile soils with good organic matter content. However, soils must be having good aeration and drainage, as they are sensitive to waterlogging conditions. The favorable soil pH for the natural growth of the plant is 5–8.5. The plant needs to be grown in full sunlight, though in hotter regions it can be planted in areas with partial shade. Sufficient windbreaks should be provided around as the tree is shallow-rooted and gets easily uprooted by strong winds.

Agronomic Practices

Propagation

Seeds are generally used for propagation. However, cuttings/grafting can also be practiced for propagation to avoid cross-pollination with different cultivars, especially in the open field. Propagation by seed is ideal under protected environmental conditions. Propagation is done from cuttings that should be free from viruses, taken from basal and aerial shoots. Cuttings may be taken from a 1–2-year-old shrub having 10–30 mm thick shoot diameter and 45–100 cm long.

Nursery Raising

Seedlings are raised in the nursery under protection until a height of about 1–1.5 m, as they are non-tolerant to frost conditions. Seedbeds should be incorporated with a good amount of organic manures and provided with partial shade. A seed rate of

200–250 g is sufficient to raise seedlings for transplanting one-hectare area. Seed sowing can be done from April to May.

Transplanting

The seedlings are ready for transplanting between 5 and 6 weeks of seeding in the nursery or 0.5–1 m height. Seedlings should be hardened off before transplanting to enable them to withstand transplanting shock and reduce the mortality rate.

Planting Distance

Plant spacing depends upon regions and planting systems. In New Zealand, single row planting at a spacing of 4.5–5 m × 1–1.5 m is recommended. In the Andean region, planting is done more densely, with a spacing of 1.2–1.5 m between plants. In Indian conditions, slightly closer spacing of 3 m × 1.5 m is practiced.

Plant Management

Proper plant management in tree tomato cultivation is important for improving the performance and acquiring its maximum potentiality.

Pruning

Pruning is necessary to regulate plant growth and simultaneous maturity. Initially, removal of the tip of young plants leads to the development of the desired plant stature. Later, once the tree had retained its proper height, pruning should be done just to remove old/dead branches and previously fruited branches, as it produces smaller fruits with inferior fruit quality. Light pruning results in bearing medium-sized fruits and heavy pruning results in large-sized fruits. At about 1–1.5 m plant height, the roots on one side may be cut and the tree may be leaned to the opposite side to allow more fruiting branches all along the trunk rather than just at the top.

Mulching

The plant being shallow-rooted and having a sensitive root system is non-tolerant to drought stress. Mulching is an important intercultural operation to preserve soil moisture, suppress weeds, and improve soil conditions.

Shelter

During full fruiting, the plant being shallow-rooted does not provide adequate support and lateral branches being fragile are easily breakable. Therefore, plants need to be protected against strong winds by planting windbreaks around the planting area.

Irrigation

Irrigation water should be supplied continuously as they are shallow-rooted and prone to water stress conditions, resulting in a decreased plant growth, fruit size, and productivity.

Nutrient Management

The recommended dose of nutrients is 170:45:130–190 kg N:P:K/ha. P and K are applied in the initial growing season, while N should be splitted in several doses equally throughout the year.

Harvest and Yield

The tree is fast-growing and it comes to bearing fruits after 1.5–2 years. Several harvestings are necessary as the ripening of fruits does not occur simultaneously. Flowering and fruit set occurs throughout the year in regions with a little annual variation. However, in regions with pronounced seasons, fruits ripen in the autumn season. The peak production of fruit is achieved after 4 years of planting and the life span of the crop is about 12 years. On average, a single tree yields more than 20 kg of fruits per year, and the yield per hectare is about 15–17 tons.

Postharvest

The fruit is climacteric in nature, so ripening can be achieved with minimal loss in fruit quality by harvesting premature fruits and inducing ripening using ethylene in a controlled atmosphere chamber. These fruits can be stored for up to 20 days at ambient temperature. However, in cold storage at a temperature of about 3.5 °C ± 1 °C, the fruit can be stored for up to 8 weeks or more. According to Reddy (2013), the fruits that require ripening can be left at room temperature for further ripening process or they can be placed in perforated packaging materials and can be stored in the refrigerator for up to 2 weeks.

Diseases and Pests

Tree tomato mosaic, cucumber mosaic The tree tomato, even though susceptible to CMV (cucumber mosaic virus) and potato virus, is known for its resistance to TMV (tomato mosaic virus). Aphids are probably the main vectors for the spread of diseases, leading to reduced productivity of the crop.

Management Use of healthy planting materials, avoiding planting near older plants, and control of vectors.

Root-Knot Nematodes (*Meloidogyne* spp.) High temperature and moisture stress conditions encourage the incidence of root-knot nematodes in the plant. The disease results in stunted growth and unproductive plants.

Powdery Mildew *Erysiphe* spp. and *Oidium* sp. causes the older leaves to shed prematurely and severe defoliation if not controlled.

Management Maintaining a good vegetative growth to overlook the loss of leaves and treatment with sulfur at regular intervals.

Mites and Aphids Controlled by spraying Malathion (0.05%) or Endosulphan (0.05%).

Husk Tomato/Tomatillo (*Physalis philadelphica* and *Physalis ixocarpa*)

The tomatillo, also known as "Husk tomato," "Mexican ground cherry," and "Mexican husk tomato," is an annual plant of the solanaceous crop. The fruits are tart in taste, having small and more or less round-shaped with green or green-purple colored.

Origin and Distribution

Tomatillos or husk tomatoes are known to have been originated from Central America and Mexico. In the United States, it was known to have been cultivated since 1863; further, it was distributed in the Bahamas, Jamaica, Puerto Rico, and Florida. In India, it was introduced in the 1950s, where the fruit was added to a variety of traditional dishes and is locally cultivated. Later on, it was distributed to Australia, South Africa, and Kenya by the middle of the twentieth century. Currently, countries such as the United States, Mexico, Canada, and Spain are the main contributors in the wholesale tomatillo market (Tridge Company 2020).

Nutritional Importance and Uses

Husk tomatoes are a rich source of dietary fibers, vitamins, minerals, and phytonutrients or antioxidants. According to USDA National Nutrient Data, tomatillos per 100 g of berries provide 32 kcalories, 5.84 g carbohydrates, 1.9 g dietary fiber, 11.7 mg vitamin C, 1.85 g niacin, 114 IU vitamin A, 39 mg phosphorus, 20 mg magnesium, 7 mg calcium. It has antioxidant phytochemicals known as withanolides (*Ixocarpa lactone-A*), which have been found to have antibacterial and anticancer properties. The fruit contains antioxidants viz. β-carotene 63 μg and lutein–zeaxanthin 467 μg. It is also a rich source of Cu, Fe, P, Mn, and other minerals.

The fruits can be consumed raw when ripen, processed into jams or chutneys, and soups. In Mexico and Guatemala, tomatillos along with hot peppers are processed to make a sauce, which is used to add taste to meats and other foods by roasting and grounding them together.

The plants are widely known as folk medicine, treating various ailments like malaria, dermatitis, asthma, hepatitis, rheumatism (Wu et al. 2005), worm complaints (Ahmad et al. 1999). Husk tomato has the potential of anti-carcinogenic effects and antioxidants due to its high β-carotene content. It is also used for bile juice secretion and activating the liver function and antibiotic activity (Perry and Matzger 1980).

Botany

The plant grows to about 30–60 cm in height, bears typically yellow color flowers with dark spots toward the base after 60 days of sowing. Unlike any other solanaceous fruits, the berry of tomatillo develops inside a thin, semitransparent calyx or thin papery husk which later on upon maturity, dries out and splits opens to expose the berry inside. The fruit size varies from 2.5 to 4 cm in diameter weighing about 30–50 g with numerous seeds.

Varieties

Mayan Husk Tomato, Purple Coban, Tiny and De Milpa (purple), Amarylla (green), Cape gooseberry, Golden Nugget, Mexican Husk, and Rendidora.

Climate and Soil

A plant-like tomato is a warm-season crop, sensitive to frost conditions. The area having full sunshine and is well-drained and the soil has a pH of 7.0. The best day temperature being 26–32 °C and 15–21 °C at night, low humidity, and sparse rainfall.

Agronomic Practices

Nursery and Transplanting

Seedlings are raised before the last frost is expected. The seeds are sown 1/4 inch deep at a spacing of 45–60 cm apart in rows. Seeds of 158 g is sufficient to raised seedlings of a one-hectare area. The seedlings are transplanted in the main field after 6–8 weeks of sowing.

Harden the plants when the weather reaches 10 °C by setting them outdoors gradually. The tomatillo grows well in tomato cages or on its own. If grown in cages, set the plants 60 cm apart, or if they are allowed to sprawl, set them at around 91 cm. Generally, 90 cm × 90 cm spacing is required. For the tomatillo flowers to set fruit at least two plants should be grown per hill.

Nutrient Management

Before planting, mix a good amount of compost in the soil or add 10–10–10 fertilizer. For more number of flowers initiation and fruit production, the plants should be fertilized again when the first flowers start initiating, using a 5–10–10 fertilizer. Higher fruit yield may be obtained by applying a higher dose of P and K. According to Shabana (2016), husk tomato applied with a combination of 50% mineral NPK + 50% compost + sulfur amendment is best for all the growth, yield, and quality attributes.

Irrigation

The plants do well without much water but do not tolerate drought conditions. An organic mulch of about 2–3 inches can be a great way to help retain moisture and check weed growth.

Harvesting and Yield

The fruit is ready to harvest in 65–85 DAT (days after transplanting) and continues up to 1–2 months. The harvesting indices are firm fruits, the husk become papery, and husk or calyx just begins to split. According to Heber (2019), the fruit is ready for harvesting at about 75–100 days after transplanting when they fill out their husks and the husks just begin to split. On average, the plant bears 60–200 fruits/plant/growing season. The per-hectare yield is about 22 tons. The fruits can be stored successfully at 12–15 °C with 85–90% relative humidity for up to 3 weeks.

Pests and Diseases

Aphids Infestation causes discoloration of the leaves as they suck the cell sap from the plants and stunted growth.

Control Growing varieties tolerant to the pest and application of insecticides in case of heavy infestation.

Cutworms Causes severe damage to the roots of the plants; the stems of young plants are cut at the soil line.

Control Soon after harvesting the crop debris should be removed from the soil, spraying azadirachtin, carbaryl, deltamethrin, flubendiamide at stem collar region covering at least 7 cm above the soil line.

Slugs Feeds on the leaves at night making large holes on foliage.

Control Handpick the slugs or use a trap made of cornmeal or beer.

White Fly Spraying azadirachtin/pyrethrin.

Anthracnose Causes circular lesions on the fruit.

Management Use of disease-free seen seeds, practice crop rotation with non-related plants, seed treatment with hot water before planting, collection and dumping of crop debris in the soil after harvesting of the crop to prevent the presence of inoculum in the soil.

Bacterial Leaf Spot Damages can be translucent spots with yellow edges that slowly enlarge and may further become irregularly circular with a red-like center.

Management Practice crop rotation with non-related species and destruction of infected plant debris.

Root-Knot Nematodes Mainly infects the roots of the plants causing galls to form on the root and plants wilt or become stunted.

Management Planting varieties of Nema-Gone marigolds around the field and use of nematicide.

References

Ahmad, S., Malik, A., Yasmin, R., Ullah, N., Gul, W., Khan, P. M., Nawaz, H. R. & Afza, N., (1999). Withanolides from *Physalis peruviana*, *Phytochemistry*, 50, 647–651.

Heber, G. (2019). How to grow and harvest tomatillos (January 30, 2019). https://gardenerspath.com/plants/vegetables/tomatillos/.

Jaenicke, H. & Höschle-Zeledon, I. (2006). Strategic framework for underutilized plant species research and development: With special reference to Asia and the Pacific, and to Sub-Saharan Africa. *Bioversity International.*

Lewis, D. H. & Considine, J. A. (1999). Pollination and fruit set in the tamarillo (*Cyphomandra betacea* (Cav.) Sendt.). *New Zealand Journal of Crop and Horticultural Science,* 27,113–123.

Morton, J. F. (1987). Fruits of warm climates. Creative Resource Systems, Winterville, N.C., USA. 437–440.

Perry, L. M., & Matzger, J. (1980). Medicinal plants of East and Southeast Asia MIT Press. *Cambridge MA*, 620.

Reddy, C. (2013). The earth of India, All about tamarillo. http://theindianvegan.blogspot.com/2013/03/all-about-tamarillo.html

Shabana, A. I. (2016). Organic Husk Tomato (*Physalis peruviana* L.) Production for Exportation. *Journal of Plant Production*, *7*(8), 843–850.

Shaiyelee, (2016). Tree tomato: An exotic fruit of Nilgiri Hill region at South India. https://shaileyeesphotographyblog.wordpress.com/2016/11/11/tree-tomato-a-exotic-fruit-of-nilgiri-hill-region-at-south-india/

Tridge Company (2020). Tomatillo. https://www.tridge.com/intelligences/tomatillos/production.

Verhoeven, G. (2016). ***Cyphomandra betacea*,** Plant Resources of South-East Asia. https://uses.plantnet-project.org/en/Cyphomandra_betacea_(PROSEA)

Wu, S. J., Ng, L. T., Huang, Y.M., Lin, D.L., Wang, S. S., Huang, S. N. & Lin, C. C. (2005). Antioxidant of Physalis peruviana, *Biol. Pharm. Bull.* 28, 963–966.

Chapter 9
Production Technology of Underutilized Vegetables of Apiaceae Family

Sophiya Bhatta

Introduction

Apiaceae family also called as Umbelliferae is known as the 16th largest family of flowering plants. It is a family of 3700 species and 434 genera (Gorvett 2017) and is commonly known as carrot, celery, parsley, or fennel family as it includes major and minor vegetables such as carrot (*Daucus carota*), celery (*Apium graveolens*), coriander/cilantro (*Coriandrum sativum*), cicely (*Myrrhis odorata*), cow bean (genus *Oxypolis*), cow parsnip (genus *Heraceleum*), cumin (*Cuminum cyminum*), dill (*Anethum graveolens*), lovage (*Levisticum officinale*), parsley (*Petroselinum crispum*), parsnip (*Pastinaca sativa*), fennel (*Foeniculum vulgare*), anise (*Pimpinella anisum*), asafoetida (*Ferula assa-foetida*), caraway (*Carum carvi*), smallage (*Apium graveolens*), water hemlock (genus *Cicuta*), water parsnip (genus *Sium*), turnip-rooted chervil (*Chaerophyllum bulbosum*), turnip-rooted parsley (*Petroselinum tuberosum*), and skirret (*Sium sisarum*). Several vegetables of this family are toxic such as hemlock water-dropwort (*Oenanthe crocata*), poison hemlock (*Conium maculatum*), and water hemlocks (*Cicuta* spp.). Research has shown that a few Apiaceae species can cause dermatitis when damp skin is exposed to bright sunlight (*Heracleum*, *Pastinaca*). Apiaceae plants are herbaceous and aromatic, and different parts of these plants such as roots, leaves, and stem are extensively used as food, flavor, repellents, spices, and also for medicinal purposes throughout the world. The essential oil derived from these vegetables is rich in antioxidants and aroma and also possesses antimicrobial properties. Along with that, these plants also possess anti-cancerous, hypoglycemic, hypolipidemic, hepatoprotective, and other activities because of which these plants are widely used as an alternative and healthy food for the prevention and treatment of many disorders (Aćimović et al. 2018). These vegetables are also a good source of phytochemicals, which are used for the prevention,

S. Bhatta (✉)
School of Agriculture, Lovely Professional University, Phagwara, Punjab, India

Savita et al. (eds.), *Production Technology of Underutilized Vegetable Crops*,
https://doi.org/10.1007/978-3-031-15385-3_9

treatment, and curing of diseases as the plants belonging to this family possess ethno-medicinal properties.

S. No.	Name	Botanical name	Chromosome number
1	Turnip-rooted parsley	*Petroselinum tuberosum*	22
2	Turnip-rooted chervil	*Chaerophyllum bulbosum*	22
3	Skirret	*Sium sisarum*	40

Origin and Distribution

Since ancient times, the Apiaceae family was known to mankind. The family originated in Australasia (region including Australia, Tasmania, New Zealand, New Guinea, New Caledonia, and several island groups) (Nicolas and Plunkett 2014). Later, subfamilies of the Apiaceae family started to diverge along the southern hemisphere, Apioideae and Saniculoideae in Southern Africa, Azorelloideae in South America, and Mackinlayoideae in Australasia (Calviño et al. 2016). The Apiaceae family is scattered worldwide but is most prevalent and diverse in areas having temperate climate such as Eurasia and North America. It is rarely seen in areas with the tropical climatic conditions. Areas having arid climate and Mediterranean region favor the diversification of species. Asia has maximum number of genera (289) and the largest generic endemism (177) of the Apiaceae family, and Europe has 126 genera and only 17 are endemic. One hundred and twenty-one genera are found in Africa, whereas North Africa has 82 genera, 13 of which are endemic. Northern and Central America possess huge diversity with 80 genera and 44 endemics, whereas South America has less diversity with 35 genera and 15 of which are endemic. Twenty-seven genera with18 endemics are found in Oceania.

Turnip-Rooted Chervil

Turnip-rooted chervil originated from Eastern Europe, and during the first part of twentieth century, it was very popular in France (Bois 1927). Due to lower production of *C. bulbosum*, it almost disappeared in 1975. Since 1978, the National Institute of Horticulture of France (NIH) carried out a research program which helped in its rapid economic development (Péron 1989).

Turnip-Rooted Parsley

Europe and Western Asia are the native places of turnip-rooted parsley. It is cultivated for its aromatic leaves which also possess pharmaceutical properties.

Skirret

The word skirret has been derived from a Dutch word *suikerwortel*, which means "sugar root." Skirret is also known as skirwort and is mainly cultivated for its sweet, edible roots. This member of carrot family originated in Asia and is widely consumed in China and Japan.

Nutritional and Medicinal Uses

The Apiaceae family is known for its phytochemical diversity which was observed by humans from its odor and flavor; therefore, these vegetables are used for various purposes such as foods, beverages, flavorings, remedies, and industrial uses.

Turnip-Rooted Chervil

Turnip-rooted chervil has very high starch content (37% dry matter, containing 76% starch). It also contains other nutrients such as carbohydrates, sucrose, and reducing sugars. Starch is consumed as a staple food by majority of the people throughout the world. It is used in food and beverage industries and also plays a major role in paper and textile industry (Slattery et al. 2000). To meet the current demand of starch, only a few commercially available crops like potato, maize, and wheat are used (Martin and Smith 1995); therefore, cultivation and commercialization of turnip-rooted chervil can be a solution to the limited source of starch. The roots of turnip-rooted chervil are consumed both in raw and cooked form. The raw root has a starchy flavor and is aromatic and tough. Upon cooking, the roots are transformed into sweet and floury forms. Turnip-rooted chervil is called as gourmet vegetable due its biochemical properties and chestnut-like flavor (Imbault et al. 1985).

Turnip-Rooted Parsley

Turnip-rooted parsley, also known as Hamburg root parsley, is one of the most popular vegetables in central Europe (Holland, Germany, Poland, Hungary, and Austria). Its roots are rich in vitamin A, C, and K and also contain a good amount of copper, iron, and iodine. It is also rich in potassium, calcium, phosphorous, sodium, and folic acid. The roots can be consumed as both raw and cooked. Apart from roots, its leaves are also used to garnish soups and stews. The roots contain oxalates, which can get concentrated and crystallized in our body fluids; therefore, people with kidney- or gallbladder-related disorders should not consume turnip-rooted parsley.

Skirret

Skirret is mainly cultivated for its long white tuberous roots and etiolated sprouts. The tuberous roots of skirret are rich in dry matter (16%) and sugars especially sucrose (65% of dry matter). The sprouts of skirret have 8% dry matter and are a great source of sugars (12%) and proteins (25%). The sprouts are richer in vitamins than the roots. Sprouts that are cultivated under greenhouse have more vitamins as compared to those grown under open field conditions (Leclerc and Peron 1988). The roots can be consumed either raw as salad or cooked. It can be consumed after baking, roasting, and frying or as stew. Roots are sweet, firm, and floury but have a woody core. The woody core present in the root should be removed before cooking. Skirrets are good for people having digestion problems, loss of appetite, and chest complaints.

Botany

Turnip-Rooted Chervil

Turnip-rooted chervil is a biennial crop which grows up to a height of 4 feet. The leaves are alternate, spirally arranged, and pinnately compound. Its inflorescence is umbel and produces around 1000–36,000 flowers per plant. Flowers of turnip-rooted chervil are protandrous, and the style elongates only after pollen dehiscence; therefore, there is no self-pollination. It is a hermaphrodite plant and is pollinated by insects (entomophilous).

Turnip-Rooted Parsley

Turnip-rooted parsley is a herbaceous plant which is biennial in nature. It grows up to a height of 2 feet and spreads up to 15–23 cm. Vegetative growth of the plant is completed during the first year, whereas it produces yellow flowers and small dry fruits during the second year. The lower leaves are bi- or triternately divided. It has umbel inflorescence and forms slender roots with white flesh.

Skirret

Skirret is a herbaceous, perennial plant which grows up to a height of 3–4 feet and spreads up to 30 cm width. Leaves of skirret are large, shiny, dark green, compound, and pinnate. It has umbel inflorescence which produces small white flowers. The roots of skirret are cylindrical, grayish white, and 6–8 inches long formed in clusters from the base of the stem like sweet potatoes. It is a hermaphrodite plant and is pollinated by insects (entomophilous).

Production Technology of Underutilized Vegetables

Climate and Soil Requirement of Underutilized Vegetables

Soil

Turnip-rooted chervil, turnip-rooted parsley, and skirret grow best in soil with 5.5–7.0 pH. Loamy soils which are loose, friable, well-drained, and rich in organic matter are best for the cultivation of these crops. Cultivation in heavy and clayey soils should be avoided. Deep to medium ploughing must be done to bring the soil to a desired physical condition as compaction of the soil affects the yield adversely.

Climate

Turnip-rooted chervil, turnip-rooted parsley, and skirret are cool season crops. The roots of these crops develop the best flavor, texture, and size at a lower temperature ranging from 15 to 20 °C. Rise in temperature can result in stunting, forking, bolting, and death of the plants.

Irrigation and Intercultural Operations in Underutilized Vegetables

Irrigation

Turnip-rooted chervil, turnip-rooted parsley, and skirret do not produce good roots under dry conditions; therefore, frequent irrigation is necessary. Drip irrigation system is highly efficient and recommended for their production. A light irrigation should be facilitated before the sowing of seeds for better germination.

Intercultural Operations

Hoeing is practiced during the initial stages of crop development to prevent weed growth, at least two hoeings should be done to prevent the weeds from growing. Chemicals such as influtalin/ethafluralin (1.1 kg/ha), linuron (1 kg/ha), or thiobencarb (6–8 kg/ha) can also be used to control the weeds.

Propagation Material, Edible Parts, and Planting Time of Underutilized Vegetables of Apiaceae Family

Plant name	Propagation material	Edible part	Planting time
Turnip-rooted chervil (*Chaerophyllum bulbosum*)	Seeds (biennial root vegetable)	Roots	October–December
Turnip-rooted parsley (*Petroselinum tuberosum*)	Seeds (annual, biennial root vegetable)	Roots and leaves	August–November
Skirret (*Sium sisarum*)	Seeds or root division (perennial root vegetable)	Roots	October–November

Postharvest Handling

Harvesting

Harvesting should be done at proper stage. Especially, Apiaceae plants grown for roots should be harvested during the cool weather as it will help in extending their storage life and maintaining their postharvest quality. Temperature above 27 °C should be avoided during harvesting. Suitable time for harvesting is early morning or late evening. Harvesting during late evening is suitable for distant markets, whereas morning hour harvesting is suitable for the local market. Size is one of the best maturity indices to harvest most of the Apiaceae family crops. Plants grown both for leaves and roots should be harvested when the leaves and roots have attained a good size but are still tender and juicy. Careful handling must be done to avoid any bruises and cuts.

Harvesting Methods

Plants of the Apiaceae family can be harvested both manually and mechanically. Crops grown for roots are mowed or the roots are pulled out from the ground. Apiaceae plants grown for roots are separated from the leaves or green tops after harvesting the plant with the help of knife.

Cleaning and Washing

Washing or cleaning is done to remove dirt, dust, extraneous matter, and pathogenic load from the surface of vegetables. To ensure proper washing of the commodity, chlorine solution of 100–150 ppm can be used to prevent the buildup of inoculum during the storage. The pH of washing solution should be 6.5–7.5 as it gives the best results.

Sorting/Grading

Plants of the Apiaceae family grown for roots are sorted to remove diseased, damaged, misshapen, overmature, insect-infested, and rotten roots. Roots that are damaged by insect attack should be immediately removed and discarded for preventing further spread to the normal and healthy roots.

Roots lacking firmness, having roughness, poor color, sunburn, splitting, or cracking are also considered as of lower quality.

Storage and Packing

Apiaceae plants grown for roots can be stored at a temperature of 0–1 °C with relative humidity of 98–100% for up to 6–9 months. High humidity is important to maintain the crispness and also controls the desiccation of roots. To store the roots for a long period, artificial cooling must be done before storage. The best method of artificial cooling for the roots of Apiaceae family is hydrocooling. The roots of Apiaceae family are packed in crates, trays, sacks, plastic bags, or nets (Paltrinieri and Staff 2014).

Conclusion

Lack of good quality seeds, marketing, and proper postharvest facilities are the main reasons behind the lesser cultivation of these vegetables. Underutilized vegetables (skirret, turnip-rooted parsley, turnip-rooted chervil) of the Apiaceae family have been neglected for a long time, but proper utilization of such vegetables can contribute to the rural and national economy as these underutilized vegetables require lesser inputs to produce the desired amount of yield. These vegetables can easily withstand harsh and unfavorable climatic conditions which make them suitable to be grown the in remote and marginal land of the country to provide food for the poor rural people. These vegetables have higher nutritional value and consist of valuable nutrients like carbohydrates, proteins, vitamins, and minerals as well as some bioactive non-nutrients that contribute to dietary health which makes them more valuable for cultivation. Lot of farmers can get benefit from the diversification, introduction, and cultivation of these minor crops. New industries can be established, and rural community can be strengthened by the utilization of their products. Moreover, diversification also helps in crop rotation which benefits the agroecosystem by reducing pest and pathogen problems and also improves the soil fertility and texture (Van Soest 1993; Fritz and Myers 2002; Poincelot 2004).

References

Aćimović, M. G., Rat, M. M., Tešević, V. V., & Dojčinović, N. S. (2018). Anticancer Properties of Apiaceae. *Phytochemicals in Vegetables: A Valuable Source of Bioactive Compounds*, 236.

Bois, D. (1927). *Les plantes alimentaires chez tous les peuples et à travers les âges.: Histoire, utilisation, culture. 1. Phanérogrames légumières*. Lechevalier.

Calviño, C. I., Teruel, F. E., & Downie, S. R. (2016). The role of the Southern Hemisphere in the evolutionary history of Apiaceae, a mostly north temperate plant family. *Journal of Biogeography*, *43*(2), 398–409.

Fritz, M., & Myers, R. (2002). Diversifying cropping systems. In: Bulletin 401. Sustainable Agricultural Network, University of Vermont, Burlington VT, USA.

Gorvett, Z. (2017). The Mystery of the Lost Roman Herb.

Imbault, N., Joseph, C., & Billot, J. (1985). Biochemical study of the root reserves of tuberous chervil (*Chaerophyllum bulbosum* L.). *In*: Activity report of the research working group on tuberous chervil: 19 p.

Leclerc, J., & Peron, J. Y. (1988, September). MINERAL, SUGAR AND VITAMIN CONTENTS OF SKIRRET (SIUM SISARUM L.). In *I International Symposium on Diversification of Vegetable Crops 242* (pp. 325–328).

Martin, C., & Smith, A.M. (1995). Starch biosynthesis. Plant Cell 7: 971–985.

Nicolas, A. N., & Plunkett, G. M. (2014). Diversification times and biogeographic patterns in Apiales. *The Botanical Review*, *80*(1), 30–58.

Paltrinieri, G., & Staff, F. A. O. (2014). Handling of fresh fruits, vegetables and root crops: A training manual for grenada. *Rome, Italy: Food and Agriculture Organization of the United Nations*.

Péron, J.Y. (1989). Les potentialités d'élargissement de la gamme des légumes dans lafamille des Apiacées (= ombellifères): l'exemple du cerfeuil tubéreux (*Chaerophyllum bulbosum* L.) et du chervis (*Sisum sisarum* L.). Acta Hort. 242: 123–131.

Poincelot, R.P. (2004). Sustainable Horticulture: Today and Tomorrow. Prentice Hall, Pearson Education Inc., New Jersey.
Slattery, C. J., Kavakli, I. H., & Okita, T. W. (2000). Engineering starch for increased quantity and quality. *Trends in plant science*, *5*(7), 291–298.
Van Soest, L. J. (1993). New crop development in Europe. *New crops*, 30–38.

Chapter 10
Production Technology of Underutilized Vegetables of Brassicaceae Family

Akshita Bisht, Vamsi Krishna, and Savita

Introduction

The Brassicaceae/Cruciferae family is widely dispersed throughout the world, with significant diversity in temperate Eurasia and North America, including California. The family has adapted to a broad range of environments, including wetlands, disturbed habitats (many of which are weeds), foothills, high mountains, and deserts, albeit just a few species may be found in marshes. Cabbage, cauliflower, mustard, radish, and turnip are some of the vegetables that belong to this family. These are the widely cultivated vegetables, but some other vegetables also fall under this family which includes sprouting broccoli, Brussels sprout, kale, and Chinese cabbage. These vegetables are highly nutritious and have medicinal importance also, but their cultivation is limited on small scale due to unawareness among the people and farmers. All the vegetables of this family have medicinal importance because they are rich in sulfur compounds. The details of underutilized cruciferous vegetables are given in Table 10.1.

A. Bisht
Department of Vegetable Science, College of Agriculture, Govind Ballabh Pant University of Agriculture and Technology, Pantnagar, India

V. Krishna
Department of Horticulture, School of Agriculture, Lovely Professional University, Phagwara, Punjab, India

Savita (✉)
Graphic Era Hill University, Dehradun, Uttarakhand, India

Savita et al. (eds.), *Production Technology of Underutilized Vegetable Crops*,
https://doi.org/10.1007/978-3-031-15385-3_10

Table 10.1 Botanical description of cruciferous vegetables

Crop	Botanical name	Chromosome no.	Center of origin
Sprouting broccoli	*Brassica oleracea* var. *italica*	18	Mediterranean
Knol-khol	*Brassica oleracea* var. *gongylodes*	18	Mediterranean
Brussels sprouts	*Brassica oleracea* var. *gemmifera*	18	Mediterranean
Collard/kale	*Brassica oleracea* L. var. *acephala* DC	18	Asia minor
Chinese kale	*Brassica oleracea* var. *alboglabra*	18	South China
Chinese cabbage	*Brassica rapa*, subspecies *pekinensis*, and *chinensis*	20	Northern China
Garden cress	*Lepidium sativum* L.	16, 32	Southwest Asia
Upland cress	*Barbarea vulgaris* R. Br.	16	Europe
Sea kale	*Crambe maritime*	60	Northwest Europe
Catran	*Crambe tatarica* Busch	60	–
Rat tail radish	*Raphanus sativus* var. *caudatus*	18	–

Sprouting Broccoli

Broccoli is a very important and nutritious vegetable, and the plant resembles cauliflower, except instead of curd, and broccoli produces a head (green, purple, or white) with a strong fleshy flowering stem. However, it is very commonly grown in India. In India, broccoli cultivation has not yet been exploited commercially and is limited to kitchen and market gardens located near the big cities and tourist areas. However, its popularity is increasing gradually because of awareness among the educated masses about its high nutritive value, cancer prevention properties, and its potential in the developing tourism industry and also as a relatively new crop.

Origin and Distribution

Broccoli was introduced in the United States and Europe. It is originated from wild cabbage (*Brassica oleracea* var. *sylvestris*). In India, it was introduced during the second half of the twentieth century. Broccoli production ranks fourth in India, behind cauliflower, cabbage, and knoll-khol, though no figure for area utilization and volume of the crop is available. The United States is the main growing country, but broccoli is also grown in Northern Europe, Italy, and cooler regions of the Far East. In India, it is cultivated in Jammu and Kashmir, Himachal Pradesh, Uttar Pradesh, and Nilgiri Hills.

Nutritional Importance and Uses

Sprouting broccoli is a highly nutritious vegetable because of its relatively higher content of minerals (phosphorus, potassium, calcium, sodium, iron, and other trace elements), vitamins (A, B, and C), and proteins whereas low in carbohydrates and fat. Broccoli head exhibited higher levels of vitamin B complex (vitamin B2, B3, B5, B6, B9, and K), amino acids, glucoraphanin, neoglucobrassicin, and some amount of omega-3 fatty acids, while leaves contain three times more carotenoids, chlorophylls, vitamins (E and K), total phenolic content, and antioxidant activity (Liu et al. 2018). Sprouting broccoli is consumed fresh or processed (frozen). Both its heads along with fleshy stems and spears are cooked and consumed like cauliflower. It is consumed as a vegetable, salad, and soups. Because of their high nutritional value, the leaves can also be used for human nutrition as a substitute for kale and the stems for the feeding of animals. Sulforaphane is a well-known anticancer compound present in broccoli (Lee et al. 2017; Vanduchova et al. 2019). It is also effective against COVID-19 (Bousquet et al. 2021). The precursor of sulforaphane is glucoraphanin; however, the amount of sulforaphane produced is determined by the myrosinase enzyme (Jeffery et al. 2003; Moreno et al. 2006; Ares et al. 2013). It is rich in glucosinolates (40–80 mg/100 g of edible portion) and quinone reductase detoxifying enzyme-inducing activity (Lee et al. 2017). Excessive consumption of broccoli causes goiter disease because certain glucosinolates are associated with toxic effects.

Botany and Taxonomy

Sprouting broccoli plant grows erect about 60 cm high with large flower heads. Flowers are yellow and seeds seriate. The inflorescence is a raceme. Flowers are actinomorphic and bisexual (complete). The calyx is gamosepalous with four sepals and the corolla gamopetalous with four petals and cruciform. Pollination is entomophilous and takes place through honeybees. Because of sporophytic self-incompatibility, it is a cross-pollinated crop. Anthesis occurs early in the morning, from 7:30 to 8:00 a.m., and continues until 11:30 a.m., depending on weather conditions. On the day of anthesis, pollen fertility is at its peak. Stigma receptivity is a maximum of 2–3 days before anthesis due to the protogyny and remains receptive even up to 4 days after anthesis. Pollen germination is hampered by glycoproteins found in the stigma and penetration of pollen tubes through the styler tissue. Cytoplasmic male sterility system in which male sterility is manifested due to interaction of cytoplasm and nuclear gene is generally utilized in hybrid seed production of broccoli. The cytoplasm of Ogura male-sterile radish has successfully been utilized in developing cytoplasmic male sterility in broccoli (Rana 2018).

Varieties

The varieties of sprouting broccoli are grouped according to their maturity, i.e., early-, medium-, and late-maturing types. The early-maturing cultivars are annual, requiring no vernalization to initiate inflorescence and flowering, while the late-maturing cultivars are biennial types requiring vernalization to produce inflorescence. The open-pollinated varieties released from the public-funded institutes are as follows:

Pusa Broccoli KTS-1

A variety developed (at IARI, Vegetable Research Station, Katrain, India) through selection from segregating exotic germplasm has medium-tall (40–50 cm) plants. It has waxy, dark green foliage with somewhat wavy margins. The head is green in color and has little beads that are somewhat elevated in the center. Its average main head weight is 350–450 g. It takes 90–105 days after transplanting to be mature in a temperate climate, which may be 5–10 days earlier in the tropical climate. Its average yield is 160 q/ha.

Palam Samridhi

A variety is developed (at Himachal Pradesh Krishi Vishva Vidyalaya (HPKVV), Palampur, Himachal Pradesh, India) through selection from an exotic material that produces compact green heads clear of bracts and yellow eyes. The average weight of the head is 300–400 g. The average yield varies from 150 to 200 q/ha. It is ready to harvest 85–90 days after transplanting.

Palam Haritika

It is a late variety and has dark green leaves with a purple reddish tinge. It gives green, soft, and crisp heads that are appropriate for salad and cooking. It attains marketable maturity in 145–150 days with 175–225 q/ha average yield.

Palam Vichitra

A variety with open dark leaves developed (at HPKVV, Palampur, Himachal Pradesh, India) through recurrent breeding produces purple heads. A purple tinge is prominent on the stem at seedling as well as on leaf margins at the young plant stage. The head is medium-sized, purple-colored, compact, and attractive, and attaining marketable maturity in 115–120 days after transplanting, and gives a 225–250 q/ha yield.

Palam Kanchan

A heading type variety (developed at HPKVV, Palampur, Himachal Pradesh) has long, broad, bluish-green upright leaves with prominent white midrib and veins. The heads are compact, attractive, and yellowish-green, attaining marketable size in 140–145 days after transplanting and giving 250–275 q/ha yield.

Punjab Broccoli-1

A variety was developed at Punjab Agricultural University, Ludhiana (India), through selection. The leaves and head are dark green with a bluish tinge. It gives 70 quintals per hectare average yield.

The hybrids are also being sold in India by the private seed companies, viz., Fiesta and Lucky by Bejo Sheetal Seeds Private Limited, Green Beauty by Doctor Seeds Private Limited, CBH-1 and CBH-2 by Century Seeds Limited, Pushpa and Prema by Seminis Vegetable Seeds India Limited, Premier, Premium Crop and Titleist by Pahuja Takii Seed Limited, Tahoe and Monfalis by Rizwan Seeds Company, and Royal Queen by Rajendra Hybrid Semences Private Limited.

Climate and Soil

It is a vegetable crop grown throughout the winter season, but different cultivars of the same crop require different climatic conditions, especially temperature, for the development of their edible portion and conversion from vegetative to reproductive phase. Broccoli is relatively tolerant of environmental stress. Being cool hardy, it thrives well in the winter season in the plains, low and mid-hills of the north (up to 1500 m above mean sea level), and during spring-summer months in higher hills. It thrives well in a cool humid climate and can withstand light frost. The low-temperature injury causes browning, freezing, and ultimately rotting of bud clusters. High temperature is responsible for the loosening of heads, premature opening of buds, and the yellowing of petals exposed over the head surface. Broccoli grows in nearly all types of soil but thrives best in well-drained sandy loam soil rich in organic matter. The plant is slightly tolerant to acidic soil, but the optimum soil pH for its cultivation is 6.0–6.5. Less fertile soils should be supplemented with manure and fertilizers so that the crop may be able to make rapid and uniform development immediately after transplanting. Dry soil conditions result in increasing fiber content in the shoots. Its plants in acidic soils with pH less than 5.5 demand relatively higher molybdenum, which can be supplied through the foliar spray of sodium molybdate or liming the soil. The crop raised in alkaline soil requires the addition of boron (addition of borax or foliar sprays of boron).

Agronomic Practices

Seed Rate and Sowing

The seed rate may vary depending upon the variety, plant spacing, and real value of the seed. Generally, 300–400 g seed for a 1-hectare area is required to raise the seedlings. A quantity of 1 g contains 350 seeds. For direct sowing, the seed rate is about 1.2–1.6 kg/ha. It is propagated by seed, which is sown on well-drained finely prepared raised nursery beds, which are of 3.0 × 1.0 × 0.15 m size. The land selected for nursery should have sandy loam soil and plenty of sunlight and should be ploughed and prepared thoroughly. Well-decomposed farmyard manure or compost at 20–25 kg and single super phosphate at 200–250 g per bed are incorporated in the top 5 cm soil at the time of bed preparation. Alternatively, 50 g of (19:19:19) complex fertilizer, 200 g powdered neem cake, and 10 g Furadan 3G per bed are added and mixed thoroughly in the soil. Soil drenching is done with captan, thiram, mancozeb at 0.3% solution or Bavistin as 2% solution at 5 l/m^2. Chlorpyrifos at 2.5 ml/l of water may be used to protect the nursery from soil-borne insect pests. After the sowing of seeds, the nursery beds are covered with straw or dry grass. Polyethylene sheets can also be used to cover the beds except when protection from sun heat is required. Water is applied every day with a fine rose can for the first few days, and thereafter, light irrigation is provided to ensure uniform germination and good growth. Grass or polyethylene mulch must be removed as and when the seedlings start emerging. Light hoeing and weeding are needed. The nursery beds may be drenched with mancozeb, Ridomil MZ 72, or captan at 0.25–0.3% at weekly intervals. The nursery beds are protected from hot sunshine and heavy rains. After 4–6 weeks of sowing, the seedlings are ready for transplanting depending upon the prevailing weather conditions.

Sowing Time

The time of sowing depends on variety, growing season, and location. In north Indian plains, the seeds are sown in nursery bed from August to September. In hills where snowfall is rare or mild, seed sowing is carried out from July to August. In higher hills where snowfall is severe, nursery sowing is done from March to April. Mid-September at Bajaura (Kullu valley) and the second fortnight of October to November at Dhaulakuan (low hills) have been reported to be the best period for transplanting broccoli in Himachal Pradesh.

Transplanting

Before transplanting, hardening off of seedlings must be done to acclimatize the plant. Withholding irrigation 4–5 days before uplifting of seedlings and adding 4000 ppm sodium chloride or spraying 2000 ppm Cycocel (CCC) are the various methods used for hardening. The nursery beds are made thoroughly moist 2–3 h before lifting the seedlings to avoid root damage. Although the yield is higher, closer spacing results in smaller heads, less production of spears, and delays the horticultural maturity. The optimum spacing for planting broccoli in the field is 45 × 45 cm. Spacing in light soils can be reduced to 45 × 30 cm. At Indian Agriculture Research Institute (IARI), New Delhi, 50 × 40 cm spacing has been found best for the variety Pusa Broccoli Kt Sel-I. Depending upon the spacing, about 33,000–45,000 plants can be adjusted in a hectare area.

Irrigation

Being a shallow-rooted crop, broccoli requires light irrigation frequently. An even supply of moisture is needed for broccoli seedlings to be established. The soil should have good water holding capacity to obtain the maximum yield. Moisture scarcity in the soil during the vegetative growth and head development stage results in small-sized heads and reduced yield. Depending upon soil type and weather conditions, irrigation should be done at 7–10 days intervals to ensure optimum moisture for proper vegetative growth and head formation.

Nutritional Requirement

Broccoli is a heavy feeder crop and requires farmyard manure at 20–25 t/ha, nitrogen at 150 kg, phosphorus at 100 kg, and potassium at 50 kg/ha. Full P and K along with one-third quantity of N is applied as basal application. The rest two-third quantity of nitrogenous fertilizer is top-dressed in two equal splits within 4–6 weeks of transplanting and just before the initiation of heading. A weekly spray of 2% urea 20 days after transplanting increases the biomass of whole plants, marketable yield, and total antioxidant activity. Like cauliflower, broccoli is also sensitive to the deficiency of micronutrients, especially molybdenum, in highly acidic soils and boron in alkaline soils. Application of 1.0 kg sodium or ammonium molybdate and 10–15 kg borax is recommended in such soils. Lime is used to make molybdenum available in acidic soils (pH 5.5), as soils with a pH greater than 7 limits the availability of boron.

Intercultural Operations

Weeding, hoeing, and earthing up are very important as weeds create a serious problem in the cultivation of broccoli crops due to wider spacing, higher fertility, and frequent irrigations. It is a shallow-rooted crop. Hence, light hoeing and weeding are required to protect the roots from damage. In the initial stage of crop growth, about two to three shallow hoeings are required to remove weeds and facilitate good aeration into the soil around the root system for proper growth of the plants. Two earthing-up operations are sufficient for optimum growth. Manual weeding and hoeing with hoe are done after 2 weeks of transplanting. Top-dressing with nitrogenous fertilizer followed by earthing up is carried out after 4–5 weeks of transplanting, and thereafter, spot weeding can be done if required. Preplant soil incorporation of weedicides like fluchloralin (Basalin 48 EC) 1.0 kg/ha and nitralin 0.75–1.0 kg/ha or preplant surface application of pendimethalin (Stomp 30 EC) 1.0 kg/ha and oxadiazon 1.0 kg/ha or postplant application of alachlor (Lasso) 1–2 kg/ha controls the weeds effectively (Rana 2017).

Harvesting and Yield

Harvesting starts when the heads have reached the right size. The central head constitutes the major portion of the yield and thus should be removed along with a 15–20 cm fleshy stem, when green buds are small and tightly closed. Delay in harvesting causes opening of buds or emergence of yellow petals and loosening of heads, thereby making them unfit for consumption. Broccoli yield ranges from 7 to 20 t/ha depending on the cultivar, maturity period, and harvesting method (Rana 2018).

Postharvest Management

The respiration rate of broccoli is very high so it cannot be stored for long period due to its perishable nature. The harvested heads are required to be hydro-cooled. At room temperature, the harvested produce can be kept for 2–3 days only but remains in good condition at 0 °C temperature and 95% RH. It should never be stored along with climacteric fruits, which can cause yellowing of green buds due to the production of a large quantity of ethylene. Broccoli heads or spears are tied in bunches of about 250–500 g and packed in waxed boxes or plastic crates. Deterioration occurs quickly in non-wrapped heads due to yellowing, weight loss, stem hardening, and chlorophyll degradation. Exogenous application of phytosulfokine (PSK) delays yellowing and retains the nutrition of broccoli during cold storage (Aghdam et al. 2020). The edible covering of chitosan and tea polyphenol reduced yellowing, improved sensory quality, and increased the nutritive value of broccoli (Fang et al. 2022).

Physiological Disorders

The physiological disorders have symptoms similar to certain diseases, but no pathogen or casual organism is involved in these disorders. These may be due to either deficiency of essential elements or climatic factors.

Hollow Stem

This is a nutritional disorder resulting from the deficiency of boron and excessive use of nitrogen. It occurs more at wider spacing. The water-soaked spots appear on the stem, while the one caused due to excessive use of nitrogenous fertilizers is distinguished by the clear white stem with no sign of disintegration. Pinkish or rusty brown spots appear on the head surface in the advanced stage, known as brown or red rot and gives bitter taste. If the deficiency is noticed during plant growth, spray the crop with borax 0.25–0.50%.

Internal Tipburning

The disorder may be caused due by the prevalence of high temperature; calcium deficiency; nitrogen, potassium, and manganese excess; soil moisture scarcity; and high endogenous level of indole acetic acid (IAA). Broccoli requires a low temperature for its normal growth, but if it is grown at a slightly higher temperature, the water requirement of the crop is increased, which is not fulfilled due to poor water movement within the plant; as a result, the leaf margins start burning. It can be controlled by spraying 0.5% calcium chloride.

Heat Injury

In many parts of the world where the temperature is higher than the normal requirement, broccoli produces rough heads with elongated peduncles and uneven head size. This disorder occurs only in heat-sensitive varieties. Hence, only heat-resistant varieties should be grown in areas with high temperatures.

Bolting/Loose Heads

Due to delayed planting, the young plants are exposed to low temperatures, leading to bolting without forming heads or producing loose heads. Growing bolting-resistant varieties, an adequate supply of nutrients and following proper plant protection measures are some of the management practices.

Blindness

The blind plants fail to produce any head. In such cases, the terminal growing point of the plants is damaged by mechanical injury, insects, low temperature, or frost in the early stages of plant growth. Such plants produce many shoots at ground level.

Floret Yellowing

The broccoli head's florets are the most sensitive part of the plant. Yellowing of floret is caused by delay in harvesting, storage of produce at high temperature, and ethylene exposure. It is also called marginal yellowing.

Leafy Heads

Small green leaves protrude out through the head due to high temperature, temperature fluctuations (day and night), or nitrogen imbalance.

Broccoli Buttoning

Button size (2.5–10 cm diameter) heads are formed, and it occurs after the transplanting of seedlings. Affected plants show decreased foliar growth, thereby resulting in too few nutrients to nourish the head to a marketable size.

Knol-Khol

The "knol-khol" is a cool-season vegetable cultivated for its edible knob (*the swollen portion above the ground*). It is also referred to as Kohlrabi, Navalkol and *Ganth gobhi*, etc. The knob is consumed as raw or cooked vegetables. The edible part is the knob which develops entirely aboveground.

Origin and Distribution

North European is the native place of knol-khol or kohlrabi and then spread all over the world including South Asian countries. It is widely distributed in the Northern state of Kashmir in India. The major kohlrabi growing states in India are Kashmir, West Bengal, Maharashtra, Assam, Uttar Pradesh, Punjab, and some parts of South India, but on a small scale. The whole of Afghanistan is well suited for knol-khol cultivation.

Nutritional Importance and Uses

Kohlrabi or nulkol is an excellent source of various nutrients. Its flavor and texture are similar to broccoli or cabbage, but milder and sweeter. Likewise, Brussels sprouts and knol-khol are also rich in vitamins (A&C), folic acid, and dietary fiber. Moreover, it is an excellent source of vitamin B complex and minerals (copper, calcium, potassium, manganese, iron, and phosphorus). It in fact can meet up to 95% of daily vitamin C and match 17% of digestive fiber requirements. It contains sulforaphane, which has anticancer properties. Kohlrabi is a powerhouse of antioxidants including anthocyanins, isothiocyanates, and glucosinolates. It has antidiabetic, antioxidant, hypolipidemic, antihyperglycemic, cardioprotective, and anticancer properties (Chauhan et al. 2016). A knol-khol juice is beneficial for diabetes patients to reduce their blood glucose levels (Selvakumar et al. 2017).

Botany and Taxonomy

Knol-khol is an herbaceous, erect, branched annual or perennial herb and has a taproot system. The leaves are simple, alternate, radical, or cauline, usually entire, sometimes lobed, petiolate, and exstipulate reticulate venation. The floral morphology is similar to sprouting broccoli.

Varieties

There are primarily two varieties: white and purple. The details of some important varieties are given below:

Large Green The knobs are green, round, and large-sized with small tops. These are usually tender and delicately flavored with white flesh. Harvesting is done after 75 days of transplanting and gives 225–250 q ha^{-1} average yield. It is very much suitable for western Himalayas.

White Vienna It is an early-maturing variety that takes about 55–65 days to mature after transplanting. The plants are small and have medium green stems and leaves. The knobs are globular in shape, light green in color, silky, and soft. Its yielding potential is 175 q ha^{-1}.

Purple Vienna The knobs are purplish-blue with greenish-white flesh. It takes about 70 days to mature and has more yield potential as compared to White Vienna.

King of North It is an early-maturing variety. The plants are dark green with about 25 cm in height and bear flattish-round knobs. It matures in 60–65 days after transplanting.

Climate and Soil

Knol-khol is a winter season crop grown in temperate and moist climate. It is tolerant to low temperatures and frost. Fifteen to 30 °C temperature is required for seed germination. The optimum temperature for crop growth ranges from 15.5 to 18 °C. It may be cultivated in a wide range of soils but thrives best in a well-drained sandy loam soil with enough organic matter content. It does well in the pH range of 6.0–6.8.

Agronomic Practices

Land Preparation

Nursery soil is prepared to a fine tilth by ploughing before the preparation of beds. Soil drenching with 1% formalin is done to sterilize the soil at least 15–20 days before sowing to protect the seedlings from fungal attacks. After soil drenching, light irrigation is given to the beds. Then beds are covered with transparent polythene sheets for 1 week. All disease-causing organisms and weed seeds are killed as a result of the increased temperature and relative humidity. This is known as soil solarization, and it may be feasible during the hotter months. Soil drenching can also be done with captan or thiram at 2–3 g per liter of water, and seed treatment is done with captan or thiram at 2–3 g per kg of seed. The standard procedure for raising the nursery should be followed. After sowing of seeds, the beds are covered with a straw or other organic mulches which help in the quick germination of seeds. Apply water over the mulch during the initial stage, i.e., 15–20 days of sowing, while during the later stage, watering should be done through furrows. The mulch should be removed immediately after seed germination. The seedlings are ready for transplanting after 1 month of seed sowing.

Seed Rate and Sowing

Around 1–1.5 kg seeds are sufficient for the 1-hectare area to cultivate "knol-khol." The optimum time for raising knol-khol's nursery is during August. The seedlings are transplanted at a spacing of 25 × 25 cm, 25 × 30 cm, 25 × 40 cm, or 30 × 45 cm depending on the soil and climate. The yield is more in close spacing, but the size of knobs is reduced. Bairwa et al. (2017) reported that in respect of knob diameter and volume, 45 × 30 cm spacing is ideal, whereas 30 × 30 cm spacing is best in terms of overall knob yield (q/ha). In comparison to late cultivars, early cultivars require less spacing.

Nutrient Management

Manure and fertilizer requirements in "knol-khol" depend upon the fertility status of the soil. Apply 20 tonnes well-rotten FYM, 90 kg N, and 80 kg P & K per hectare. Before transplanting, apply the half dose of N and the full doses of P and K. After 1 month of transplanting, the remaining half dose of N is applied in two split doses, first at the time of knob formation and second at the time of knob formation. For acid soils low in Ca and Mg, the application of "dolomite" as lime should be followed. The use of organic manures along with biofertilizers improves the growth and yield of knol-khol (Shah et al. 2019; Hussain et al. 2020). Islam et al. (2020) reported the highest yield of knol-khol using 50% vermicompost +50% poultry manure.

Irrigation

Likewise other cole crops, knol-khol also requires a sufficient moisture supply for proper growth and development of the plant. First and light irrigation is done immediately after transplanting, and subsequent irrigation can be given at 15 days intervals. The critical stage of irrigation is during knob formation.

Intercultural Operations

The intercultural operations are similar to that of other cole crops. Two to three hand weedings and hoeing are required to kill young weeds. In the early stage of plant growth, the field should be kept weed-free. Earthing up is done after 4–5 weeks of transplanting. Foliar spray of Stomp 30 EC (Pendimethalin) at 2.5 l per ha is applied to control the weeds.

Harvesting and Yield

The harvesting of knol-khol is done when the knobs attain about 5–8 cm diameter. For marketing purposes, the roots are removed from the plants, and plants are tied in bunches along with the leaves. In general, yield ranges from 200 to 250 q/ha.

Brussels Sprout

Brussels sprout is a hardy cool-season slow-growing crop and gained popularity in the nineteenth century. It is commonly grown in the United Kingdom and northwest Europe but grown in the United States of America (California), Eastern Europe, and Australia on small scale. In India, the cultivation of Brussels sprout has not been exploited commercially and is limited to kitchen gardens, places of tourist importance, and near big cities. Being a nutritious vegetable, it is gaining importance by the high living people.

Origin and Distribution

Brussels sprout has also evolved from wild cabbage in 1750, and the plant was first appeared in Brussels, Belgium, and by 1800, it had spread to England and France; however, in the late eighteenth and early nineteenth centuries, the crop became widespread in Europe and California.

Nutritional Importance and Uses

Nutritionally, it is quite comparable with broccoli for vitamins and minerals except for vitamin A and calcium. It has more protein and carbohydrates content in comparison to other cole crops. It is an excellent source of vitamin K (177 mg/100 g edible portion). Brussels sprout is a nutritive green vegetable grown for fresh consumption as well as for processing (freezing and canning). Its taste is much like cabbage but has a slightly milder flavor and a denser texture. Sprouts are stir-fried, boiled, steamed, or cooked as a vegetable. Sprouts are eaten raw as salad and liked in soups. The leaves can also be used as a substitute for kale. The frozen product of brussels sprouts is popular in the USA and the UK.

Brussels sprouts contain 160–250 mg/100 g edible portion of glucosinolates, which are known to combat cancers such as breast, ovarian, lung, bladder, colon, and prostate cancer. It prevents constipation, maintains blood sugar, and checks to overeat. Sulforaphane also protects the stomach lining by obstructing the overgrowth of *Helicobacter pylori*, a bacterium that can lead to peptic ulcers. Brussels sprout has been used to determine the potential importance of crucifers on thyroid function and provides health benefits without putting the thyroid gland at risk. Beta-carotene of Brussels sprout acts as a potent antioxidant. It is a suitable food for those who are suffering from water retention, obesity, arthritis, and constipation. It helps in healing wounds.

Botany and Taxonomy

Brussels sprouts are commonly grown as annuals, whereas for seed production, they are biennial. Its inflorescence is a raceme, and its flowers are yellow, actinomorphic, and bisexual (complete). Calyxes are gamosepalous with four sepals, and the corolla is gamopetalous with four petals and cruciform. Pods are linear and dehiscent, the beak of pods cylindrical or conical, and seeds seriate. Pollination is entomophilous and takes place through honeybees and butterflies. The presence of sporophytic self-incompatibility makes it a cross-pollinated crop, which remains inactive at the bud stage. It becomes active in stigma at the flowering stage. Anthesis occurs in the morning hour between 7:30 and 8:00 a.m. and continues up to 11:30 a.m., depending upon prevailing weather conditions. Pollen fertility is observed at its maximum on the day of anthesis. The stigma becomes receptive 2–3 days before the day of anthesis due to the protogyny condition of the flowers. Pollen germination and pollen tube penetration through the styler tissue are hampered by glycoproteins found in the stigma. Seeds are borne in siliqua. The plant for seed production needs exposure to low temperature (4–10 °C) during winter to transit to the generative phase. A temperature of 7 °C for 28 days is sufficient to induce flower initiation, whereas the induction period is 35 days at 4–10 °C. The plants at 14–23 °C do not transform to the generative phase and remain vegetative. The edible part of Brussels sprout is a swollen auxiliary bud, which is about 5–8 cm in diameter and is composed of tightly packed leaves resembling a miniature cabbage head. It is also known by several other names such as sprouts, buttons, or mini cabbage heads. The sprouts are round and white, green, or purple.

Varieties/Hybrids

The productivity, harvest period, shape, size, color intensity, and compactness of the buds are important attributes of Brussels sprouts. Its cultivars may be dwarf or tall.

Characteristics	Dwarf varieties	Tall varieties
Plant characteristics	Dwarf varieties have to stem less than 50 cm in length and sprouts are small and crowded on the stem	The tall varieties have plant stem more than 50 cm in length and are late in maturity
Suitable for cultivation	These are early and suitable for areas where the growing season is short	These are suitable for the areas where the growing season is long, especially in England, other parts of Europe, and North Indian hills
Varieties	Improved Long Island, early morn, dwarf improved, frontier Zwerg, and Kvik	Hilds ideal, red vein, and amager

The hybrids marketed by different companies in Japan and Europe are Asgard, Boxer, Crystal, Jade Pearl, Cross, Dorman, Kundry, Oliver, Predora, Rasmunda, Richard, Rider, Rovoka, Rogor, Royal Marvel, Sonara, Stephen, and Valiant. In India, the crop improvement work has been carried out only at the Temperate Vegetable Research Station of Indian Agricultural Research Institute, Katrain, and the open-pollinated variety Hilds Ideal has been found suitable for cultivation in hills as well as in Maharashtra, India. Early Dwarf, Early Morn, and Bowler are the most promising genotypes exhibiting high sprout and seed yield at Kalpa, Himachal Pradesh, India. The description of some important varieties grown in India is given as follows:

Hilds Ideal

It is an open-pollinated cultivar with a plant height of 55–60 cm and 45–50 sprouts per plant. Each sprout on average weighs 60–70 g and diameter of 7–8 cm. Sprouts are light green, globular, and solid and take 115 days for first picking after transplanting and give four to five pickings at about 10-day intervals. Its average yield is 160 q/ha.

Jade Cross

A Japanese F1 hybrid cultivar has become popular for its short-stemmed, early, high-yielding, and single-harvest quality. It belongs to a dwarf group of varieties.

Rubine

It is a high-yielding commercially grown variety that bears red sprouts.

Brussels Dwarf

Plants are dwarf (50 cm in height), early, and suitable for areas of the short growing season. It is a high-yielding variety, and its sprouts are medium-sized.

Climate and Soil

Brussels sprout requires a long cool and humid climate and grows well in regions with moderately severe winters. It can withstand frost and extremely cold weather. For the good growth and development of sprouts, an average temperature of 16–20 °C is favorable. Sunny days with light frost during the night result in the production of the best quality sprouts, which are solid, tender, and sweet. Puffy and loose sprouts are produced at temperatures above 27 °C. It can tolerate 0 °C temperature and even up to −10 °C. Being a strong biennial plant, it does not bolt and flower unless exposed to chilling temperature (<10 °C) for 6–8 weeks. Brussels sprout is a nutrient exhaustive crop, which demands highly fertile soil with a pH between 6.0 and 7.0. In acidic soil, light dusting in the planting hole with lime reduces soil-borne disease called clubroot and increases the yield of Brussels sprouts. It grows into a hard top-heavy plant with a thin base, which is easily damaged by strong winds. If the plant falls over, it will continue to develop but will not produce as much as the plant that remains upright. Light sandy soils cannot adequately anchor the Brussels sprouts plant, even when the plant is supported with stakes or the soil is mounded up around the base of the plant. Therefore, the ideal soil texture for its cultivation is dense clay loam with a few rocks clay since tight clay loam soil helps in holding the roots firm when the large plant is blown by high-speed winds.

Agronomic Practices

Seed Rate and Sowing

The seedlings of Brussels sprout are raised from seeds sown on well-drained and finely prepared nursery beds. Nursery management practices are similar to those followed in broccoli. About 300–400 g of seeds are sufficient to raise the seedlings for a 1-hectare area. For direct seeding, the seed requirement is four times higher, i.e., 1.2–1.6 kg/ha. Sowing time depends on variety, growing season, and location. In north Indian plains and low hills, sowing of seeds is done in early September. The optimum time for nursery sowing in the hills up to 1500 m average mean sea level with mild or rare snowfall is early July, whereas it is March–April in very high hills experiencing heavy snowfall. Transplanting of seedlings is done after 4–5 weeks of seed sowing. Transplanting of seedlings is done in the evening either on flat or raised beds. In agriculturally advanced countries, direct sowing of seeds is done for commercial cultivation, and thinning is done to maintain the recommended spacing. 60 × 45 or 60 × 60 cm spacing is followed for dwarf varieties, whereas wider spacing of 75 × 45, 75 × 60, or 90 × 90 cm is kept for tall varieties, especially where the growing period is relatively longer. For Hilds Ideal variety, a spacing of 60 × 45 cm, which accommodates about 33,000 plants in a hectare area, has been found appropriate for north Indian hills and Peninsular India. The sprout size increases as the distance between rows increases.

Nutritional Requirement

Brussels sprout needs a heavy intake of nutrients. Besides the application of farmyard manure at 25 t/ha, nitrogen is applied at 200 kg, phosphorus at 90 kg, and potassium at 80 kg/ha, depending upon the varieties and soil fertility. The entire quantity of farmyard manure, phosphorus, and potassium with one-third dose of nitrogenous fertilizer is given as a basal dose before transplanting. The remaining two-third dose of nitrogen is applied in two split doses where the first split dose is given after 1 month of transplanting and the second dose after 2 months of transplanting.

Irrigation

An even moisture supply is needed to ensure continuous and optimum plant growth to get a higher yield. Surface irrigation or repeated spot watering manually is carried out until the seedlings are established. Optimum moisture level at the time of head formation is very essential. In general, the Brussels sprout field is irrigated at 8- to 10-day intervals. Every week, it needs 2–3 cm irrigation water. Following a low tunnel system of cultivation reduces the water requirement and improves the growth and yield (Acharya et al. 2019).

Intercultural Operations

Deep weeding and hoeing can easily damage the roots because it is a shallow-rooted crop. Weeds create a serious problem in the cultivation of crops due to wider spacing, higher fertility, and frequent irrigations. Hoeing within 2–3 weeks of transplanting is done to remove weeds and to facilitate good aeration into the soil. Top dressing with nitrogenous fertilizers followed by earthing up is carried out after 6–7 weeks of transplanting. Weedicides recommended for cabbage and other cole crops prove effective in Brussels sprout as well. Before transplanting seedlings, weedicides like fluchloralin 0.5 kg/ha, trifluralin 1 kg/ha, nitrofen 2 kg/ha, oxyfluorfen 1 kg/ha, alachlor 0.2 kg/ha, or butachlor 2 kg/ha 2–3 days can be applied to control the broad-leaved weeds. Removal of the terminal bud of the plant about 3 weeks before harvest is done to ensure uniform sizing at maturity of sprouts in areas having a short growing period, e.g., north Indian plains. Pruning of lower leaves before the start of the harvest of sprouts leads to yield reduction because of a reduction in photosynthetic area.

Harvesting and Yield

Harvesting of crops starts when sprouts attain proper size and firmness and before they start loosening and bursting. Harvesting is done at least three to five times at 7- to 10-day intervals. The harvest period varies according to the agroclimatic zone. Harvesting is done from January to February in the North Indian plains, whereas in hills and very high hills of northern India, the harvesting period is November to March, and July to September, respectively. Generally, the leaves are removed from the plants before harvesting, which starts from the bottom upward and continues for several weeks. In agriculturally advanced countries, a single harvest with a machine is carried out for industrial processing. The entire crop is harvested after about

4–6 weeks of topping (removal of apical or terminal vegetative bud). Delay in harvesting results in poor-quality sprouts. In India, a yield of 150 q/ha has been reported. A single harvested crop generally gives a yield of 30–50 q/ha.

Postharvest Management

Brussels sprout is highly delicate as the respiration rate of freshly harvested sprouts is very high. After removing the leaves from the cut stem, the harvested Brussels sprout is required to be hydro-cooled. After grading, the fresh sprouts are packed in cardboard boxes with proper aeration and sent to the market at the earliest possibility. The entire plant can also be stored in a cool underground room, and sprouts are removed as per the requirement of the family, especially in high hills. Brussels sprouts can be kept for 2–3 days at room temperature, while at low temperature (°C) and 95% relative humidity, they can be kept in good condition for 6–8 weeks. Sprouts can be stored for 16 weeks at 0 °C temperature and only for 11 days at 20 °C temperature if they are packed in polythene ceramic film, whereas if polythene packed sprouts are stored at 0–10 °C in a modified atmosphere containing 2–5% CO_2 and 14.5–16.8%, O_2 reduces the losses and improves their quality in comparison to open storage at the same temperature. Application of sorbic acid at 0.05% and benzoic acid at 0.2% solution hinders the development of harmful microorganisms and preserve the quality of Brussels sprout heads (Pusik et al. 2020).

Physiological Disorders

The physiological disorders have symptoms similar to certain diseases, but no pathogen or casual organism is involved in these disorders. These may be due to either deficiency of essential elements or climatic factors.

Loose Sprouts

Affected sprouts are large, leafy, open, and with poorly formed heads. This may be due to several causes such as moisture and temperature fluctuation. Excessive use of nitrogenous fertilizer is another cause of loose sprouts in Brussels sprouts. The use of a balanced amount of nitrogenous fertilizers can control this disorder.

Tipburn

A collapse of tissue and cell death in plant leaves is symptoms of this disorder caused by a calcium deficiency. High humidity, soil moisture scarcity, and excess use of potassic and nitrogenous fertilizers are other reasons for tipburn and also aggravate calcium availability. The application of an adequate amount of calcium fertilizers and a balanced amount of nitrogen and potassium helps to control this disorder.

Internal Browning

Yellowing of leaves at distal ends within the sprout, often along a line across the middle part of the sprout and midway between its growing tip and outer surface, is caused by a calcium deficiency in the sprouts. Application of calcium nitrate, maintenance of soil moisture, and use of a balanced dose of nitrogenous fertilizers help in controlling this disorder.

Blindness

This is one of the common disorders of all the cole crops. In this disorder, plants remain without terminal buds, and leaves become large, thick, leathery, and dark green. Low temperature at the initial growth stage, terminal bud damage during transplanting, and injured by insects and pests are the main causes. It can be controlled by protecting the crop from low-temperature injury, careful handling at transplanting, and timely control of insect pests.

Kale

Kale is a very important salad vegetable crop also called "Portuguese Galega kale" in Spain (Dias et al. 1993). It is mainly a winter vegetable when other vegetables are scarce. It is mainly grown in countries like China, Canada, Korea, the Netherlands, Taiwan, Vietnam, and the Middle East countries. In India, kale is grown in the northern parts, especially on the Chinese border (Chinese kale). Similar to this in Kashmir, Karam saag is grown, which is closely related to Chinese kale. The only difference is that Chinese kale is grown in tropical regions and Karam saag requires a chilling temperature. Among different kales, curly kale is cultivated for its rosette leaves, which bear at the top of the stem, and is harvested as individual leaves or in a bundle.

Origin and Distribution

Asia Minor is the native place of kale and was carried to Europe in approximately 600 B.C. by tribes of Celtic wanderers. People began producing this first ancient cabbage plant in the Mediterranean region, where huge leaves were consumed as part of the plant. The continuous predilection for bigger leaves resulted in the evolution of the vegetable known as kale (*Brassica oleracea* var. *acephala*) which translates to "vegetable garden cabbage without a head." For thousands of years, kale has been grown as a leafy vegetable and is still being cultivated. With the progression of time, people started to express a preference for plants with a dense cluster of central young tender leaves. Therefore now kale is propagated more frequently.

Nutritional Importance and Uses

Kale is a highly nutritious vegetable that contains a high amount of vitamins (A, K, and C) and calcium. Sulforaphane, a chemical with potent anticancer properties, is found in the leaves of kale. It also contains glucosinolates that are glucobrassicin and sinigrin (Jeon et al. 2018). Flavonoids quercetin and kaempferol flavonoids are also found in kale (Akdaş and Bakkalbaşı 2017). It is used in salads and soups. It's widely used in the Netherlands in a traditional winter stamp pot dish called boerenkool. In Ireland, it's combined with mashed potatoes to make colcannon, a traditional dish. Japanese kale is primarily used for decorative or ornamental purposes. Dried kale, also known as "kale chips," has recently become very popular, although drying significantly reduces the nutritive and phytochemical content (Oliveira et al. 2015). In the development of novel functional foods and beverages, including apple juice with the addition of frozen and freeze-dried kale leaves, which are high in minerals and phytochemicals (Biegańska-Marecik et al. 2017). The nutritional composition of kale juice fermented with Lactobacillus strains is excellent (Kim 2017).

Apart from relieving the symptoms of gastric ulcers (Šamec et al. 2017; Šamec and Salopek-Sondi 2019), kale has been used to treat diabetes, rheumatism, bone weakness, ophthalmologic problems, hepatic diseases, anemia, obesity, and other conditions (Lemos et al. 2011; Gonçalves et al. 2012; Kuerban et al. 2017). It also stimulates the immune system, lowers the risk of colon cancer, reduces blood sugar and cholesterol levels, and improves insulin sensitivity (Dey 2017).

Botany and Taxonomy

Kale is a biennial non-heading with curled succulent leaves but grown as an annual crop for vegetable production. A flowering stalk arises in the center from the base, which is thickened at the base and narrows toward the tip with small yellow-colored flowers. The floral morphology is similar to the other brassicaceous vegetables. The plants may be dwarf or tall, but dwarf types are most preferred. There are different varieties of kale word kale such as collard (*B. oleracea* L. var. *Viridis* L.), marrow stem kale (*B. oleracea* L. convar. *acephala* (DC.) Alef. var. *medullosa* L.), palm kale (*B. oleracea* L. convar. *acephala* (DC.) Alef. var. *palmifolia* L.), scotch kale (*B. oleracea* L. convar. *acephala* (DC.) Alef. var. *sabellica* L.), thousand-head kale (*B. oleracea* L. var. *ramosa* DC.), and Portuguese Tronchuda cabbage (*B. oleracea* L. var. *costata* DC.) (Diederichsen 2001).

Varieties

A wide range of varieties are available in kale, but they are typically classified according to the type of leaves such as curly leaved kale, plain leaved kale, rape kale, leaf and spear (a hybrid between curly leaved and plain leaved kale), cavolo nero also known as Black cabbage, Tuscan cabbage, Tuscan (Tuscano) kale, Lacinato kale, and Dinosaur kale. In these leaf shapes, there are many varieties with varied growth times from transplanting to harvesting, viz., Blue Armor (45–75 days, hybrid), Blue Curled Scotch (65 days), Blue Knight (55 days, hybrid), Dwarf Blue Curled (55 days), Dwarf Blue Scotch (55 days), Dwarf Green Curled (60 days), Dwarf Siberian (65 days), Greenpeace (65 days), Hanover Late Seedling (68 days), Konserva (60 days), Red Russian (40–60 days), Squire (60 days), Verdura (60 days), and Winterbor (60–65 days, hybrid). Depending upon stature, kale is grouped into two types as given below:

- Dwarf types: Dwarf Green Curled Scotch and Dwarf Moss Curled
- Medium tall types: Hamburger Market and Moss Curled

However, in India, there are no released varieties by the public sector. Only local varieties and some imported varieties are grown. In Kashmir, a famous variety named Karam Saag is cultivated on a large scale.

Scotch Kale

Dwarf Blue Curled Scotch A squat plant, which is good for container culture, produces curly blue leaves that are good in salads when young, or cooked when mature. It is a very cold hardy plant and become ready for harvest in 55 days.

Redbor A plant with deepest red-purple leaves whose color enhances with cold looks gorgeous in a flowerbed or as an edging. Its flavor is sweet. It takes 28 days for baby kale and becomes ready for harvest in 55 days.

Winterbor Its blue-green plants are 60–90 cm tall, extremely hardy, and very productive. It takes 28 days for baby kale and 60 days for mature kale.

Siberian (Russian) Kale

Red Russian It has blue-gray, flat, deeply cut leaves. Veins and stems are blue-green in warm weather, turning red with cold. It is one of the most tender kale, delicious raw in salads, and can be harvested at 25 days for baby kale and normally matures 50 days after sowing.

Blue-Curled Vates Its great flavored leaves can be used like lettuce. It is the best cold weather kale, medium green in color, and matures in 60 days.

White Russian Its taste is mild and sweet and is excellent for cool weather salads. It matures in 58 days.

Heirloom Kale

Lacinato An Italian heirloom also known as Nero di Tosca, Tuscan Black, or Dinosaur produces 30–60 cm long, 7 cm wide, slightly crinkled, and deep blue-gray leaves, which are excellent for cooking. It is heat and cold tolerant, and it takes 30 days for baby kale and normally matures in 65 days (Rana 2018).

Climate and Soil

Kale is a winter-season vegetable crop. Being highly resistant to frost, it can withstand relatively unfavorable conditions, and thus, it is a very reliable crop. A temperature range of 12–24 °C is required for seed germination. It grows quite well even in Southern Greenland, but it is not commercially grown in India. A form of kale known as Karam Saag is grown in Jammu and Kashmir, India. At high temperatures and relative humidity, the glucosinolate content in kale leaves is decreased but increases as CO_2 concentration rises (Chowdhury et al. 2021). It thrives well in medium-heavy, fertile soil with plenty of water, but it also prefers light soil. It can tolerate salt more than cabbage and can also be grown at the C8 salt index (Pavlović et al. 2019). Soil pH ranging from 6.0 to 6.8 is ideal for its cultivation.

Agronomic Practices

Seed Rate and Sowing

Direct seeding as well as transplanting are practiced for growing kale. About 3.5–4.5 kg is required for direct seeding crop and 350–400 g for transplanting a hectare area. Seeds can be sown in September–October in the plains of north India and August–September in the hills. For spring and early summer harvest, the seeds are sown during the first week of April in Virginia and warmer southern areas, and in Pennsylvania and cooler parts, seeds are sown in the third week of April.

Direct Seed Sowing

The field is prepared after adding well-decomposed farmyard manure and fertilizers to the soil. The seeds are sown at 3.5–4.5 kg per hectare in the rows spaced 45–60 cm apart. The seeds are usually sown after any danger of hard frost, and for a fall harvest, sowing is done in such a way that the crop would mature in cool weather. The seeds are thinly sown at a 3.5 cm distance and 1.2 cm depth. Thinning is required to maintain 15–30 cm between plants when the seedlings attain three to four leaves. The soil is kept moist during germination to prevent the formation of a hard crust on the soil surface, as it will hamper germination.

Transplanting

The seeds of kale are sown in raised nursery beds of 3.0 × 1.0 × 0.15 m in size for raising seedlings. The soil should be loose and friable added with farmyard manure supplemented with inorganic fertilizers for better growth of seedlings in the nursery. The seedlings in the nursery are kept free from biotic and abiotic stresses during seedling growth. The seedlings become ready for transplanting when seedlings attain two to four true leaves. Transplanting of the seedlings is done at 45–60 × 15–30 cm between rows and plants. The seedlings are transplanted slightly deeper than they were previously sown in the nursery.

Nutritional Requirement

The requirements of organic manure and fertilizers will depend upon the type of soil, climate, and cultivar used. Kale can successfully be grown with farmyard manure 18–23 t/ha supplemented with 120–180:75–80:125 kg N:P: K per hectare. The application of half quantity of N along with full P and K as the basal dose is done during field preparation. The remaining half quantity of the nitrogen is applied in a split dose, i.e., 4 weeks after transplanting. It is beneficial to spray urea at 0.5–1.0% every 3 weeks. Kale grows and yields better with intermediate doses of bovine manure and rock powder (Pereira et al. 2020). In kale, the plant growth-promoting bacteria improve vegetative growth, yield, and quality (Helaly 2017).

Irrigation

Crop irrigated immediately after sowing seeds in the main field in the direct sown crop for better germination and soon after transplanting of the seedlings. Then irrigation should be given at 10- to 15-day intervals. The total number of irrigations may be six or seven during the crop season, and it may again vary with the variety, type of soil, and season. There should be adequate soil moisture for optimal growth and development of the plant.

Intercultural Operations

The plant is not able to compete with weeds so the field must be weed-free at the initial stage of plant growth. Thus, at least two shallow hoeings are beneficial and necessary for keeping the weeds down and providing a good soil environment for plant development. One or two hand weedings are enough to remove weeds. Thinning is also essential to reduce competition among plants for space, light, nutrients, and moisture for the direct-sown crop. Mulching is also done with a 2.5–5.0 cm thick layer of organic matter, keeping the mulch 2.5 cm away from the plant stem. Mulching helps in conserving soil moisture and provides nutrients to the plants and controls weeds. Faba bean as a cover crop increases the mineral, protein, and carbohydrate content, but ryegrass increases the biomass production of kale (Siva et al. 2019).

Harvesting and Yield

Kale is harvested with a knife while young and tender. For fresh market, the whole rosette tops are frequently harvested, while in tall cultivars, the lower leaves if adequately curled are harvested and packed individually. Frost favors the development of flavor and color in kale. After harvesting, it is tied in small bundles. Since the leaves have a high respiration rate, bunches are subjected to hydrocooling, and after proper drying, they are kept in the refrigerator. The yield of kale varies from 10 to 25 t/ha. Yield may also vary with crop variety, environmental, and soil factors.

Postharvest Management

Kale is a highly perishable crop, so after precooling, bunches are packed in polyethylene-lined crates and protected by crushed ice and kept at 0 °C temperature and 95–98% relative humidity. Under such conditions, it can be held for up to 10–15 days in excellent condition without wilting and loss of quality. Air circulation is necessary to remove heat produced by respiration.

Collard

Collard (*Brassica oleracea* L. var. *acephala* DC) is a very important crop due to its various uses, high nutrition, and short duration (Chang et al. 2019; Abellán et al. 2019; Jeon et al. 2018). It also originated from wild cabbage (*Brassica oleracea* var. *sylvestris*). It is a non-heading open leafy type cabbage that is consumed in various parts of the world and is closely related to kale and cabbage. Although both kale and collard are of the same variety, acephala, the main difference between them is the leaf shape. The leaf shape of kale is curly, whereas the leaf shape of collard is smooth. The leaves lack frilled edges, which make them different from kale. In Kashmir, it is known as Haak and can be found in almost every meal in Kashmiri households, where both leaves and roots are consumed. It is also grown as an ornamental plant.

Origin and Distribution

Eastern Mediterranean or Asia Minor is the native place of collard. It's been cultivated since the time of ancient Greek and Roman civilizations and might have been introduced into the United States earlier through settlers from the British Isles probably in the late 1600s, though it was first mentioned in the late seventeenth century as collard greens. Leafy collard is the earliest vegetable cultivated by man, and it is thought that it has been consumed as food since prehistoric times. It is mainly grown in North America, and Central and Northern Europe (Thomson et al. 2007; Schmidt et al. 2010). It is also cultivated in Brazil, Bosnia, and Herzegovina, India, Northern Spain, Portugal, the Southern United States, different parts of Africa, and Southern Croatia.

Nutritional Importance and Uses

Collard contains good amounts of vitamins (A, C, E, and K), proteins, and minerals along with soluble fibers but is low in calories. However, collard greens carry antioxidant properties due to the presence of antioxidant phytonutrients such as ferulic acid, caffeic acid, kaempferol, and quercetin. It also contains a significant amount of phytonutrients called glucosinolates such as glucobrassicin, glucoraphanin, gluconasturtiin, and glucotropaeolin, which help in activating the detoxification of enzymes by regulating their activity (Giorgetti et al. 2018; Akram et al. 2020; Cartea and Velasco 2008). Glucosinolates content is lower in vegetative tissues than in reproductive tissues like flowers and seeds (Holst and Fenwick 2003).

Both the roots and the leaves, which may be cooked together or separately, are consumed. Haak-rus, which is a soup prepared by cooking whole leaves in water, salt, and oil, is usually consumed with rice in Kashmir. The leaves are also cooked by mixing them with meat, fish, or cheese. Consuming raw or cooked collard greens also helps in lowering cholesterol. It contains a high amount of folate (a critical vitamin B), supporting cardiovascular health, which is 50% higher than broccoli, 100% higher than Brussels sprouts, three times more than cabbage, and seven times more than kale. The fiber in collard (about 4 g per 100 g edible portion) is a natural choice for a healthy digestive system. However, it contains a high content of oxalate so should be carefully consumed by individuals prone to kidney stones.

Botany and Taxonomy

Collard is a biennial dicotyledonous herbaceous plant, but commercially, it is cultivated as an annual. It has a rosette form of vegetative growth on the stem but does not form a head. Collard has large, spoon-shaped smooth, and succulent leaves. It can grow up to 120 cm tall with a loose crown of leaves on a hard stalk. In addition, collard, being self-incompatible, is a cross-pollinated plant, which produces viable seeds through insect pollination. Its inflorescence is the terminal raceme. Flowers are hermaphrodite and bright yellow. Anthesis and dehiscence take place in the morning, and the pollen grains remain available until the early afternoon hours, depending upon the prevailing temperature. The fruit dehisces longitudinally at maturity into distinct equal parts. Seeds are small, globular, and smooth, having a uniform brown color.

Varieties

The important varieties of collard are as follows:

Open-Pollinated Varieties

Georgia

The oldest open-pollinated variety produces slightly savoyed leaves of a blue-green color. The plant gives low to moderate yield as per the growing environment. It tolerates cool weather effectively but has a severe problem of bolting among the open-pollinated varieties.

Morris Heading

An old open-pollinated variety produces medium green, slightly savoyed, broad, waxy leaves. In late planting, plants form loose heads. It is an old favorite variety with good flavor and nutrition. The leaves give a sweeter taste because of frost. Its yielding ability varies from low to medium.

Vates

A non-heading variety with compact growing habit bears dark green, glossy, flat to wrinkled leaves on 75 cm plants. The variety with long-standing is fairly uniform with moderate yield potential and has resistance against bolting and frost. This has been the standard variety for many years, especially for processing. It matures in 75–80 days.

Champion

A selection from Vates does not bolt readily in spring. Smooth, deep green, flat to slightly savoyed leaves with short internodes characterize this open-pollinated variety. It gives a moderate yield and takes 78 days from planting to harvesting.

Georgia Blue Stem

A darker blue-green variety with a good yield matures 60 days after planting.

Hybrids

Bulldog

A variety with upright growth habits bears dark green and slightly savoyed leaves.

Hi-Crop

A variety previously sold in the US market as Blue Max has very attractive, blue-green, semi-savoyed leaves and requires 75 to 85 days from planting to harvesting. It has a small intermodal distance, which makes it a little more difficult to bunch than other varieties. It is prone to bolting.

Flash

A variety similar to Vates has a uniform, smooth, dark green, glossy, flat to wrinkled leaves. It has excellent yield potential and does not bolt readily in spring. Unlike Vates, it has an upright growth habit and, thus, is good for bunching.

Tiger

The hybrid has a straight plant growth habit like Georgia, having slightly savoyed, blue-green leaves developed by the private sector. It is suitable for the fresh market.

Top Bunch

A Georgia-type variety with a uniform growth bears slightly Savoy-type medium bluish-green leaves. It has a semi-erect growth habit that gives rise itself well to bunching.

Climate and Soil

Collard is a winter season crop, and 10–29 °C temperature is required for seed germination, while 15–18 °C is good for vegetative growth. It can tolerate severe cold weather in comparison to other members of the cole group. The mature plants can tolerate frost and light to medium freezing temperatures. For higher yield, collard requires plenty of sunshine and warm weather. It is sensitive to environmental conditions. When the temperature goes below 10 °C for more than 10 days, bolting takes place. Collard can be grown on a wide range of soils, but well-drained sandy loam soils are good for an early spring crop, while higher yield can be obtained from heavy loam soils rich in organic matter. pH 6.0–6.5 is ideal for its cultivation (Rana, 2018).

Agronomic Practices

Seed Rate and Sowing

Seed is the most critical and costly input, so optimum seed quantity should be used to raise the crop. The seed rate varies with the method of planting, which may be direct seeding or planting seedlings. For a direct-seeding crop, 1.5–2.5 kg/ha seed is required, whereas 350–400 g/ha seed is required for transplanting of seedlings. Optimum sowing time is most important to exploit the yield potential of any vegetable crop. The planting time varies from region to region according to prevailing climatic conditions. The seeds can be sown in September–October in the northern Indian plains, August–September in hills, and April–May in higher mountainous hills.

Direct seeding of collard can be done manually, but it requires more seed quantity followed by thinning of plants to avoid overcrowding. Seed should be placed in shallow and moist furrows at a depth of 1.5 cm, but it should not be deeper than 2.5 cm. For harvesting young collard plants like mustard greens, direct sowing of seed is done at 5–10 cm spacing with an interrow spacing of 30–45 cm. The seeds may germinate in 6–12 days. The plants are grown at 90 × 40–45 cm between rows and plants to carry harvesting at the full-grown stage. Transplanting of the seedling is done after 4–6-week of seed sowing.

Nutritional Requirement

For obtaining optimum growth and yield, it is essential to apply only the required quantity of chemical fertilizers. During field preparation, well-decomposed farm-yard manure is applied at 20–25 t/ha. The crop needs nitrogen 100–150 kg, phosphorus 40–75 kg, and potassium 50–100 kg/ha where one-third of the quantity of nitrogen and full P and K is applied during field preparation, while the two-thirds quantity of nitrogen is applied in two split dose at an interval of 1-month interval. The critical stage of irrigation is the leaf enlargement phase.

Irrigation

Collard is a cool-season crop and needs comparatively less water for its better growth and development. In direct seeding crops, pre-sowing irrigation is required to have optimum moisture at sowing time to get better seed germination. Depending upon the weather conditions and availability of soil moisture, irrigation should be given at 10–15 days intervals. It is advisable not to irrigate the field very frequently, but it is essential to maintain proper moisture in the soil as it is a shallow-rooted crop. Frequent irrigation is important to obtain a good plant stand during hot weather.

Intercultural Operations

Shallow hoeing is required to remove weeds and avoid root injury. Alachlor at 2 kg a.i./ha is applied immediately after sowing to control annual and broad-leaved weeds. Trifluralin is also widely used at 0.60–1.0 kg a.i./ha as preplant incorporation (PPI) to control many weeds such as grass, pigweed, and purslane.

Thinning

Thinning, which is done to maintain optimum spacing and to reduce competition among plants, is an important operation in the direct-seeded crop. In the direct-seeded crop, plants are left to grow to a height of 10–15 cm after the emergence or till there is crowding within the row. Then, thinning of plants is done gradually by maintaining a distance of 30–45 cm between plants. Crowding results in smaller and light green leaves.

Earthing Up

Being a shallow-rooted crop, it needs earthing up essentially to cover the roots, which are exposed due to the application of irrigation water. This helps in preventing the plants from lodging and provides a better soil atmosphere to the developing roots and thereby to the plants. Earthing up should be done 30–40 days after planting.

Harvesting and Yield

Harvesting starts when the development of a large rosette of leaves occurs at the crown of the plant. It is appropriate to harvest collard greens, which have firm, tender vividly deep green leaves with no sign of wilting, yellowing, or browning. Collard is harvested in different ways depending upon the method of planting, which are given as follows: (i) Collard is harvested at a very young age when the stalks are tender, clean, and free of discoloration and wilted leaves. (ii) The whole plant is cut at a very young stage just like mustard greens followed by successive cutting. (iii) The entire plant is cut when it is about half grown. (iv) The entire plant is cut when it is full-grown. (v) The tender leaves are harvested from full-grown plants and marketed by tying in 1/2 or 1 kg bunches. It is necessary to trim the plants to remove damaged leaves before they are tied into bunches and are sent to the market. It is essential to make bunches of collard leaves or plants having a uniform size. Only tender stems of plants should be packed. The average yield of collard green varies from 100 to 250 q/ha, depending upon crop variety, soil fertility, environmental conditions, and cultural practices adopted by the growers.

Postharvest Management

Cleaning of collard greens should be done before marketing. Quality is best achieved if collard greens are precooled by hydro-, vacuum, or forced air cooling before transportation as rapid air movement is needed to remove field heat from leaves. Collards are top iced at 0 °C for sale, which is necessary for maintaining quality, as they are perishable. Plants are either bulk loaded into trucks or should be packed into bushel baskets, wire-bound boxes, crates, or cartons for distant shipment and supermarket sales. Collard greens, in general, are sold in bunches of two to three plants, depending on market demand. Greens are not usually stored but will remain in good condition for 10–14 days at 0 °C temperature and at least 95% relative humidity. Collard greens should be top-iced to maintain crispness. These greens should not be stored or shipped with ripening tomato, musk melon, or other ethylene-producing fruits because it is sensitive to ethylene.

Physiological Disorder

Bolting

It is characterized by the development of flower stalks at an immature stage. It occurs when the crop is subjected to 2–10 °C temperature for more than 10 days. The degree of temperature-induced bolting response depends upon the type of variety.

Chinese Cabbage

Chinese cabbage is a quite popular vegetable in China, Taiwan, Korea, Japan, and Mongolia, and recently, it has become popular in India. It includes both non-heading and heading types. The non-heading group (*Brassica rapa* L. var. *chinensis*) includes pak choi, Chinese mustard, celery mustard, and chongee, and its plant is characterized by several thick white petioles and glossy green leaf blades that form a celery-like bunch. The heading group (*Brassica rapa* L. var. *pekinensis*) includes pet sai, celery cabbage, Chinese white cabbage, Peking cabbage, Won Bok, Napa, Hakusai, Pao, and Bow Sum. Within the heading group, there are chihili types that grow erect and develop cylindrical heads and chefoo types that form compact round heads of green-bladed and white petiole leaves. These are two distinctly different groups often used as leaf vegetables. It is known as Chinese Leaf in the United Kingdom, Wong Bok or Won Bok in New Zealand, and Wombok in Australia and the Philippines. Now, its cultivation has spread to many countries because of its high tonnage capacity, nutritional value, and low oxalate content.

Origin and Distribution

Brassica is one of the most important genera containing around 37 different species. *Brassica compestris* emerging in Near East reached China through several pathways about 4000 years ago, and pak choi in Southern China developed separately. The earliest known records suggest that Chinese cabbage arose from a cross between pak choi from South China and turnip from North China. Heading type Chinese cabbage has its center of diversity in Northern China, but according to Kučera et al. (2018), it is originated from European/Mediterranean origin and was imported into China concurrently with the spread of Buddhism. It has adapted well to a relatively cool climate. Pe-tsai cabbage (heading type) and pak choi cabbage (leafy type) are now considered two different groups. Heading type cultivars can be further grouped into (i) cylindrical head forms called chichili (*B. rapa* var. *pekinensis* or *cylindrica*), (ii) round and compact heading types called chefoo (*B. rapa* var. *pekinensis* or *cephalata*), and (iii) open-headed form (*B. rapa* var. *pekinensis* or *laxa*).

Nutritional Importance and Uses

Chinese cabbage is a highly nutritious vegetable that contains a high amount of vitamin C than lettuce and is also rich in calcium, potassium, and magnesium. It also contains citric acid and anticancer compound glucosinolates. The total glucosinolates content varies from 0.097 to 0.337 and 0.39–0.704 g/kg of the fresh edible portion in heading type and leaf types, respectively.

Like common cabbage, Chinese cabbage is used as a vegetable, a salad, in soups, and other Chinese dishes and also processed as a brined product or pickles such as kimchi. The white tender leaves midribs are sliced or coarsely shredded in salads. Being low in oxalate, the patients suffering from kidney or gallbladder stones can consume Chinese cabbage saag; otherwise, those patients are advised not to consume either common sarson saag or one from raya because of its rich oxalate content. Moreover, the leaves being succulent are easily cooked with minimum time and fuel. Its leaves are also dehydrated for off-season use.

Chinese cabbage regulates blood pressure and blood sugar. Being high in iron fights anemia and fatigue and keeps the hemoglobin level high in the blood. Chinese cabbage contains glucosinolates, which prevent cancer. Its large amount makes Chinese cabbage rich in flavonoids and vitamins (A and C), which act as antioxidants and protect the body from free radicals, degeneration, cataracts, and loss of vision. The body requires Chinese cabbage to produce collagen, a protein, which keeps the skin youthful and elastic. It boosts and maintains a healthy body immune system. Chinese cabbage is antirheumatic, antiarthritic, antiscorbutic, and resolvent (Rana 2018).

Botany and Taxonomy

Chinese cabbage is a multilayered biennial crop but grown as an annual vegetable production. The plant grows to a height of 25–50 cm. Although the Chinese cabbage head is firm but less than the cabbage head when fully matured. The outer leaves are pale green, while the inner leaves are creamish white. The head length is about 20–25 cm, and the diameter is 15–20 cm. The leaf blades of the leafy Chinese cabbage are thick, smooth, glossy, dark green, and almost round. The headed types have a cylindrical or barrel-shaped head with a broad central midrib. The leafy types grow faster than the headed forms. It comes in flowering from May to August. The floral morphology and pollination behavior are similar to broccoli and knol-khol.

Varieties

Usually, two types of Chinese cabbage cultivars, viz., pet-sai cabbage heading type (*Brassica rapa* var. *pekinensis* or *B. pekinensis*) and pak choi cabbage leafy type (*Brassica rapa* var. *chinensis* or *B. chinensis*), are cultivated in India. In the heading type, the heads are formed like cabbage, while in the leaf type, the loose leaves are formed with an open center.

Leafy Types

Chinese Sarson No. 1

A leafy type variety with semi-erect growth habit (developed at PAU, Ludhiana, India) possesses field resistance to Alternaria leaf spot and bears light green, broad, and puckered leaves. Its midrib is white, succulent, and tender. The first cutting is done 30 days after planting and gives a a total of 6–8 cuttings with a yield potential of 400 q/ha.

Saag Sarson

A leafy type variety (developed at PAU, Ludhiana) was released in 2014 where leaves are large with prominent midrib. This is most suitable for processing and preparation of good quality saag, which is better than Raya and Gobhi Sarson. It is rich in ascorbic acid and minerals but low in oxalates. This is also approved by Markfed Canaries Jalandhar, Punjab, for processing. The saag prepared from this cultivar is suitable for export. On average, it yields about 500 q/ha.

Palampur Green (PUCH-3)

A late-bolting leafy type variety (developed at CSK Himachal Pradesh Krishi Vishvavidyalaya, Palampur, India) produces tender green leaves with a creamy stem. It takes about 25–30 days for the first cutting and gives a total of five to six cuttings each at 15-day intervals.

Pusa Saag

A leafy type variety developed (IARI, New Delhi, India) through a cross between Wong Bok (Suttons) and turnip. Its taste is like that of Local Sarson.

Solan Selection

A leafy type variety (developed at Dr. Y.S. Parmar University of Horticulture and Forestry, Solan, Himachal Pradesh, India) produces light green smooth and tender leaves with fleshy petioles. Its average yield is about 150–190 q/ha.

Hybrid Shan Tung

A leafy type hybrid is relatively tolerant to heat and adapted well to the subtropical and mild climate. Light green leaves and white thin petioles are very tender and delicious. This hybrid is fast-growing and matures in 3–4 weeks. Its young leaves can be picked for salad purposes. The leaves are suitable for stir-frying and soups.

Green Seoul

A leafy type hybrid widely grown in Korea primarily for the fresh market is a fast growth, attaining maturity in about 4–5 weeks. Its leaves are very tender and well flavored and are used for cooking as well as for pickling.

Hybrid Taiwan Express

A leaf-type hybrid suitable for densely growing year-round in subtropical regions produces light green leaves with frilled edges, which can be harvested within 30 days after sowing. It can be grown very well in a mild and warm climate.

Chin Sun

A leafy type Taiwanese hybrid suitable for salad, stir-frying, and soup preparation produces very tender green leaves and petioles with excellent texture and flavor. Its plants can be harvested at any growing stage. Some of the other leafy type hybrids grown in Southeast Asia are Canton Pak-choi, Pak choy Green, Pak choy White, Green Boy, Lei Choi, Hung Chin, Shanghai Pak choi, and Joi Choi.

Heading Types

Solan Band Sarson

A heading type variety (developed at Dr. Y.S. Parmar University of Horticulture and Forestry, Solan, Himachal Pradesh, India) produces long barrel-shaped light green semi-compact heads with six to nine outer leaves. The head length is 20–25 cm, diameter 10–15 cm, and average head weight is about 700–1100 g. It becomes ready for harvesting in 120 days, and its average yield is about 350–400 q/ha. Optike, an early heading type hybrid resistant to Fusarium wilt developed by Bejo Sheetal Seed Private Limited, produces barrel-shaped short and compact heads with vigorous growth habit. The head is of light green color and good in taste. The other head type varieties popular in Southeast Asia are Chefoo, Wong Bok, Spring Giant,

Tokyo Giant, Tropical Pride, Tropical Delight, Early Top, Tip Top, China King, Winter Giant, Oriental King, Chihili, and Michihili.

Climate and Soil

Chinese cabbage is grown in a cool and warm-season crop in the tropics and subtropics. During vegetative growth, it requires about 20–23 °C temperature (Chowdhury et al. 2021), while during the head initiation stage, the lower temperature of around 15 °C is considered desirable. High temperature during head formation reduces the compactness of the heads. In the northern plains of India, the rise in temperature in the spring season transits the plants into the reproductive phase, and flower formation occurs. It can be grown in all types of soils ranging from light to quite heavy, but it prefers well-drained moisture-retentive soils with 6.5–7.5 pH. If the pH is below 5.5, the soil should be limed to raise the pH since, at low pH, the calcium and other major nutrients become unavailable to the plants. Liming also reduces the incidence of clubroot disease. It is also very much sensitive to high pH and hence not suitable for cultivation in saline soil (Pavlović et al. 2019).

Agronomic Practices

Seed Rate and Sowing

Approximately 500 g/ha seed is required to raise the seedlings, and about 2.5 kg seed per hectare is sufficient for direct seeding crop. For sowing in portrays, around 200–300 g seed is adequate for producing seedlings enough for planting a hectare land area. In North Indian plains and low hills, the seeds are sown in mid-September, and transplanting is done in mid-October. In mid-hills, the ideal sowing time is September to October, and the transplanting time is October to November. In high hills and dry cold desert areas, the crop is sown or transplanted from April to June.

Raising of Seedlings

Raised nursery beds of 15 cm high, 3.0 m long, and 1.0 m wide are prepared for the sowing of seeds. Nursery beds are usually covered with fine nylon mesh screen net to protect the nursery from heavy rain, blazing sun, and virus transmitting insects. The upper 5–7 cm soil layer of the bed is prepared by mixing garden soil, well-decomposed farmyard manure, and sand in a ratio of 2:1:1. In well-drained soils, calcium ammonium nitrate is incorporated at 40 g, superphosphate at 50 g, and muriate of potash at 30 g/m2. The seeds are sown 5 mm deep in rows spaced at 6 cm apart, and thereafter, the beds are covered with organic mulch such as paddy straw or dry grass and then irrigated with a rose can. The mulch cover is removed after the completion of seed germination. Transplanting of seedlings is done after 3–4 weeks of seed sowing or when they attain four to five leaves. Water supply is restricted 3–5 days before transplanting to harden the seedlings. The seedlings can also be raised in portrays under protected conditions, but the seed requirement is almost half. About two to three seeds per plug are sown at a depth of 0.5–1.0 cm. The seeds take 3–5 days to emerge, and thinning is done at the 2–3 leaf stage, keeping one seedling per plug. After thinning, water-soluble fertilizer is applied.

Transplanting

Direct seeding and transplanting are done in the case of leafy type varieties, while heading types are usually transplanted. The distance between rows is maintained at 30–45 cm and between plants at 30 cm in heading types, while the leafy types are planted at 25–30 × 25 cm spacing between rows and plants; however, in leafy type, the best results are obtained when transplanting is done at 45 × 45 cm spacing. In direct-sown leafy type crop, thinning is normally done after 3 weeks of sowing.

Nutritional Requirement

The crop requires the incorporation of well-decomposed farmyard manure about 30–40 t/ha at least 2–3 weeks before sowing/transplanting at the time of land preparation. Besides, nitrogen is applied at 125 kg, phosphorus at 75 kg, and potash at 75 kg/ha. The method of fertilizer application is the same as in collard. Excessive application of nitrogen can lead to increased susceptibility to insect pests and diseases. For heading types or chefoo types, some scientists recommend the use of 10 tons of well-decomposed farmyard manure along with nitrogen 75 kg, phosphorus 30 kg, and potash 30 kg in a hectare land area. *Trichoderma* biofertilizer enhances nutrient uptake, and tolerance to environmental stresses improves the head quality and flowering (Ji et al. 2020). *Bacillus amyloliquefaciens* KC-1 was very effective to control soft rot disease (Cui et al. 2019). Horse manure also gives the highest Chinese cabbage yield in comparison to cattle manure (Laczi et al. 2016). With an increase in nitrogen application, tipburn and bolting disorders are decreased (Vavrina et al. 1993).

Irrigation

Chinese cabbage water requirement usually varies from 400 to 600 mm. The crop needs a continuous supply of water with first irrigation given immediately after transplanting and subsequent at 10- to 15-day intervals as per the need of the crop because it is a shallow-rooted crop. In heading types, sufficient moisture should be available during head formation since inadequate watering can delay maturity, reduce the head size, and thus yield. Excess irrigation during head formation can cause the formation of poor-quality heads. Heavy irrigation after a dry spell especially at maturity can cause the bursting of heads.

Intercultural Operations

Intercultural operations are needed to control weeds and provide aeration to the soil around roots. Crop weed competition has been reported between the second and seventh week after direct seeding or transplanting; therefore, the weed intensity should be low during this period. The weeds are controlled by hand hoeing or in combination with the preemergence application of herbicides. Two to three weedings are sufficient to control the weeds. The time of weeding depends on the weed emergence behavior of the crop. Mounting up of soil around the base of the plants, especially in the case of early crops, favors better growth of the plants. The recommended herbicides should be used to control the weeds, especially in the transplanted crop in situations where labor is expensive. Preplant application of Stomp 30EC (pendimethalin) at 2.5 l/ha followed by one hand hoeing is effective.

Harvesting and Yield

The first cutting of Chinese cabbage is done in mid-November. In heading types, the heads are usually harvested manually when they are well developed and firm but before the development of the flower stalk. The heads should be handled carefully as they are very easily injured. Harvesting should always be done in the evening with the outer leaves trimmed off. The heading type Chinese cabbage is usually harvested 65–80 days after transplanting, while in leaf types, first picking can be done 30–45 days after sowing or transplanting. The crop gives a 20 to 50 t/ha average yield; however, a higher yield up to 100 t/ha can also be obtained using better varieties and improved production and protection techniques.

Postharvest Management

In leafy type, damage occurs due to rough handling, washing, or loading onto the trucks and transport to retail markets. The weight loss occurs during transport to retail markets and home storage, while yellowing of leaves occurs due to high temperature. In head types, the damage occurs due to rough handling at all stages in collection centers and during loading. Storage decaying is caused due to poor ventilation and freezing during storage. Immediately after harvesting, the heads should rapidly be cooled to as close to 0 °C as possible.

Physiological Disorders

The major physiological disorders developed in Chinese cabbage due to the environmental and nutritional factors are tipburn, pepper spot, bolting, and petiole freckles.

Tipburn

The tipburn disorder is caused due to calcium deficiency in young leaves, and a condition of high temperature and low relative humidity is characterized as brown to black necrotic areas on leaf margins. It occurs in both wet and dry months as plants start heading. The disorder is also strongly associated with soil salinity, nitrogen fertilization, and climatic conditions. Avoid excessive use of nitrogenous fertilizers, and spray the crop with calcium citrate (25 g/100 liters of water) twice a week to control this disorder. Spraying daminozide at 1–5 g/liter of water at the seedling stage also reduces the incidence of tipburn in Chinese cabbage.

Bolting

Bolting is a process of developing seed stalk. Chinese cabbage when grown in late spring under long-day conditions develops a seed stalk and thus undesirable poor quality heads result. Late sown crop normally tends to bolt as compared to fall season sown crop. In Chinese cabbage, the bolting response is under the control of genetics. Hence, some of the cultivars bolt more swiftly than the other cultivars. If the sowing is to be done in the spring months, slow-bolting cultivars should be selected for growing since these cultivars are tolerant to warm conditions and produce compact heads. For flowering, this crop is not only sensitive to photoperiod but also sensitive to temperature. The evidence in the literature indicates that exposure of plants to a 4.4 °C temperature for a week or 10 °C for 2 weeks or prolonged temperature below 13 °C at the seedling stage induces bolting. Growing the crop under short days and warmer temperature conditions keep the plant in the vegetative phase. Lastly, the factors that check the growth of the plants such as deficiency of nutrients or water scarcity also induce bolting.

Petiole Freckles

Corking and brown streaks are developed on the midribs due to boron deficiency. To overcome this problem, borax is applied 10–20 kg/ha at the time of sowing, depending on soil type.

Garden Cress

Garden cress is a fast-growing perennial crop also known as pepperwort, pepper grass, or garden pepper cress. It is called Asalio or Chandrasoor in India. It is commercially cultivated as an important medicinal crop for the production of seeds. In some parts of the world, it is a popular green vegetable for use as toppings or as a leafy vegetable. The leaves and seeds having a peppery taste are consumed in relatively small amounts since their peppery taste can be overwhelming.

Origin and Distribution

Garden cress is originated from Southwest Asia and later spread to various parts of Europe, Britain, Italy, France, Germany, Syria, Greece, India, and Egypt. Persian cultivated this plant as early as 400 B.C. and used it even before bread. It was introduced to the United States from China. In India, it is cultivated in southern states on a commercial scale.

Nutritional Importance and Uses

Garden cress is nutritious vegetable which contains calcium (81.0 mg), potassium (606.0 mg), phosphorus (76.0 mg), magnesium (38.0 mg), iron (1.3 mg) as well as vitamin A (6917 IU) and C (69.0 mg), folic acid (80.0 µg), thiamine (0.08 mg), riboflavin (0.26 mg), and niacin (1.0 mg). Its seeds are rich in protein (2.6 g), dietary fiber (1.1 gm), fat (0.70 g), and photochemical. Garden cress is usually cultivated for its leaves and seeds. Its leaves are used in salad, baby greens, and garnish. It is also eaten with cottage cheese, boiled eggs, or mayonnaise and chutney with bread. In Arab countries, its powder is consumed with honey or hot milk and boiled seeds with drinks. The seeds are bitter and considered an abortifacient, diuretic, expectorant, aphrodisiac, antidiabetic, antibacterial, antispasmodic, anticancerous, antiasthmatic, hypocholesterolemic, analgesic, hepatoprotective, coagulant, gastrointestinal stimulant, gastroprotective, laxative thermogenic, depurative, galactagogue, emmenagogue, and ophthalmic. It also has cytotoxic effects on different carcinoma cell lines (Alkahtani et al. 2020). Its seeds are ground into a paste that can be used to relieve pain and swelling in rheumatic joints. The ethanolic seed extract is effective in treating inflammatory bowel disease.

Botany and Taxonomy

Garden cress is an erect, small glabrous annual herb. Its leaves are ovate or ovate-lanceolate and glabrous, and flowers borne in terminal and axillary paniculate cymes are brownish-purple, bisexual, regular, and tetramerous. There are six stamens, and the ovary is superior, flattened, apex emarginated, and stigma capitate. Fruit is round or ovate and usually two-seeded. Seed germination is epigeal, seedling cotyledons are trifoliolate, and leaflets are spatulate.

Varieties

Dadas

A newly registered variety for cultivation in the Eastern part of Turkey produces curly leaves.

Bahar

A variety for cultivation in Turkey produces plain leaves.

Wrinkled

A variety with extremely slow bolting developed through a cross between curled and Persian cress is recommended for cultivation in the United States of America. Its leaves with ruffled edges look like curly parsley.

Curled

A variety of dark green and curly leaves and slow bolting nature is recommended for cultivation in the United States of America. Some other important varieties of garden cress recommended for cultivation in the United States of America are Crinkled, Crumpled, and Persian.

Climate and Soil

Garden cress is a winter season crop that requires low temperature and high relative humidity. It can successfully be cultivated in areas with full sunshine as well as partial shade. The ideal mean temperature for its cultivation is 12–22 °C. The seed germination is slow and poor when the soil temperature is less than 8 °C. Under long-day and warm temperature conditions, the plants start bolting, which spoils the quality of the produce. It is mostly grown on all types of soils but thrives best in well-drained fertile and organic matter-rich soils and has a good moisture-holding capacity. It prefers slightly acidic to neutral soils with a pH of 4.9–7.0 for its successful cultivation.

Agronomic Practices

Seed and Sowing

In subtropics, September last week to October first week is the sowing time whereas March–April in temperate areas. Seed rate usually varies with viability, sowing time, and growing condition. A seed rate of about 3.0–4.0 kg is sufficient to sow a hectare land area. Garden cress is propagated by seed. Sowing is done either by broadcasting or in lines at 15–20 cm spacing and 0.5–1.0 cm depth. The seeds after sowing are covered lightly with fine soil or compost. Seed germination takes place after 4–6 days of sowing, depending on the temperature and humidity. It can be grown year-round in pots, bowls, boxes, or flat plates, putting or hanging on a windowsill.

Irrigation

Garden cress is a shallow-rooted crop; thus, it performs best if moisture in the soil is kept at field capacity. In summer, irrigation is applied at 4- to 6-day intervals, and in winter, at 10- to 12-day intervals to keep the soil moist. The field is kept moist during dry periods since water-deficit stress during the growth period reduces the yield greatly.

Nutritional Requirement

Garden cress requires a very low amount of fertilizer due to its very short growing period. However, to harvest a good yield, it is necessary to fertilize the crop periodically with soluble liquid fertilizers. The roots and shoots of young seedlings of garden cress contain heavy metals like cadmium and lead which are toxic to human health (Souri et al. 2019). Humic acid and fulvic acid can help plants absorb less cadmium by stabilizing cadmium in the soil and preventing cadmium from moving from the plant's roots to its shoots and leaves (Yildirim et al. 2021).

Intercultural Operations

At the initial stage of crop growth, hoeing and weeding are required, when the plants are established since the crop plants at this stage compete poorly with weeds. Hoeing is done to prevent the uprooting of shallow roots, which slows down plant growth. The crop needs continuous moisture supply, which is impossible to provide all the time; however, the moisture can easily be conserved by covering the soil with straw, shredded newspaper, dry grass, or polyethylene sheet. The weeds, which deplete field soil moisture, can also be checked with the use of mulch.

Harvesting and Yield

Garden cress is a fast-growing crop, and it is harvested in the same month of sowing, i.e., 2 or 3 weeks after emergence. The leaves as and when attain a height of 5–8 cm and turns green are harvested with the help of scissors. Only the older leaves without touching the younger ones are harvested for immediate use so that the plants may not stop growing. The average yield of garden cress leaves is 60 q/ha.

Postharvest Management

Garden cress cannot be stored for long period, so it should be used as soon as possible after harvesting to preserve its freshness and flavor.

Upland Cress

Upland cress (*Barbarea verna*) is also called American cress, dryland cress, cassabully, and creasy salad. In form and flavor, the upland cress resembles the watercress. The word Lepidium has been derived from the Greek word Lepidium, which means a little scale, as the pod shape is like scales, and the species verna simply means spring. Normally, it is cultivated in hot areas of the world. Upland cress is occasionally found in the gardens of Florida.

Origin and Distribution

Upland cress is native to the parts of Central, Southeastern, and Southwestern Europe, as it has been grown there as a leafy vegetable since the seventeenth century. Thereafter, this species has been widely introduced to the rest of the Europe, Africa, America, and Asia. The species is widespread in France, Italy, Spain, and Belgium; however, in Malaysia, it is grown on a small scale.

Nutritional Importance and Uses

Upland cress is rich in calcium and iron as well as in vitamins, chiefly vitamins A, B, C, E, and K. It is also high in phytonutrients, having twice the amount of vitamin A than broccoli and twice the amount of vitamin C than oranges. It also has a high concentration of lutein, which helps to prevent macular degeneration and β-carotene (Xiao et al. 2019). It has a high protein and calorie content in its seeds. Upland cress calcium is comparatively less soluble than that from skim milk and deprives the availability of calcium from spinach. The leaves of the upland cress having a radish-like sharp spicy flavor are cooked as a grilled vegetable and as a garnish in fish dishes. It is also used raw in salads, sandwiches, and in any recipe in which watercress is used. Its freshly harvested leaves, shoots, or very young flower buds are used to add vigor, especially to soups and egg preparations. Besides, its leaves may also be used as a substitute for parsley, and young sprouts may be added together with mustard sprouts to give a spicy taste. In the old world, it was used for the healing of wounds. During colonial times, it was consumed in the winter to ward off scurvy disease. The addition of dark green leafy vegetables that are innately rich in beta-carotene (provitamin A) is strongly recommended by nutritionists and dieticians to cure vitamin A deficiency. Being fat and cholesterol-free is useful for weight loss and good for cardiovascular health. It is a mild diuretic, helping to get rid of excess water from the body.

Botany and Taxonomy

Upland cress is a biennial herb, but it is cultivated annually for leaf purposes. It has a diverging taproot system. The foliage of the plant is not so dense. The leaves are dark green, very short, stocky, and almost square-shaped, and with a zesty tenderness. The leaf petiole is 15–20 cm long, and margins are slightly notching, resembling curled parsley. In the first season, it forms a rosette of leaves, and in the second season, it develops inflorescence. Its small yellow flowers are hermaphrodite but cross-pollinated in nature. Pollination is performed by flies, bees, and beetles. Flowering takes place from December to January, and seeds mature from March to April. The seeds remain viable for 3–5 years.

Varieties

Upland cress is an underexploited wild plant of the brassica family; thus, improvement work has not yet been done in India to develop varieties or hybrids. However, several local selections without proper nomenclature are available. Farmers usually grow the locally available varieties.

Climate and Soil

Upland cress is a winter season crop that requires low temperature and high humidity. The seed germination is better at lower (12–23 °C) than at higher (25 °C) temperatures. 15 °C is the optimum temperature required for seed germination while for vegetative growth is 18–20 °C; however, the growth is slow at 5–7 °C. Low temperature tends to increase leaf thickness but decreases leaf size and smoothness. It thrives in the open conditions as it can withstand frost, although it needs protection in colder months at night and during very heavy frost, which can be provided using cloches with sacks. If it is grown at a high temperature and as the plant matures, its flavor becomes very spicy, but the young leaves' taste is just fine. Under a condition of dry soil and high temperature, its leaves and stems become unpleasant and bitter. Upland cress can be grown in a site with full sunshine as well as partial shade. The crop in summers can successfully be grown by providing some shade, while the winter crop succeeds in sunny locations, needing 6–8 h of light for vegetative growth, but the increasing light intensity and atmospheric temperature increase the production of biomass since environmental conditions influence the leaf number and size. It can be grown almost round the year, but its production is highest under short-day and mild temperature conditions. Being a long-day plant for flowering, upland cress under long days and warm weather conditions tend to develop seed stalks; therefore, it is not possible to grow it during spring-summer in plains, especially under long-day conditions. Upland cress is mainly grown on a wide range of soil from light to heavy provided the soil has a proper drainage facility and good water-holding capacity. However, sandy loam organic matter-rich soil is preferred over heavier soils since the crop is susceptible to waterlogging conditions. The ideal soil for its cultivation is fertile, moist, and properly drained with adequately decomposed compost. Although it can be grown at a soil pH of 4.5–7.5, the soil pH of 6.0–6.8 keeps the plants healthy and nourish the plant better. Proper land preparation is done to ensure satisfactory seed germination and for further plant growth.

Agronomic Practices

Seed and Sowing

Generally, 1.7–2.25 kg seed is enough for a 1-hectare area. Upland cress can be propagated by root divisions or cuttings taken during the spring months. However, propagation by vegetative parts is less adopted. Therefore, it is usually propagated by seeds, and the direct sowing method is used for its cultivation either by broadcasting or by line sowing. However, sowing seeds in lines is always better than the broadcasting method, as it facilitates intercultural operations and harvesting. 15–20 cm spacing is maintained between the lines and 8–10 cm between the plants. Spacing between plants influences the number of leaves and their size. Lower plant density facilitates hand-harvesting for the bunching of leaves, and on the contrary, high plant density results in more upright leaf growth. The optimum plant density can be obtained by sowing disease-free healthy seeds. Based on soil type and moisture availability, the depth of seed sowing should be 0.6–1 cm. The sowing depth is maintained a little greater in light soils and a little lesser in heavy soils. The seeds after sowing are covered with light soil. Its seeds take 2–3 weeks to germinate. In subtropical climate, the crop is sown from August last week to September first week, and in temperate areas, the sowing time varies from early spring to late summer, but under cloches, it can be grown round the year without any difficulty except in January and February. In mild climates, it can be grown during any month for a succession supply of the leaves, and for a steady supply, the sowings are made after every 3–4 weeks; however, during August–September, the plants are protected from extreme blazing sun and heat using shading net.

Nutritional Requirement

Farmyard manure at 8–10 t/ha is given during land preparation. Lack of phosphorus and potash results in dry leaf tops and leaf necrosis, respectively, whereas poor foliage growth due to nitrogen deficiency. Nitrogen, phosphorus, and potash are applied at 80:60:40 kg per ha. In silt and clay loam soils, the amount of nitrogen, phosphorus, and potash may be reduced to 100, 40, and 30 kg/ha, respectively. One-third quantity of N along with full phosphorus (P) and potassium (K) should be given as basal dose, and two-third amount of nitrogen should be given in splits after each cutting. In severe yellowing of leaves in later stages, nitrogen 40 kg/ha may be applied as a side dressing. The application of nitrogenous fertilizer normally increases the leaf production of the upland cress by increasing the leaf size. Moreover, potash and phosphorus improve the leaf quality by increasing leaf density.

Irrigation

Upland cress is a moisture-loving crop; thus, it needs to be watered regularly. The soil is kept consistently moist after sowing seeds but not fully soaked since waterlogging as well as drying of soil is detrimental to this crop. Irrigation is applied only at the base of the plant; thus, it should not be applied through a sprinkler system since it is essential to keep the foliage dry. It is not necessary to irrigate the crop in winter since the irrigation requirement of upland cress is usually not too high in the cool season due to the low rate of evapotranspiration during this period, but during summers, it needs a considerable amount of water for proper growth and development since the evapotranspiration rate is high during summer months. In very hot weather, if the crop is grown in dry soil, its nips become unpleasant and bitter. The first light irrigation is given immediately after sowing, and subsequent irrigations are given as per the requirement of the crop. The plants grow best if regular watering is done. For incessant harvesting, the crop should be watered regularly.

Intercultural Operations

The intercultural operations adopted in upland cress are similar to other cruciferous vegetables. Keeping the field weed-free is essential, especially for the crop grown for vegetable purposes since the weeds, particularly those which are similar in appearance to upland cress and difficult to separate, contaminate the crop; therefore, removing such weeds from the upland cress field is most crucial. Regular hoeing after each cutting or mulching with organic material can be done to suppress the weeds. Mulching will also help in keeping the soil moist, which is essential for the production of the best quality upland cress. The crop grown for the fresh market hardly requires thinning since this operation is labor-intensive. However, thinning is essential when the crop is sown by broadcasting. When the seedlings become large enough to handle, the extra seedlings are removed, keeping the plants at a spacing of 20 cm. Pruning may also be done to obtain quality produce. To promote new growth, the sprouts 10 days after germination are cut at ground level with the help of scissors when tender leaves of the plant just begin to have a green color.

Harvesting and Yield

Harvesting of upland cress depends on the purpose of its use. Usually, the crop becomes ready for harvest at the five to six leaf stage, which is attained 7–8 weeks after sowing under favorable environmental conditions. Leaves, shoots, flower buds, and seeds all are delicious, and their seed oil is edible. Young tender leaves are best for use. Harvesting should not be done by pulling out the plant; it is done by cutting the outer leaves. When the plant begins to develop flowering stalks, harvesting of leaves should be stopped since the leaves by this time become extremely bitter. For the harvesting of flower heads, it is always advisable to cut the flower heads fresh as and when needed, and they should be consumed soon after harvesting. Upland cress average yield is about 37.5 quintals per hectare.

Postharvest Management

Being a leafy vegetable, the upland cress cannot be stored for long periods. The leaves, however, can be stored in low-density polyethylene film bags in the refrigerator for 2–3 days.

Sea Kale

Botanically, sea kale is known as Crambe maritime. The word Crambe comes from the Greek words Krambe, which means cabbage or crucifer, and maritime means seaside. Since the plant can tolerate salt, it can be raised in regular gardens and on beaches. It is nearly a perfect primitive food, and it is difficult to imagine that it would not be on a primitive menu of man's food. It is very similar to leaf cabbage, though its leaves are comparatively thicker. Its shoots are very crisp with a fresh, nutty flavor and slightly bitter taste and are used like asparagus. Its roots, which are very starchy and a little sweet, can be eaten after boiling and roasting. Sea kale, with its large leaves and fragrant flowers, has recently become a more popular ornamental plant. It has gained a Garden Merit Award from the British Royal Horticultural Society. Sea kale in general is an insect nectar plant, so bees and parasitoid wasps frequently visit the plant, which is grown as a groundcover.

Origin and Distribution

Sea kale is a wild species found on both Mediterranean and Atlantic Ocean seashores and cliffs in Northwest Europe. It was first grown in the sixteenth century, and by the eighteenth century, it had become a popular garden vegetable in Europe and North America. It could not find a place in modern agriculture, as it was difficult to store or ship it well.

Nutritional Importance and Uses

Fresh sea kale contains an adequate amount of mineral salts, viz., potassium and sulfur. Like other greens, sea kale contains nitrates levels (>17 mg/100 g). Every part of the sea kale plant is edible. Very young leaves and thin stems are used either raw or cooked like kale, collard, or asparagus since the older leaves become tough and bitter. Its roots are very nutritious and eaten raw; however, more calories are available when eaten after cooking or roasting. Like broccoli, the flower heads along with their peduncles can also be eaten in raw or cooked form. This plant is also used as fodder for cattle, sheep, swine, and poultry. Like other Brassicas, its high-fiber content makes it very beneficial for digestion, which helps in preventing colon cancer. It has diuretic properties; as a result, its consumption aids in the removal of toxins from the body and improves kidney function and aids the elimination of excess salts, water, and fat from the body. It makes the body's immune system strong and protects the body from common colds and cancer due to its high ascorbic acid content.

Botany and Taxonomy

Sea kale, an herbaceous plant, is about 60–90 cm tall with deciduous leaves. Its deeply lobed leaves with long stalks, purple veins, and wavy edges are usually over 30 cm long, smooth, and fleshy. The leaves are alternate. The lower leaves contain blades lobed with an entire toothed margin. The newly formed leaves are purple, which later become green. A thick waxy layer covers the leaves, which acts as an effective water repellent. The plant has a strong taproot system. After completing its basic vegetative phase, the plant develops hundreds of 0.8–1.5 cm large white flowers, which open in spring over several weeks and release a honey-like fragrance. The inflorescence is a corymbose raceme, containing several flowers. The flower is botanically regular, i.e., actinomorphic, having four petals, four sepals, and six stamens, out of which four are long and two are short. The gynoecium is fused with a single carpel. The pollination is performed by honeybees and wind; thus, the sea kale has large genetic variability. Its indehiscent fruit usually contains one seed, which germinates very slowly with a low germination power due to seed dormancy and slow water imbibition because of the thick and hard integument.

Varieties

Lily White

It is the most widely grown variety, producing pure white heads with good flavor. It takes 5–14 days for germination and 45–50 days to maturity after seed sowing and gives a 200–300 q/ha yield.

Ivory White

It is mostly similar to Lily White in most of its characteristics.

Climate and Soil

Sea kale usually requires little maintenance; thus, it is quite easy to grow it in gardens. The basic requirement of sunlight and water results in healthier plant growth. Its strong taproot system makes it a drought-tolerant plant. Being a hardy plant, its cultivation is possible under any environmental condition. Although it is resistant to frost, temperatures above 30 °C are most preferable. High-velocity winds seem to cause it a little problem. Sea kale can be grown on all types of soil. Its natural environment is organic matter-rich sandy soil with good water-holding capacity.

However, it should be cultivated in well-drained deep heavy soils rich in plant nutrients but low in organic matter for getting a higher yield. It grows well in soils with a pH of around 7.0.

Agronomic Practices

Seed and Sowing

Sea kale is planted from mid to late spring in the hilly tracks of India. Sea kale can be propagated by root cuttings known as thongs and seeds. Propagation by seed is more affordable than thongs, which take more time to start. Its propagation by seed is much similar to cabbage. The plants through thongs look like second-year plants. When it is propagated by seed, the plants start to deteriorate after about 7 years. Therefore, the best way is to replace a few plants each year. Root cutting is a quicker method of propagation. The cuttings can be obtained from well-established healthy plants at least 3 years old, with no sign of rotting on the crown. Taking root cuttings from a 1-year-old plant is not desirable. Thongs should be lifted carefully in October–November after the shedding of leaves. The thongs should be straight, clean, 15 cm long, and of pencil thickness. To avoid planting them upside down, a straight cut across the bottom and a slanting cut across the top are made. The cuttings are tied into bundles and kept straight in a box containing sand in a cool shed until sprouting in March. These thongs are soaked in water for 1 h before planting. The thong is placed vertically in the soil during planting. The top end of the thong starts thickening in the soil and then produces sprouts, which later reach the soil surface, but the sprouting of thongs can be quite slow. While preparing for planting, leaving the strongest central bud all the buds are rubbed off. A hole is made with a dibber, and the cuttings are planted 2.5 cm below soil level, keeping a 37 cm distance each way. The seeds are very corky, which enables them to float at sea for several years, which is why the germination is very slow. This corky case can be removed carefully to speed up the germination process, as the seed can imbibe water more easily. The seeds are sown 2.5 cm deep in moist soil or portrays in late May. The seeds take 2 weeks to germinate. The seedlings in seedbeds are thinned at the three- to the four-leaf stage to make them strong. As and when the seedlings become ready for transplanting, they are picked out and transplanted in well-prepared seedbeds in the last week of May. Only the healthiest seedlings are planted in the field. Plants can be cropped once they are more than 12 months old. The young seedlings are very attractive to slugs; thus, the seedlings must be protected against slugs.

Nutritional Requirement

Sea kale responds well to soil amendments like farmyard manure. Thus, the crop needs regular application of fertilizers, which are to be applied from March to October. The slow-release granular fertilizers or easily water-soluble fertilizers should be applied after every 15–20 days. During the autumn months, the nutritional requirement can be fulfilled with some slow-releasing fertilizers, or with some vermicompost.

Irrigation

The crop needs irrigation at 10–15 days intervals depending on season and soil moisture. It is advisable not to irrigate the field very frequently; however, care should always be taken that the field may not dry and become too compact; otherwise, the root development is affected adversely. The crop is sensitive to waterlogging conditions; thus, water stagnation should be avoided. The irrigation requirement of sea kale varies from region to region. In coastal areas, no irrigation is required. During the coldest months, irrigation should be applied only when the soil dries out completely, and in the case of low temperature, irrigation frequency is reduced. For better growth and performance of plants, maintaining adequate humidity is also very necessary.

Intercultural Operations

Sea kale is a shallow-rooted crop; hence, shallow weeding and hoeing are required to protect the roots from damage. Weeds are a problem during the first month after transplanting on account of more vacant space. Manual weeding and hoeing with a hoe (Khurpi) are done after 2 weeks of transplanting. Top dressing with nitrogen fertilizer followed by earthing up is carried out after 4–5 weeks of transplanting, and thereafter, spot weeding can be done if required. In India, the chemical weedicides have not been experimented with; however, preplant soil incorporation of weedicides is recommended for crucifer crops, viz., fluchloralin (Basalin 48 EC) 1.0 kg/ha, or alachlor (Lasso 50 EC) 2.0 kg/ha or pendimethalin (Stomp 30 EC) 1.0 kg/ha should prove effective.

Blanching

Blanching is one of the most important operations, which is to be done in January or early spring. The purpose of blanching is to protect the plant from sunlight exposure. This operation makes the shoots milder in flavor but decreases the nutrient content. The crowns of 2-year-old plants are covered either with special earthen pots, large plastic buckets, or with a mixture of well-decomposed garden compost and leaf mold. Dry grass, sugarcane tresses, or paddy straw can also be scattered loosely over the plants for blanching purposes, but pots are preferred over any other types of cover.

Winter Protection

Sea kale plants can be forced in situ, in which the same plants can be grown for several years, or they can be lifted and forced indoors in warmer conditions. After dying back of crowns, the plant refuses to be removed from the field, and the crowns are covered with an 8 cm thick layer of dry grass, traditional clay seakale pot, or black polythene attached to a wooden frame during winter months, which keeps the temperature a bit high. Such covering will also exclude light. Whatever covering is used, it should be at least 37 cm high and firmly held down so that it may not blow away. This process makes the stems ready for cutting within 3 months. Using a sharp knife, the stems are cut with a little piece of root attached. Cutting is stopped in May, and the plants are allowed to resprout. They can then be blanched again in the following year.

Forcing Indoors

The roots are dug after the first frost or lifted before the frost but left on the ground to expose them to frost. After trimming off the side roots, they are packed in boxes or large pots, and the frame is covered with black polythene attached to a wooden frame to exclude light. Roots can also be forced into a greenhouse or cool room.

Harvesting and Yield

Sea kale leaves and shoots at the small and tender stage are harvested in spring since the older leaves are tough enough; thus, they need a long cooking period. The flowering heads are ready for cutting in summer when they are 15–18 cm long. They are harvested as per the need. If they are not consumed at once after harvest, they become poor in color and deteriorate quickly. When the plant has become dormant, its roots are lifted from the field. The roots take about 100 days to reach maturity. For obtaining good-size roots, 3-year-old plants are harvested. Only the blanched roots are stored for an extended period, as the stalk does not stay fresh for a longer time. After blanching, the stems having a length of 15–25 cm should be harvested by cutting them below the crown. Improved varieties also have greater yield potential as compared to local varieties. On average, sea kale gives a 200–300 q/ha yield.

Postharvest Management

The main crop of sea kale is the spring shoots. The blanched shoots like asparagus are cut when they attain a length of 15–20 cm and have a slight hazelnut flavor. In its first and second years, sea kale is slow to grow; thus, it should not be harvested until the third year. Sea kale does not store very well; therefore, it should be used within a day.

Physiological Disorders

Brown Head

The browning of heads occurs due to excessive temperature, poor growth, and disease. The intensity of this disorder can be controlled by applying an adequate amount of nutrients.

Cracking

A sudden increase in soil moisture can cause cracking of heads. Cracking is particularly a problem in sea kale when heavy irrigation is applied near maturity. Inside the head, the young leaves expand quickly with the sudden increase in soil moisture; as a result, the heads develop cracking. There will be no head cracking with a regular and uniform supply of moisture.

Tipburn

Tan or light brown tissues are the characteristic symptoms, which may later turn dark brown or even black. The tissues that are affected lose moisture and appear papery. The causes of tipburning are the application of high doses of nitrogen and high humidity in the atmosphere. This disorder occurs due to calcium deficiency in soil and plant tissues, particularly in the margins of inner leaves. High temperature and high levels of nitrogen are other reasons for this disorder. Under such conditions, the plants are unable to supply sufficient calcium to actively grow inner young leaves at a critical point. Neither soil nor the foliar application of calcium will be effective in reducing tipburning because calcium is fixed by the outer leaves and not further translocated to the young growing leaves. The only solution to tipburning is to grow resistance to tolerant cultivars.

Hollow Stem

Hollow stem in sea kale occurs due to boron deficiency and excessive use of nitrogenous fertilizer. The hollowness due to boron deficiency appears in water-soaked areas, while the hollowness due to the excessive use of nitrogenous fertilizer is recognized as a lucid white stem with no tissue disintegration sign. The incidence occurs more in crops planted at wide spacing. The plants that produce larger heads are prone to this disorder. The hollow stem is also infected with fungi, which enter through the openings on the stem. To avoid this problem, grow those varieties that are resistant to the hollow stem. Spraying borax 0.25–0.3% can cure the hollowness caused due to boron deficiency.

Black Petiole

This disorder has been noticed in the recent past. Being internal, this disorder cannot be seen on outer leaves. As the cabbage heads tend to mature, the dorsal side of inner leaves midribs or petioles turns dark or black at or near the point where the petiole is attached to the core. This complex form of the physiological disorder is entirely due to certain unfavorable environmental factors.

Plant Protection

Diseases

Important diseases of underutilized cruciferous vegetables along with their symptoms and control measures are discussed below (Table 10.2):

Insects-pests

Important insect pests of underutilized cruciferous vegetables along with their damaging symptoms and control measures are discussed below (Table 10.3):

Conclusion

All the cruciferous vegetables have high nutritional and medicinal value due to glucosinolate content and their hydrolysis products. They are also good for preventing and treating inflammation, as well as a variety of gastrointestinal and digestive issues, and chronic diseases. Although there is no doubt that including underutilized brassicaceous vegetables in one's diet can provide health-promoting compounds, their cultivation is limited to a small scale. Therefore, the underutilized cruciferous vegetable cultivation can be improved by developing high-yielding cultivars and commercialization of their production.

Table 10.2 Important diseases of brassicaceous vegetables

Disease	Symptoms	Control
Damping-off (*Pythium* spp., *Phytophthora* spp., *Rhizoctonia solani*, and *Fusarium* spp.)	Two types of damping-off, viz., pre- and postemergence; seeds are unable to germinate in preemergence, whereas in postemergence damping-off, the stem tissue becomes soft and water-soaked, and the plant collapses at the collar region near the soil surface	Adopt long crop rotation, avoid waterlogging, and provide good drainage; treat seed with Bavistin, thiram, captan, mancozeb, or Apron 2.5 g/kg of seed; Drench (5 liter of water/m^2 area) the soil with fungicide captan or thiram 2.5 g/l of water
Downy mildew (*Peronospora parasitica*)	Yellowish, irregular, or angular lesions on the lower leaf surface are soon covered with downy growth of fungus, the upper surface becomes chlorotic and leaves may drop off prematurely, rotting infected heads	Follow long crop rotation; eradicate cruciferous weeds from the field; avoid overhead irrigation and excess watering; seed treatment with hot water at 50 °C for 30 min; spray Indofil M 45 0.2%, Indofil Z-78 0.2%, or Ridomil MZ 0.25% at 7-day intervals
Powdery mildew (*Erysiphe polygoni*)	White powdery masses appear on both sides of the leaves, which gradually turn completely yellow, curl, dry up, and fall off; dry weather, moderate temperature (15–25 °C), and shady conditions are favorable for the development of this disease; spores are carried to new hosts by the wind; spores and fungal growth are sensitive to extreme heat (above 32 °C) and direct sunlight	Follow long crop rotation with non-host crops; provide good air circulation and sunlight; avoid excessive use of nitrogenous fertilizer; avoid overhead sprinklers for irrigation; spray the crop with Bavistin 0.1%, wettable sulfur 0.3–0.5%, Sulfex 0.2%, or Karathane 0.5%
Stalk rot or white rot or watery soft rot (*Sclerotinia sclerotiorum*)	Yellowing and falling off prematurely of lower leaves; small irregular dark brown to black necrotic lesions on leaf petioles and stems; stem, twigs, and inflorescence become straw-like and dry in March	Follow crop rotation; maintain good sanitation and good drainage; keep the field weed-free; seed treatment with 50 ppm aureofungin and 50 ppm streptocycline for 30 min; seedlings treatment with 0.25% Benlate for 5–8 min before transplanting; spray the crop with carbendazim 0.05% at 10 days intervals
Black leaf spot or blight (*Alternaria brassicae* or *Alternaria brassicicola*)	Small concentric dark spots on lower leaves; rooting of the head; brown to black spots on pods and seeds; disease spreads through infected seeds, crop residues, cruciferous weeds, infected seedlings, and transmission of spores by wind and water	Adopt crop rotation; use disease-free seed; maintain good sanitation in the field; remove and destroy all affected leaves and heads, seed treatment with captan, thiram, or ceresan at 2.5 g/kg of seed; spray the crop with mancozeb or Ridomil MZ-72 at 10–15 days intervals

(continued)

Table 10.2 (continued)

Disease	Symptoms	Control
Club root (*Plasmodiophora brassicae*)	It is caused by a root parasite and soil-borne fungus, which enters through root hairs; yellowing of foliage, flagging and wilting of plants; small spindle-like spherical club-shaped swellings develop on roots and rootlets; more prevalent on acidic soils having pH below 7.0 and less serious on heavy soils and those containing low organic matter	Follow long crop rotation with non-cruciferous crops; grow resistant cultivars; raise soil pH above 7.0 by liming; maintain good sanitation and eradicate weeds; dip the roots in 0.05% solution of PCNB (pentachloronitrobenzene) at transplanting
Black rot (*Xanthomonas campestris*)	Initially, chlorotic lesions appear along the leaf margins, which progress toward forming V-shaped blotches followed by blackening of veins and death of the whole plant. Production of small and poor-quality heads. The head formation is restricted if the attack is early and severe	Follow crop rotation; removal and destruction of infected parts of the plant; grow resistant varieties; hot water treatment of seeds at 50 ± 2 °C for 30 min followed by 30 min' dip in streptocycline 100 ppm; spray the crop with streptocycline 0.01% or 100 ppm at monthly intervals; planting on raised beds to facilitate drainage
Broccoli head rot (*Pseudomonas spp.*)	It is a soil-borne disease; biosurfactant is released by bacteria in water-soaked areas on heads; development of small black lesions on water-soaked florets; mainly due to high temperature (28 °C)	Grow resistant cultivars; avoid using high doses of nitrogenous fertilizers; avoid pesticides application during head formation; transplant the seedlings at wider spacing
Fusarium wilt (*Fusarium oxysporum*)	Dull yellow-green discoloration of leaves and drooping of the lower leaves; older leaves usually die first followed by the death of the entire plant; young infected plants often die; masses of spore-bearing stalks appear on dead tissue and may look like small pink cushions.	Deep summer ploughing; grow resistant varieties; raising of seedlings on raised beds; avoid excessive use of nitrogenous fertilizer; avoid overwatering and provide good drainage facilities.

(continued)

Table 10.2 (continued)

Disease	Symptoms	Control
Ring spot (*Mycosphaerella brassicicola*)	Seed-borne disease; circular light brown to black spots appear on leaves; yellow halo may surround the ringspot; entire leaf becomes yellow and leads to early senescence and defoliation in cases of severe infection; small spherical fruiting structures develop within the leaf spots; diseased leaves fall off prematurely in severe cases	Follow crop rotation; use disease-free seed; avoid growing crop adjacent to brassica fields; hot water seed treatment at 50 °C containing 0.2% captan or thiram for 20 min; spray chlorothalonil 0.005% or a mixture of penconazole and mancozeb 0.2%
Anthracnose (*Colletotrichum higginsianum*)	Small circular to irregular-shaped grey to straw-color dry spots appear on leaves; lesions may coalesce to form large necrotic patches causing the leaves to turn yellow and wilt; lesions may split or crack in dry centers	Follow long crop rotation; provide good soil drainage facilities; maintain phytosanitary conditions in the field; remove all cruciferous weeds; hot water seed treatment at 50 °C for 30 min; treat the seed with thiram or captan before sowing at 2.5–3 g/kg of seed; foliar spray with Dithane M 45, Topsin M, or Bavistin 2 g/liter of water
Root-knot nematode (*Meloidogyne* spp.)	The areas of irregular and stunted plant growth; infected plants wilt during hot afternoons in periods of water-stress; galls are formed on roots	Follow 2 years of crop rotation; fumigate the nursery beds; treat nursery bed soil with methane sodium at 25 ml/m^2 or carbofuran at 2 g a.i./m^2; dip the seedlings bare roots in dimethoate at 500 ppm or carbosulfan at 1000 ppm at transplanting time
Wire stem rot (*Rhizoctonia solani*)	Plants infected with wire stem show reddish-brown discoloration of the stem near the soil surface; the outer skin of the stem tissues is worn out, leaving a dark, wiry, and woody inner stem; plants may be twisted but do not split, and hence the name wire stem; stunted and weak growth of the plants	Discard the infected seedlings at the time of transplanting; follow crop rotation; drench the soil with 0.2–0.3% Brassicol solution; spray the crop with Dithane-M-45, Difolatan 80, or Daconil at 0.2% at 15-day intervals

(continued)

Table 10.2 (continued)

Disease	Symptoms	Control
Black leg (*Leptosphaeria maculans*)	Black leg occurs in Chinese cabbage growing areas; black pycnidia are visible as small dots on hypocotyl and cotyledons of the young plants. In the advanced stage, gray necrotic lesions with dark purple or black margins appear on the stem, which may extend to the entire surface and damage the cortex of the root	Plough the field deep in hot summer months; use disease-free seed; follow 3 years of crop rotation; use raised beds for the planting of the crop; provide hot water seed treatment at 50 °C for 30 min; collect and destroy crop debris after harvesting, seed treatment with benomyl slurry combined with thiram to eradicate black leg disease
Collard green mosaic virus	The disease is transmitted by aphids and can be recognized by poor growth of plants with a mosaic pattern of light green on the normal green color of leaves, crinkling, and distortion of leaves	Destroy wild and volunteer plants; kill the overwintering aphids
Turnip mosaic virus (TUMV)	Viral diseases in Chinese cabbage spread through aphids; symptoms include mosaic mottling, black speckling, or stippling of heads, ring spot, malformation, discoloration, chlorotic lesions, and puckering of leaves; causes lumpy or warty growths on veins of leaf undersurface and vein clearing in the later stage; black sunken spots develop on leaves all over the head during storage	Grow resistant varieties; use a reflective mulch to deter the aphids; use biological predators to suppress the population of aphids; protect the seedlings from aphids in the nursery; spray the crop with malathion 50 EC 0.1%, imidacloprid 0.04%, or pymetrozine 50% 0.025%

Table 10.3 Important insect pests of underutilized brassicaceous vegetables

Insect-pest	Damaging symptoms	Control
Cutworm (*Agrotis* spp.)	Cutworms cut the stem of newly set seedlings at ground level and bite the foliage, which leads to uneven crop stand and too much gap filling	Use well-rotten farmyard manure; plough the field to expose the caterpillars to predators; pick the caterpillars manually at night by torch or very early morning; use pheromone traps; foliar spray with aqueous *neem* seed and *neem* leaf extracts three times at weekly intervals; add chlorpyrifos 20 EC at 2.0 l/ha after mixing in 25 kg of sand before transplanting the seedlings
Cabbage butterfly (*Pieris brassicae*)	The female adult lays eggs in clusters on the upper surface of the leaves; young caterpillars scrape the leaf surface, and older caterpillars damage the leaves by feeding on leaf margins, and they proceed to the center and skeletonize them; infested plants do not develop heads	Follow mustard trap cropping; pick the infested leaves and destroy them; foliar spray with malathion 0.1%, dimethoate 0.05%, monocrotophos 0.05%, or Nuvan 0.05%
Diamond back moth (*Plutella* spp.)	The adult female of this insect lays eggs on the undersurface of the leaves; grayish green caterpillars feed on the leaves, thereby leaving intact the parchment-like epidermis and holes in the leaves; leaves turn brown and fall; growth of young plants is greatly inhibited and such plants do not form heads	Follow mustard or marigold trap cropping; pick the infested leaves manually and destroy them; spray neem seed kernel extract (5%) at regular intervals; spray the crop with malathion 0.1%, Nuvan 0.05%, fenvalerate 0.01%, or quinalphos 0.2% at 10-day intervals
Aphids (*Brevicoryne brassica* and *Myzus persicae*)	Aphids get shelter within the dense bud clusters and leaves and suck the cell sap from tender portions resulting in poor growth of young plants, yield, and quality of the produce	Grow tolerant varieties; set yellow pan traps in the field; wash the crop with plain water at high pressure; remove and destroy the affected plant portion in mild infestation; spray malathion 0.1%, monocrotophos 0.03%, or Nuvan 0.05%
Flea beetle (*Systena blanda*)	The beetles feeding on leaves result in small holes in leaves, giving the foliage a characteristic shot-hole appearance and reducing plant growth	The application of mulch helps in preventing the beetles from reaching the surface; spray the crop with spinosad 0.02%, bifenthrin, or permethrin 0.4% to control the beetles

(continued)

Table 10.3 (continued)

Insect-pest	Damaging symptoms	Control
Leaf Webber (*Crocidolomia binotalis*)	The insects lay eggs on the undersurface of the leaves in clusters; green caterpillars web up the leaves and live inside the knotted mass; flowering and pod formation is adversely affected	Removal and destruction of infested leaves; spray with Carbaryl (4%) or Malathion (0.05%) is effective
Cabbage looper (*Trichoplusia ni; synonym Plusia orichalcea*)	The adults lay eggs on under the surface of leaves closer to the leaf margin; caterpillars make small or large holes in leaves and damage often them extensively	Pick the larvae by hand and destroy them; application of *Bacillus thuringiensis* effectively kills younger larvae; spray neem seed kernel extract at 5% at regular intervals; spray the crop with any of the chemicals like fipronil 0.15%, Thiodan 0.2%, or fenvalerate 0.05% at 10- to 15-day intervals
Thrips (*Frankliniella occidentalis* and *Thrips tabaci*)	The small minute insect sucks the sap from leaves; leaves distorted very badly if its population is high. Leaves are covered in coarse stippling and appear silvery	The use of reflective mulches early in growing season; spray the crop with Confidor 0.03–0.05%, carbosulfan 0.2%, regent 0.2%, or dimethoate 0.2%
Beet armyworm (*Spodoptera exigua*)	The young larvae are present on the leaves; they make singular or closely grouped circular to irregularly shaped holes in leaves, and leaves become skeletonized due to heavy feeding	The natural enemies parasitize the larvae; apply *Bacillus thuringiensis* to kill the larvae; spray the crop with fipronil 0.15%, Thiodan 0.2%, Polytrin-C 0.1%, cartap hydrochloride 0.05%, or fenvalerate 0.05%
Cabbage webworm (*Hellula undalis*)	The cabbage webworm, also known as cabbage borer, is the most destructive pest of sea kale; young caterpillars mine the leaves, while old caterpillars feed on the undersurface of the rolled leaves as well as on stems and growing points; feeding on young seedlings causes their death, especially when the growing point is attacked; older plants due to its attack produce new shoots and several small heads of little commercial value	Follow crop rotation of at least 2–3 years; plant only vigorous insect-free seedlings; remove and destroy the affected plant leaves; uproot and burn the infected sea kale stalks; introduce parasitic wasps such as braconid, ichneumonid, and chalcidoid in the sea kale field; spray 5% neem seed kernel extract

(continued)

Table 10.3 (continued)

Insect-pest	Damaging symptoms	Control
Harlequin or stink bug (*Murgantia histrionica*)	Both adults and nymphs damage the crop by eating all parts of the plants, including stems, fruits, and seeds. It is a very destructive pest of almost all cruciferous plants, feeding voraciously on juice inside the leaves, causing the plant to wilt, turn brown, and often die if its population is high; the females feed greedily before laying eggs. Its damage appears on stems and leaves, depending on the plant species. Old plants may survive with very poor growth	Grow resistant varieties; cover the plant rows with a lightweight floating row cover; grow a few rows of strong-scented plants such as basil, celery, chamomile, garlic, mint, rosemary, and sage around the crop; grow a few lines of mustard, broccoli, and kale as a trap crop to attract the bugs; handpick the bugs and egg masses and drop them into a bucket of soapy water; burn the plant refuges and plant debris at the end of the crop; spray the crop with insecticidal soap or spinosad at 0.1%
Striped flea beetle (*Phyllotreta striolata*)	In early spring, beetles attack the seedlings and young plants; small shiny black adult beetles because primary damage by feeding on the undersides of the leaves, creating numerous round shot holes; the seedlings may also be killed if severe damage occurs; beetles may act as a vector of plant diseases	Keep the field weed-free; protect the seedlings with fine mesh netting; plough the field after harvesting the crop to expose the larvae to predators; spray the crop with methyl parathion 8 EC 0.05–0.1%
Snails (garden brown snail – *Cornu aspersum*; white garden snail – *Theba pisana*) and slugs (gray garden slug –*Deroceras reticulatum*; banded slug – *Lehmannia poirieri*; three-band garden slug –*Lehmannia valentiana*; tawny slug – *Limacus flavus*; greenhouse slug – *Milax gagates*)	They chew succulent leaves leaving irregular holes with smooth edges behind; they leave characteristic silvery mucous trails, which distinguish them from other chewing insects; the mucous trails deteriorate the quality of the produce	Pick and destroy manually on regular basis; traps can effectively be used with baits; baiting with metaldehyde and bran (1:25 in 12 liters of water); alum may also be sprayed at 2% solution

References

Abellán, Á., Domínguez-Perles, R., Moreno, D. A., & García-Viguera, C. (2019). Sorting out the value of cruciferous sprouts as sources of bioactive compounds for nutrition and health. *Nutrients*, *11*(2), 429.

Acharya, T. P., Welbaum, G. E., & Arancibia, R. A. (2019). Low tunnels reduce irrigation water needs and increase growth, yield, and water-use efficiency in Brussels sprouts production. *HortScience*, *54*(3), 470–475.

Aghdam, M. S., Sayyari, M., & Luo, Z. (2020). Exogenous application of phytosulfokine α (PSKα) delays yellowing and preserves nutritional quality of broccoli florets during cold storage. *Food Chemistry*, *333*, 127481.

Akdaş, Z. Z., & Bakkalbaşı, E. (2017). Influence of different cooking methods on color, bioactive compounds, and antioxidant activity of kale. *International Journal of Food Properties*, *20*(4), 877–887.

Akram, W., Saeed, T., Ahmad, A., Yasin, N. A., Akbar, M., Khan, W. U., Ahmed, S., Guo, J., Luo, W., Wu, T. & Li, G. (2020). Liquiritin elicitation can increase the content of medicinally important glucosinolates and phenolic compounds in Chinese kale plants. *Journal of the Science of Food and Agriculture*, *100*(4), 1616–1624.

Alkahtani, J., Elshikh, M. S., Almaary, K. S., Ali, S., Imtiyaz, Z., & Ahmad, S. B. (2020). Antibacterial, anti-scavenging and cytotoxic activity of garden cress polysaccharides. *Saudi Journal of Biological Sciences*, *27*(11), 2929–2935.

Ares, A. M., Nozal, M. J., & Bernal, J. (2013). Extraction, chemical characterization and biological activity determination of broccoli health promoting compounds. *Journal of Chromatography A*, *1313*, 78–95.

Bairwa, R. K., Singh, S. P., Mahawar, A. K., & Das, K. K. (2017). Influence of sulphur and spacing on growth and yield attributes of Knol-Khol (*Brassica oleracea* Var. Gongylodes L.) Var. early white Viana. *Int. J. Curr. Microbiol. App. Sci*, *6*(5), 2438–2447.

Biegańska-Marecik, R., Radziejewska-Kubzdela, E., & Marecik, R. (2017). Characterization of phenolics, glucosinolates and antioxidant activity of beverages based on apple juice with addition of frozen and freeze-dried curly kale leaves (Brassica oleracea L. var. acephala L.). *Food chemistry*, *230*, 271–280.

Bousquet, J., Le Moing, V., Blain, H., Czarlewski, W., Zuberbier, T., de la Torre, R., Lozano, N.P., Reynes, J., Bedbrook, A., Cristol, J.P. & Anto, J. M. (2021). Efficacy of broccoli and glucoraphanin in COVID-19: From hypothesis to proof-of-concept with three experimental clinical cases. *World Allergy Organization Journal*, *14*(1), 100498.

Cartea, M. E., & Velasco, P. (2008). Glucosinolates in Brassica foods: bioavailability in food and significance for human health. *Phytochemistry reviews*, *7*(2), 213–229.

Chang, J., Wang, M., Jian, Y., Zhang, F., Zhu, J., Wang, Q., & Sun, B. (2019). Health-promoting phytochemicals and antioxidant capacity in different organs from six varieties of Chinese kale. *Scientific reports*, *9*(1), 1–10.

Chauhan, E. S., Tiwari, A., & Singh, A. (2016). Phytochemical screening of Knol-Khol (Brassica caulorapa) owder and juice–A comparative study. *International Journal of Home Science*, *2*(3), 123–126.

Chowdhury, M., Kiraga, S., Islam, M. N., Ali, M., Reza, M. N., Lee, W. H., & Chung, S. O. (2021). Effects of Temperature, Relative Humidity, and Carbon Dioxide Concentration on Growth and Glucosinolate Content of Kale Grown in a Plant Factory. *Foods*, *10*(7), 1524.

Cui, W., He, P., Munir, S., He, P., He, Y., Li, X., Yang L, Wang B, Wu Y, & He, P. (2019). Biocontrol of soft rot of Chinese cabbage using an endophytic bacterial strain. *Frontiers in microbiology*, *10*, 1471.

Dey, M. (2017). Toward a personalized approach in prebiotics research. *Nutrients*, *9*(2), 92.

Dias, J. S., Monteiro, A. A., & Lima, M. B. (1993). Numerical taxonomy of Portuguese Tronchuda cabbage and Galega kale landraces using morphological characters. *Euphytica*, *69*(1), 51–68.

Diederichsen, A. (2001). Cruciferae: Brassica. In: Hanelt P, Institute of Plant Genetics and Crop Plant Research (eds) Mansfeld's encyclopedia of agricultural and horticultural crops Springer, Berlin, pp 1435–1446.

Fang, H., Zhou, Q., Yang, Q., Zhou, X., Cheng, S., Wei, B., Li, J., & Ji, S. (2022). Influence of Combined Edible Coating with Chitosan and Tea Polyphenol on the Quality Deterioration and Health-promoting Compounds in Harvested Broccoli. *Food and Bioprocess Technology*, 1–14.

Giorgetti, L., Giorgi, G., Cherubini, E., Gervasi, P. G., Della Croce, C. M., Longo, V., & Bellani, L. (2018). Screening and identification of major phytochemical compounds in seeds, sprouts and leaves of Tuscan black kale Brassica oleracea (L.) spp acephala (DC) var. sabellica L. *Natural product research*, *32*(14), 1617–1626.

Gonçalves, Á. L. M., Lemos, M., Niero, R., de Andrade, S. F., & Maistro, E. L. (2012). Evaluation of the genotoxic and antigenotoxic potential of Brassica oleracea L. var. acephala DC in different cells of mice. *Journal of ethnopharmacology*, *143*(2), 740–745.

Helaly, A.A. (2017). Enhancement growth, yield production and quality of kale plants by using plant growth promoting bacteria. *J Nat Sci*, *15*, 120–130.

Holst, B., & Fenwick, G.R. (2003). Glucosinolates. In Encyclopedia of Food Sciences and Nutrition, 2nd ed.; Caballero, B., Ed.; Academic Press: Oxford, UK, pp. 2922–2930. ISBN 978-0-12-227055-0.

Hussain, A., Ganai, M. A., Shan, M. A., Jehangir, I. A., Wani, S. H., & MA, A. (2020). Growth, yield and nutrient uptake of knol-khol (brassica oleracea var. gongylodes) as influenced by organic manures, inorganic fertilizers and biochar. *Indian Journal of Ecology*, *47*(3), 681–685.

Islam, M. A., Kabir, M. Y., Shuvra, N. T., Islam, M. A., & Hera, M. H. R. (2020). Effect of different organic manures and fertilizers on growth and yield of knol-khol (var. L.). *Malaysian Journal of Halal Research*, *3*(2), 56–62.

Jeffery, E. H., Brown, A. F., Kurilich, A. C., Keck, A. S., Matusheski, N., Klein, B. P., & Juvik, J. A. (2003). Variation in content of bioactive components in broccoli. *Journal of food composition and analysis*, *16*(3), 323–330.

Jeon, J., Kim, J. K., Kim, H., Kim, Y. J., Park, Y. J., Kim, S. J., Kim, C., & Park, S. U. (2018). Transcriptome analysis and metabolic profiling of green and red kale (Brassica oleracea var. acephala) seedlings. *Food chemistry*, *241*, 7–13.

Ji, S., Liu, Z., Liu, B., Wang, Y., & Wang, J. (2020). The effect of Trichoderma biofertilizer on the quality of flowering Chinese cabbage and the soil environment. *Scientia Horticulturae*, *262*, 109069.

Kim, S. Y. (2017). Production of fermented kale juices with Lactobacillus strains and nutritional composition. *Preventive nutrition and food science*, *22*(3), 231.

Kučera, O., Šamajová, K., Čižmárová, R., & Faltýnek, D. (2018). On the origin of Chinese cabbage.

Kuerban, A., Yaghmoor, S. S., Almulaiky, Y. Q., Mohamed, Y. A., Razvi, S. S. I., Hasan, M. N., Moselhy, S.S., Al-Ghafari, A.B., Alsufiani, H.M., Kumosani, T.A., & Al-Malki, A. (2017). Therapeutic effects of phytochemicals of Brassicaceae for management of obesity. *J. Pharm. Res. Int*, *19*, 1–11.

Laczi, E., Apahidean, A., Luca, E., Dumitraş, A., & Boancă, P. (2016). Headed Chinese cabbage growth and yield influenced by different manure types in organic farming system. *Horticultural Science*, *43*(1), 42–49.

Lee, Y. S., Ku, K. M., Becker, T. M., & Juvik, J. A. (2017). Chemopreventive glucosinolate accumulation in various broccoli and collard tissues: Microfluidic-based targeted transcriptomics for by-product valorization. *PLoS One*, *12*(9), e0185112.

Lemos, M., Santin, J. R., Júnior, L. C. K., Niero, R., & de Andrade, S. F. (2011). Gastroprotective activity of hydroalcoholic extract obtained from the leaves of Brassica oleracea var. acephala DC in different animal models. *Journal of ethnopharmacology*, *138*(2), 503–507.

Liu, M., Zhang, L., Ser, S. L., Cumming, J. R., & Ku, K. M. (2018). Comparative phytonutrient analysis of broccoli by-products: The potentials for broccoli by-product utilization. *Molecules*, *23*(4), 900.

Moreno, D. A., Carvajal, M., López-Berenguer, C., & García-Viguera, C. (2006). Chemical and biological characterisation of nutraceutical compounds of broccoli. *Journal of pharmaceutical and biomedical analysis*, *41*(5), 1508–1522.

Oliveira, S. M., Ramos, I. N., Brandão, T. R., & Silva, C. L. (2015). Effect of Air-Drying Temperature on the Quality and Bioactive Characteristics of Dried G alega Kale (B rassica oleracea L. var. A cephala). *Journal of Food Processing and Preservation*, *39*(6), 2485–2496.

Pavlović, I., Mlinarić, S., Tarkowská, D., Oklestkova, J., Novák, O., Lepeduš, H., Bok, V.V., Brkanac, S.R., Strnad, M. & Salopek-Sondi, B. (2019). Early Brassica crops responses to salinity stress: a comparative analysis between Chinese cabbage, white cabbage, and kale. *Frontiers in plant science*, *10*, 450.

Pereira, M. B., Dias, T. J., Lima, N. R., Justino, E. S., Oliveira, D. S., & Martins-Veras, M. L. (2020). Plant growth and yield of butter kale (*Brassica oleracea* L. var. *acephala*), as influenced by the combined application of bovine manure and rock powder. *Acta Agronómica*, *69*(1), 38–45.

Pusik, L., Pusik, V., Bondarenko, V., Gaevaya, L., Kyruchina, N., Slobodyanyk, H., Shchetyna, S., Shchetyna, M., & Kononenko, L. (2020). Studying the Preservation of Brussels Sprout Depending on its Treatment With Antimicrobial Preparations Before Storage. *Восточно-Европейский журнал передовых технологий*, *6*(11-108), 52–59.

Rana, M. K. (Ed.). (2017). *Vegetable crop science*. CRC Press.

Šamec, D., & Salopek-Sondi, B. (2019). Cruciferous (brassicaceae) vegetables. In *Nonvitamin and nonmineral nutritional supplements* (pp. 195–202). Academic Press.

Šamec, D., Pavlović, I., & Salopek-Sondi, B. (2017). White cabbage (Brassica oleracea var. capitata f. alba): botanical, phytochemical and pharmacological overview. *Phytochemistry reviews*, *16*(1), 117–135.

Schmidt, S., M. Zietz, M. Schreiner, S Rohn, L.W. Kroh and A. Krumbein (2010): Genotypic and climatic influences on the concentration and composition of flavonoids in kale (*Brassica oleracea* var. *sabellica*). Food Chemistry (119): 1293–1299.

Selvakumar, G., Shathirapathiy, G., Jainraj, R., & Paul, P. Y. (2017). Immediate effect of bitter gourd, ash gourd, Knol-khol juices on blood sugar levels of patients with type 2 diabetes mellitus: a pilot study. *Journal of traditional and complementary medicine*, *7*(4), 526–531.

Shah, K. N., Rana, D. K., Vivek, S., & Chaudhary, I. J. (2019). Growth, yield and quality of knol-khol (Brassica oleracea var. gongylodes) as affected by fertilizer management. *Fundamental and Applied Agriculture*, *4*(3), 959–969.

Siva, N., Johnson, N., McGee, R., & Thavarajah, P. (2019). Effect of cover crops on the yield and nutrient concentration of organic kale (*Brassica oleracea* L. var. *acephala*). *Scientific reports*, *9*(1), 1–8.

Souri, M. K., Hatamian, M., & Tesfamariam, T. (2019). Plant growth stage influences heavy metal accumulation in leafy vegetables of garden cress and sweet basil. *Chemical and biological Technologies in Agriculture*, *6*(1), 1–7.

Thomson, C.A., T.R. Newton, E.J. Graver, K.A. Jackson, P.M. Reid, V.L. Hartz, E.C. Cussler, and I.A. Hakim (2007): Cruciferous vegetable intake estimate. Journal of American Dietetic Association, 107:631–643.

Vanduchova, A., Anzenbacher, P., & Anzenbacherova, E. (2019). Isothiocyanate from broccoli, sulforaphane, and its properties. *Journal of medicinal food*, *22*(2), 121–126.

Vavrina, C. S., Obreza, T. A., & Cornell, J. (1993). Response of Chinese cabbage to nitrogen rate and source in sequential plantings. *HortScience*, *28*(12), 1164–1165.

Xiao, Z., Rausch, S. R., Luo, Y., Sun, J., Yu, L., Wang, Q., Chen, P., Yu, L. & Stommel, J. R. (2019). Microgreens of Brassicaceae: Genetic diversity of phytochemical concentrations and antioxidant capacity. *Lwt*, *101*, 731–737.

Yildirim, E., Ekinci, M., Turan, M., Ağar, G., Dursun, A., Kul, R., Alim Z, & Argin, S. (2021). Humic+ Fulvic acid mitigated Cd adverse effects on plant growth, physiology and biochemical properties of garden cress. *Scientific reports*, *11*(1), 1–8.

Chapter 11
Production Technology of Underutilized Vegetables of Chenopodiaceae (Amaranthaceae) Family

Teshu Kumar and Vinod Jatav

Introduction

The family of Chenopodiaceae (Amaranthaceae) has about 165 genera and 2040 species, making it the most diverse of its parent species, Caryophyllales. Most Amaranthaceae species are annual, perennial, or subshrubs; others are shrubs; very few are vines or trees. Pigweed and french spinach are the two essential crops of the Amaranthaceae family. Pigweed (*Chenopodium album*) is considered a weed globally but is generally used for human consumption as a leafy vegetable. The details of some underutilized crops of the Chenopodiaceae family are given below:

S. No.	Common name	Botanical name	Origin	Chromosome no. (2n)	Edible part
1.	Pigweed	*Chenopodium album* L.	Western Asia	36	Leaves and tender twigs
2.	French spinach	*Atriplex hortensis* L.	India	12/18	Leaves and immature shoots

Pigweed (*Chenopodium album*)

Chenopod is a member of the family of Chenopodiaceae and is one of the main crops considered as a weed in wetlands of Asia, Europe, and North America; it can also be cultivated in varied soil and climatic conditions. The Chenopods are also

T. Kumar
Ch. Chotu Ram PG College, Muzaffarnagar, Uttar Pradesh, India

V. Jatav (✉)
College of Agriculture- GLA-University Mathura, Mathura, Uttar Pradesh, India

Savita et al. (eds.), *Production Technology of Underutilized Vegetable Crops*,
https://doi.org/10.1007/978-3-031-15385-3_11

called as pigweed, lambs' quarter, fat hen, white goose-foot, frost-blite, muck weed, dung weed, and wild spinach (Holm et al. 1977). It is known by different vernacular names in different regions of India, viz., bathua in Hindi, chandan betu in Bengali, parupukkirai in Tamil, pappukura in Telugu, and katu ayamoddakam/vastuccira in Malayalam. The name *Chenopodium album* L. has been derived from the Latin word "cheno," meaning goose, and "podos," meaning footed, whereas the species "album" is about its white powdery bloom. It is an annual summer season weed in wastelands, cultivated fields, and other disturbed habitats (Reed 1971). It is considered a popular wild vegetable in southern Africa, while in most parts of India, it is being grown as a traditional leafy vegetable. This is a polymorphic plant and is more commonly found in north India from Sikkim to Kashmir, but it is also available in southern parts of the country. Usually, it is not grown commercially, but in parts of Kullu Valley and Rajasthan, it is being cultivated on a small scale. The crop is chiefly raised for its young, tender leaves and stems. Quick growing in habit can be fitted well in multiple cropping systems. However, in many parts of the world, it is also consumed as a vegetable.

Origin and Distribution

Pigweed originated from Western Asia, but its exact location is unknown. It is found worldwide from sea level to 3600 m in elevation and from 70°N to 50°S latitude except in extreme desert climates. Most botanists believe it originated in Europe, and evidence suggests that hunter-gatherers consumed bathua during the Bronze and Iron Ages. The edible species *Chenopodium gigantium* D. Don (Syn. *C. amaranticolor*) is thought to be its nearest relative and probably derived through its cultivation. This species is commonly known as purple goosefoot or tree spinach. In Madagascar and Zambia, *Chenopodium giganteum* is considered an excellent cooked vegetable, and it occurs widely in India. The genus *Chenopodium* is large, having about 100 to 150 species mostly occurring in temperate zones of the world. It consists of leafy vegetables and pseudocereals, which are abundantly consumed in the Andean region of South America, Africa, and some parts of Asia and Europe.

Nutritional Importance and Uses

Chenopod leaves are a cheap source of nutrients because they provide large amount of calcium, phosphorus, sodium, and vitamin C and are an excellent source of carbohydrates and protein. Magnesium, calcium, and sodium content is high in mature leaves, while copper and iron content is higher in young shoots. The nutritional composition of chenopod leaves and seeds is given in Table 11.1:

Table 11.1 Nutritional composition of chenopod leaves and seeds

Constituents	Content
Water (g)	85.7
Iron (mg)	6.6
Carbohydrates (g)	7.3 (leaves), 66.0 (seed)
Zinc (mg)	0.3
Protein (g)	4.3 (leaves), 16.0 (seed)
Vitamin A (IU)	7817
Fat (g)	0.8 (leaves), 7.0 (seed)
Thiamine (mg)	0.15
Calcium (mg)	258
Riboflavin (mg)	0.4
Potassium (mg)	288
Niacin (mg)	1.3
Sodium (mg)	265
Vitamin C (mg)	90.0
Magnesium (mg)	23
Energy (kcal)	44 (leaves), 395.0 (seed)

Uses

In the past, leaves were consumed as a vegetable like spinach; the seed was ground to flour; and fruit was consumed as poultry (Mabberley 1997). The young leaves are used as a salad. The plant has diuretic, laxative, and sedative properties, and leaves as poultices are used. Chenopod is medicinally effective and helps cure anorexia, cough, dysentery, and diarrhea. It kills small worms and is also helpful in treating heart and blood diseases. It is also managed biliousness, hepatic disorders, throat troubles, spleen enlargement, burns, ulcers, abdominal pain, and eye diseases. Its seeds have long been used as appetite stimulants, anthelmintic, laxative, aphrodisiac, and tonic. Ethanol extract of its fruits has shown antinociceptive and antipruritic properties. In India, the plants have been in use since ancient times, as it is evidenced in Rig Veda and Atharva Veda that it cures all diseases, acts as a laxative, and is beneficial in piles and clearing worms. The leaves contain high levels of ascaridole, which is a potent anthelmintic and is used to treat infestations of roundworms and hookworms. It is mentioned in Charaka Samhita that it enhances digestive power. At the same time, it is reported in Sushruta Samhita that it improves memory, appetite, digestive capacity, and body strength and destroys all worms. As mentioned in the textbook of Indian medicinal plants, most of the chenopods possess anthelmintic, laxative, and blood purifying effects. Its leaves have antibacterial properties against human pathogenic bacteria, including *Salmonella typhimurium*, *Staphylococcus aureus*, and *Pseudomonas aueruginosa*. The plant produces the toxic substance of oxalates and saponins, usually in minimal quantities, which may be harmful to human beings and livestock when consumed in large amounts in a short period under certain conditions. These toxic substances are absorbed poorly by the intestine but are denatured by prolonged cooking.

Owing to its medicinal properties, it is used as a laxative, anthelmintic against hookworms and roundworms, for enlargement of spleen, for blood purification in hepatic disorders, burns, and intestinal ulcers (Agrawal et al. 2014). It possesses allelopathic potential that affects growth of the plants. The extract obtained from its plants contains bioactive compounds, e.g., phenols and flavonoids (El-Rokick Kowther et al. 2018). Many authors reported the fungicidal properties of the *Chenopodium album* (Pagno et al. 2015). *C. album* extract can be effectively used as a biological agent for controlling Fusarium root rot in sunflower (Abu-Tahon et al. 2021).

Botany and Taxonomy

Chenopodium album is a dicotyledonary herbaceous annual for tender leaves and succulent shoot production (Abu-Tahon et al. 2021). It is diploid ($2n = 12$), growing during the summer months (March–September) and a hexaploid ($2n = 54$), increasing during the winter months (November–April) (Ramakrishnan and Kapoor 1974). It is a polymorphic, mealy white, erect annual herb, which may extend to a height of 1.0–2.5 m, depending upon fertility and moisture in the soil. The whole plant is covered with varying amount of waxy substance which gives it light green appearance. The stem is soft and varies in color from reddish light or bluish-green to striped with purple or green, and single or have a few rigid angled branches. The young stems have a mealy white covering of small white hairs, whereas older stems are smoother, stout, grooved, or ribbed. Chenopod produces simple, alternate, pinnately lobed, and oblong to triangular leaves of variable shape and size, 2.0–6.0 cm long and 0.5–2.0 cm broad, having thin stalk about half of the leaf blade, the shape of which is usually ovate to lanceolate. The leaves are dull green with a pale pink center with a waxy coating and wavy margin. The leaves appear opposite to one another along the stem and are almost equal or somewhat more significant than the cotyledons. The inflorescence stands terminal and lateral, and flowers are dimorphic, bisexual, or pistillate, leafy panicle, gray to green in color, and maybe cross- or self-pollinated. Grain chenopod is predominantly self-pollinating, but up to 8–10% of cross-pollination may also occur mainly by wind. The 10–40 cm long inflorescence consists of spicate or paniculately arranged in dense heads of small rounded clusters or glomerules and having five stamens with yellow anthers tips. The ovary is superior (hypogynous) in nature, depressed globose, short-styled, and one-celled with two stigmas. Flowers are symmetric, regular bearer, pentamerous, without petals, and have sepals slight to sharply ridged, nearly covering the mature fruit, which is a nut having a thin-walled indehiscent single seed which is polymorphic (Williams and Harper 1965). Chenopod produces two types of seeds, i.e., (i) smaller black and (ii) larger brown, which is glossy, disc-shaped with a notch, and 1.2–1.6 mm diameter. It has a thin papery cover which gives the seed a dull appearance. The five-segmented perianth may also be attached with the seed and detaches during collection and subsequent handling. The black seeds are found abundantly and hard coated with almost smooth to finely striate, rugulose, or pitted. A smaller percentage of sources (5%) are relatively large and brown with thin, smooth seed coats. The plant produces dormant seeds that can survive in the soils for long periods (Lewis 1973; Bassett and Crompton 1978).

Varieties

Pusa Bathua No.1

A variety was developed at IARI (Indian Agricultural Research Institute), New Delhi, through pure line selection from a local collection. It is a first improved variety, containing more vitamin C and beta-carotene, about 60% and 10%, respectively, in its edible parts, than the wildly grown chenopod. Its purplish-green leaves are of 10.5 cm length and 3.0 cm width. The plants may extend to a height of 2.25 m, and the thickness of the clipped stem is 0.45 cm with an internodal length of 2.5 cm. Its branch is purplish-green and tender with a fine texture. It takes 175–180 days to flower, but its tender leaves may be picked up 45 days after sowing, and harvesting continues for 150 days. It is suitable for planting from October to March, and the average yield is 300 q/ha.

Ooty 1

A dual-purpose dwarf variety developed at Tamil Nadu Agricultural University, Coimbatore, may have a plant height up to 38–40 cm and has unfolded dark green to pinkish leaves. It is suitable for leaf and grain production. This variety has resistance against leaf spots, caused due to Cercospora, Colletotrichum, and against foot rot due to Macrophomina. It gives an average green leaf yield of 289 and 170 q/ha in 55 days crop duration in hills and plains. Its grain yield is 12 q/ha. Some other genotypes are Sel-2, Local-1, and Local-2.

Pusa Green

This variety having multiple cuttings was developed by the Indian Agricultural Research Institute, New Delhi, in 2016 and has dark green attractive leaves. Central Variety Release Committee (CVRC) notified this variety in December 2018 for cultivation in NCT Delhi and gives an average leaf yield of 36.8 tons per hectare. It is suitable for both direct sowing (during October) and transplanting (during November). It has luxuriant plant growth, and its leaves are smooth and large 18 cm long, 9 cm wide with medium lobing and serration. It has higher total carotenoids (91.31 mg/100 g), iron (7.6 mg/100 g), dry matter (13%), and ascorbic acid content (50 mg/100 g) on fresh weight basis. It has late bolting habit and has lesser attack of insect-pests and diseases.

Kashi Bathua 2

It has upright growth habit and plant height of 180–195 cm at 160 DAS and is an excellent source of folic acid, minerals, vitamin A and C, phenolics, and antioxidants, yield 30–32 t/ha. Seed rate: 800–1000 g/ha.

Kashi Bathua 4

Upright growth habit with purplish-green leaf and petiole and is a good source of vitamin A and C, folates, minerals, phenolics, and antioxidants, yield 35 t/ha. Seed rate: 800–1000 g/ha.

Climate and Soil

Chenopod is a fast-growing, frost susceptible annual weed, which is very commonly found in temperate regions. Being a long-day plant, it usually flowers from March to April in northern plain areas. The plants under long photoperiod (16–17 h) produce a higher seed than plants grown under 8 h photoperiod. It is readily available during the winter months at an elevation up to 4700 m and may grow as a summer crop in irrigated areas. Its wide distribution as a weed shows excellent tolerance to temperature with an average temperature ranging from 5 to 30 °C, though the optimum temperature for a higher germination rate goes from 15 to 25 °C. Red or white light induces seed germination while far-red light inhibits the germination of seed. It is also promoted by stratification and alternating temperature and light interaction. The flowering of summer crops starts during 13.5–14.0 h of light, and mean temperatures of 29–31 °C whereas, in the winter season, flowering starts during 10.8–11.6 h of light and mean temperatures of 12–15 °C (Ramakrishnan and Kapoor 1974).

Chenopod is found growing wild in pastures, agronomic fields, and vegetable croplands, gardens, orchards, vineyards, landscape areas, roadsides, and other disturbed areas. It thrives on all types of soil, varying from sandy to heavy clay. However, the best quality greens can be produced in fertile soils containing good amount of organic matter with higher moisture retention capacity. For getting a higher yield with broader leaves, the chenopod should be cultivated in well-drained, friable sandy-loam soils rich in organic matter and initial plant nutrients.

Agronomic Practices

Seed Rate

Chenopod is generally not grown commercially as it grows naturally at disturbed sites. However, it can be propagated through seeds. The brown seeds germinate earlier with higher germination rates than the black seeds, which persist longer in soil. Prechilling or treatment with either potassium nitrate or some nitrogenous radicals or chlorates can increase the germination in black-colored seeds (Williams and Harper 1965). The seed is tiny, with a test weight of approximately 1.4 g. Seed rate of about 6–10 kg for line sowing and about 20 kg of broadcasting is required for sowing 1-hectare area. *C. album* is the most diminutive light-sensitive species, has higher germination percentage in both light and dark conditions at a temperature range of 15–25 °C, but is inhibited under dark (Henson 1970). The seed dormancy can also be broken by treating the seeds with GA_3, ethylene, and KNO_3 (Saini et al. 1985).

Sowing Time

In plain areas, chenopod can be cultivated year-round due to its wider adaptability to climate except in extreme winter, as it is susceptible to frost. In temperate regions, the seedlings that emerge in the autumn rarely survive in winters. For getting higher yield with the best quality greens, it is sown from October to March in northern plain areas.

Sowing Method

Chenopod plant has a taproot system, so it may not be transplanted. The seeds, though tiny, can be sown directly in the field either by broadcasting or in lines at a spacing of 20–30 cm × 10–15 cm. The emergence of seedlings depends upon the frequency of soil cultivation. Spacing significantly influences plant growth. If seeds are sown at broader spacing, plants thrive. For grain purpose, the seeds are sown at a depth of 1–2 cm in rows that are 25–50 cm apart, depending on the soil moisture content.

Nutrient Management

Chenopod does not require too many nutrients since it is very efficient in extracting nutrients from the soil. However, the application of nitrogen, phosphorus, and magnesium in moderate amounts is essential for harvesting the best quality greens. It grows well in soils rich in magnesium, so it is used as an indicator of magnesium status in the soil. Nitrogen and phosphorus are useful fertilizers for vigorous growth of the plants and reducing the percentage of underdeveloped seeds.

Irrigation Requirement

For better seed germination and consequent growth of plants, during sowing, the soil must have ample moisture. No pre-sowing irrigation is given for preparing seedbeds if the crop is sown on conserved soil moisture. Sometimes, the seeds do not germinate due to a lack of moisture. Under such a situation, light irrigation may be given to obtaining an early and higher percentage of germination and for subsequent seedlings' growth. Depending on the season and availability of soil moisture, the crop needs irrigation at an interval of 10–15 days.

Intercultural Operations

Chenopod grows naturally as a weed in growing fields of potato, tomato, pepper, beans, lettuce, eggplant, radish, carrot, turnip, beetroot, cauliflower, cabbage, broccoli, knoll kohl, etc. However, if it is grown commercially, the field should be kept weed-free in the initial stages of crop growth as the weeds, especially *Chenopodium murrel*, which is morphologically similar, compete with the main crop for nutrients, moisture, light, and space. Thus, at least two shallow hoeings are beneficial and necessary for keeping the weeds under control and providing a good soil environment to the roots for their development.

Harvesting

Harvesting may be done by either uprooting the whole plant or clipping the leaves. The first picking may be done 40–45 days after sowing and further if is harvested for 100–150 days, depending upon variety. The shoots along with leaves are cut at ground level and when the plants attain a height of 10–15 cm and when they are tender and crisp. Harvesting of chenopod for grains is almost similar to cereal crops as it is also harvested when the seeds are ripe, and most of the leaves have turned yellow. The plants after cutting are bundled and left on the threshing floor for a few days to dry. When the plants are fully dried, the seeds are separated by either rubbing the flower spikes onto a tarp or beating them lightly with a stick, and thereafter, the seed is cleaned by winnowing. The seeds at harvest contain moisture around 20%; therefore, they are artificially dried to a safe moisture level (12–14%) so that they may be stored safely for a longer period.

Yield

The average yield of chenopod varies from 250 to 300 q/ha in hills and 150 to 200 q/ha in plains, depending on crop variety, soil and environmental factors, and cultural practices adopted by the growers. The average seed (grain) yield varies from 8 to 12 q/ha.

Physiological Disorders

No specific physiological disorder is observed in cultivated chenopod.

Insect-Pests

*Leaf Sticker (*Eurysacca melanocompta*)*

This pest attacks the plants at all the growth stages particularly at the time of grain maturation. It has a longitudinal dark band on the wings with two dark spots surrounded by light scales. Female lays 80–300 eggs into the leaf tissues. The egg-laying range mainly depends on temperature. The larvae feed on foliage, with a preference for inflorescence. The attack of this pest causes severe yield losses ranging from 20% to 50%.

Control

Leaf sticker can be controlled by (i) spraying the crop with simple water with pressure. (ii) Remove and destroy the portion of the plant heavily attacked by leaf sticker. (iii) Spray the crop with Metasystox 0.1%, malathion 0.05%, monocrotophos, dimethoate 0.03%, or phosphomidon 0.03%.

*Aphids (***Myzus persicae***)*

Both nymphs and adults suck the cell sap from the undersurface of the tender fresh leaves reducing plant vigor. The leaves curl upward and ultimately wilt. The aphids excrete honey dew like substance, due to which black sooty mold develops, which hinders the photosynthesis. The severe infestation affects the yield and quality of leaves and inflorescence adversely.

Control

Aphid can be controlled by (i) spraying the crop with simple water with pressure. (ii) Remove and destroy the portion of the plant infested with aphids. (iii) Spray the crop with malathion 0.05%, monocrotophos 0.04%, methyl demeton 0.05%, dimethoate 0.03%, or phosphomidon 0.03%.

Diseases

*Downy Mildew (***Peronospora farinosa** *f. sp.* **chenopodii** *or* **Peronospora chenopodii***)*

A most destructive disease causes much damage in chenopod growing areas throughout the world. The disease may attack the plant at any stage, even at the flowering stage. The leaves affected by this fungus show light yellow irregularly shaped areas without distinct margins on their surface. Whitish-gray mycelium grows on leaf surface at high humidity, while the mycelium is generally found absent in low humid atmosphere. The disease develops in warm and moist weather. Its infection occurs mainly in the field, but it becomes prominent after harvesting, especially at 25–28 °C temperature and >85% relative humidity.

Control

Downy mildew can be controlled by following long crop rotation with nonhost crops, using the disease-free seed to avoid the incidence of this disease, removal of the infected leaves and destroying them, and spraying the crop with Indofil M-45 or Indofil Z-78 at 0.2% at 10- to 15-day intervals.

*Leaf Blight (***Alternaria alternate***)*

The initial infection is noticed as small brown necrotic spots with concentric rings, which rapidly increase in number and size, and later on, these spots coalesce to form large irregular blotches; as a result, the leaves wilt and, ultimately, die and fall. In advanced stages, the burning effect and blight symptoms are also seen after coalescing of adjoining spots. The disease comes in the middle of February and remains throughout summer. The crop shows about 10–60% mortality in different areas, which is higher in the rainy season.

Control

Leaf blight can be controlled by irrigating the crop frequently, use of the disease-free high-quality seed, seed treatment with hot water at 52 °C for 30 min to kill the pathogen present on or below the seed coat, and by spraying the crop with Bavistin, mancozeb, Blitox, or Blitane at 0.3% three to four times at an interval of 10 days just after the appearance of the disease.

*Cercospora Leaf Spot (***Cercospora chenopodii***)*

The disease is characterized by the presence of numerous small brown circular spots on the leaves. In the beginning, the spots are small round with concentric rings, but later on, these spots increase in size, and sometimes, they coalesce. The infected leaves dry and fall.

Control

Cercospora leaf spot can be controlled by following field sanitation, growing resistant varieties like Ooty 1, and by spraying the crop with Bordeaux mixture (5:5:50) or Blitox 0.3% thrice at an interval of 15 days.

Colletotrichum Leaf Spot (Colletotrichum *spp.)*

The disease is caused by the fungus, which affects all aboveground parts of the plant and may infect at any stage of plant growth. The affected leaves show round to angular reddish-brown spots on older leaves. Spots may later dry, turn almost black and tear out, giving the leaf an irregular appearance. The severity of infection depends almost entirely on environmental conditions. A combination of high temperature and high humidity is very conducive to the spread of this disease.

Control

Colletotrichum leaf spot can be controlled by using disease-free healthy seed or resistant varieties like Ooty 1, adopting field sanitation by burning crop debris, seed treatment with 0.25% carbendazim at 2.5 g/kg of seed before sowing, and by spraying the crop with chlorothalonil 0.2% for good control of this disease.

French Spinach (*Atriplex hortensis* L.)

French spinach (*Atriplex hortensis* L.) is a hardy monoecious annual herb and belongs to the Chenopodiaceae family, which is popularly known with various names like garden orach, mountain spinach, sea purslane, saltbush, and Phaltora/Pahari Paleng/Ustak in the Ladakh region (India) (Rinchen et al. 2017). It is considered to be one of the earliest domesticated crop plants, which valued mainly for its leaves since time immemorial and still grown in a wider range of environmental conditions, as a potherb and colorful salad green (Stevens 1994). It is also used in folk medicine, and as soil erosion control, scarce scientific information is available on the chemical composition, amino acid content, or consumption of its seeds.

Origin and Distribution

The geographic origins of *A. hortensis* remain elusive, but the most probable center of origin is the Trans-Himalayan (central Asia and Siberia) and Southeast European regions, from where it has been domesticated to the Mediterranean region in around third century B.C. followed by domestication in the Americas in colonial times (Ruas et al. 2001; Sukhorukov 2014). It is distributed in diverse regions of the world, viz., Europe (France, Poland, countries of the former Soviet Union, former Serbia/Montenegro and Norway) and North America (the United States, and

Canada), central Asia, and Siberia Mediterranean region. There is also evidence of its use as a food in Switzerland as early as the Neolithic Age (Andrews 1948).

Nutritional Importance and Uses

Its 100 g leaves contain 56–62 g total carbohydrate, 17–20 g protein, 11–13 g fiber, and 5–6 g fat, along with a good amount of minerals like Ca, Mg, P, K, Fe, Cu, and Mn and amino acids (Carlsson 1975; Wright et al. 2002). It is a rich source of vitamins A and C (Siddiqui et al. 1994; Nicol 1994; Steinbach 1996). Its leaves contain a high level of bioactive compounds, i.e., sulfated flavonoids: kaempferol 3-O-sulfate-7-Oarabinopyranoside and quercetin 3-O-sulfate-7-Oarabinopyranoside (Steinbach 1996; Bylka et al. 2001).

Since time immemorial, it has been used as vegetables for local dishes with different recipes and consumed as a salad. Its leaves have a spinach-like salty taste. The leaves and seeds have diuretics, emetic and purgative properties (Polunin 1969; Duke 1983), and are used as a health tonic in traditional medicine, which helps in digestion and in the absorption of nutrients. In folk medicine, it is used for treating plethora and lung ailments (Duke 1983). Seeds can be cooked, or ground could be mixed with wheat flour for making bread. Moreover, its seeds products are also used in soups, and a valuable blue dye can be extracted from the seeds. Folk medicine is also used to treat yellow jaundice by mixing its seeds with wine. Liniments and emollients which are prepared from the whole plant are used as a remedy for indurations and tumors particularly of the throat (Polunin 1969). They also excite vomiting. It has both medicinal and food uses (Rubatzky and Yamaguchi 2012). This plant could also be used for phytoremediation to clean heavy metals like lead, nickel, zinc, and copper in contaminated areas.

Botany and Taxonomy

Atriplex hortensis is one of the hardy annual plant species grown in cold desert areas. Its stem shows erect growth habit with side branching, and plant height varies from 60 cm to 180 cm. Leaves blade is normally triangular in shapes with sagittate or cordate base in lower leaves whereas upper leaves with a rounded base. Fully developed leaves are around 4–9 cm (1.6–3.6 in) and membranaceous, reticulate (Babb and Kraus 1939). The flowers are small in size, and the color of the flower varies from greenish to reddish. Female and male flowers are borne on the same plant, hence called monoecious. The seeds are small in size, black in color, and cylindrical in shape, with 2–3 mm diameter, and the seeds are surrounded by a thin, light yellow membrane. Pollination occurs through the wind (anemophilous). The seeds dehisced after maturation and become dormant during severe winter.

Cultivated Varieties

There are four varieties commonly cultivated in the world:

1. Triumph

   Plants intermediate in size, 6–9 feet; also intermediate in many branches. Stem and leaf petioles are of same color as leaves. Leaves are dark or grass green, coarsely rugose; lower leaves ovate to cordate; large, with long-tapering, slightly rounded point; margins coarsely and irregularly sinuate two-thirds of length from base; the base of leaf broad, and angles between the base of leaf and petiole more obtuse than in Gelbe or Deep Blood Red; upper leaves becoming slightly more lanceolate; veins of the same color as leaves; more thickly covered with glandular hairs than Gelbe or Deep Blood Red, but less so than Lee Giant. Normal vertical and horizontal seeds are yellow-brown, but black seeds are also borne in both forms. Immature bracts are of same color as leaves; mature bracts are light yellow-brown, somewhat lighter than seeds.
2. Gelbe

   Plants intermediate in size, 6–9 feet, typically somewhat more branched than Triumph. Stem and leaf petioles are of same color as leaves. Leaves are pale or yellowish-green; upper surface of leaf – covered with light – grayish bloom; coarsely rugose, medium in size, with rounded point becoming progressively more lanceolate toward the top of the plant; margins of lower leaves coarsely sinuate, approaching dentate; upper leaves more nearly entire; basal angles broad, occasionally closely approximating a right angle with petiole; and sparsely covered with glandular hairs on the upper surface, intermediate coverage on the lower surface. Normal vertical and horizontal seeds are yellow-brown, but black seeds are also borne in both forms. Immature bracts are greenish-yellow, and mature bracts light yellowish-brown or somewhat lighter than seed.
3. Lee Giant

   Plants somewhat taller (7–10 feet) and considerably more branched than other varieties. Stem, leaf petioles, and leaf veins are of the same color as leaves. Leaves are dark green, flat, triangular-hastate, with lower lobes relatively much enlarged; smallest and least fleshy of any cultivated variety; margin coarsely and irregularly dentate; the base of leaf forming an obtuse angle with petiole; upper leaves very much reduced in size; and more thickly covered with glandular hairs on both upper and lower surface than any other variety. Normal vertical and horizontal seeds are yellow-brown, but black seeds of both types are also borne. Immature bracts are of the same color as leaves; mature bracts are light yellow-brown, or somewhat lighter than the seeds.
4. Deep Blood Red

   Plants are smallest and least branched of all types, which are 4–6 feet. Stem, leaf petioles, and leaf veins are dark purplish-red, darker than leaves. Leaves are striped with a lighter red and covered with grayish bloom; the upper surface of leaves is dull crimson with slight grayish bloom, whereas the lower surface is reddish-purple with a slight grayish overcast; lower leaves cordate, occasionally approaching sagittate; coarsely rugose; margins undulate; basal angles broad, sometimes almost at right angles with petioles, becoming obtuse on upper leaves; leaves toward top of the plant becoming more lanceolate and margins less undulate; and glandular hairs on leaves not so numerous as on many other varieties. The shape of seed and bracts is identical to other varieties. Normal vertical and horizontal seeds are reddish-brown; vertical and horizontal black seeds are also borne. Immature bracts are usually reddish-green, occasionally pure red, whereas the mature bracts are light yellowish-brown, identical to other varieties.

Other Varieties

Golden A rare heirloom from the old Abundant Life Seed, maintained by us since 1982. This cultivar was used for whole-genome sequencing of *Atriplex hortensis* species, and the estimated genome size was recorded around 1.1 gigabases (Hunt et al. 2020).

Climate and Soil

French spinach is a hardy plant which can tolerate various abiotic stresses like drought, soil salinity, alkalinity, and hot weather (Welbaum 2015). Young plants emerging from either early spring or a late fall seeding will withstand −3.3 °C without injury. It can tolerate a wide range of annual rainfall from 30 to 140 cm and temperatures ranging from 6 to 24 °C. However, plants thrive best in cool and sunny conditions and with a mean temperature of 21 °C. It can grow up to an elevation of 2000–2500 m and is slightly less vigorous in lower situations.

It grows well on a wide range of soil types, though well drained, fertile (rich in organic matter), clay loam to sandy loamy soils having pH of 6–7.0 are better suited for the quick growth of plants and development and high yield of leaves. It is exceptionally drought-resistant and tolerant to alkaline and saline soil conditions (Babb and Kraus 1939; Greenway 1968; Gale and Poljakoff-Mayber 1970; Handley and Jennings 1977; Wilson et al. 2000). It also thrives in any temperate climate and requires a pH of 5.0–8.2.

Agronomic Practices

Propagation

It is commercially propagated by seed, and normally direct seeding is followed, but transplanting can also be done in northern regions where a late spring frost is very common. The optimum spacing between the rows and plants should be kept around

20–30 cm and 10–15 cm, respectively. Desirable plant population per hectare (3.5–4.5 lakhs plants/ha) should be maintained by thinning to promote proper growth and development of plants and to obtain good leaves yield. This desirable plant population could be achieved from 3 to 4 kg of seed if proper soil moisture is provided at the seed sowing, which enhances seed germination and crop establishment. Sowing depth is very important in seed germination. In clay or the heavier types of soil, the seed should not be covered deeper than one-half inch, and even this may be too deep if much moisture is present, and the soil is abnormally either warm or cold. In sandy soils or light loams, the seed may be covered somewhat deeper, but not to exceed 1 inch.

Sowing

Seeds are sown early in the spring, and field preparation is done accordingly. The germination period is the most critical stage of the plant's growth, and if the soil is too cold or wet, a poor stand of plants may result. If seeds are sown in late April and during May, it gives a better stand of plants than those sown during June and July. The soil chosen for fall seedlings must be as dry as it is possible to obtain; otherwise, some of the seeds may germinate and subsequently be killed.

Nutrient Management

Although it can be produced in poorly fertile soil, additional manure treatment is necessary for best yield. At the time of field preparation or before seed sowing, well-decomposed organic manure should be applied at 25–30 tons/ha. In addition, about 1.5–2.0 kg organic manure per square meter may be applied after first harvesting to boost the growth of second leaves flush.

Irrigation

As it is a drought-tolerant crop, it requires less water in comparison to other crops. One or two irrigations are sufficient in the heavy soil, provided adequate moisture level is available at the time of seed germination, whereas more irrigations are required in the light soil to ensure maximum vegetative growth and leaf yield. First irrigation should be given at the time of first true leaves development followed by second and third irrigations which are done at about 20 days of intervals.

Intercultural Operations

Thinning is the most important operation to maintain the optimum plant population per unit area, which provides uniform sunlight, water, nutrient, and CO_2 to each plant, to maximize the yield per plant. It is done when the plants reach 4–6 inches in height, and the removed plants could be used for culinary purposes. Thinned crop (properly spaced) is allowed to grow undisturbed until it reaches the suitable size for harvesting. It serves as a regular source of greens throughout the summer.

Harvesting and Yield

Harvesting consists of stripping the young, tender leaves from the tips of the branches and the central portion of the plant. If repeated harvests are to be made, the plants should not be too severely stripped at any one harvest. Picking of leaves is done when leaves attain 2–3 inches (in breadth) size, after 25–30 days of sowing. The older, coarser leaves near the base of the plant should also be left, as they are too fibrous or tough for high quality. If the plants are not harvested at the right time at the right height, they start flowering and seed set, resulting in a low economic yield of leaves.

When the crop is raised on a commercial basis, all plants are harvested from 4 to 6 inches in height and succession plantings at 2-week intervals. The average yield of leaves is about 12–15 tons per ha depending upon the varieties and production practices (Rinchen et al. 2018; Rinchen and Singh 2015).

Insect-Pests

Flea Beetle (Systena frontalis)

Adult and larvae both are the damaging stages of a wide host range of pests. Adult infestation occurs on the leaves, in which shot hole wounds resemble the symptoms produced by fine buckshot. Whereas, the young larvae mostly feed on the roots and seldom feed on leaves. In the early stage of growing, plants are more prone to attack by this pest. Growth of the infested plants reduced significantly, and in severe cases, plants may die also. The holes in the leaves reduce the cosmetic and marketing values of these leafy vegetables.

Control

For effective management, the following practices may be followed:

- Regular scouting of plants should be done especially in the early stage of growth.
- Yellow or white sticky traps can be used until plants are well established.
- Crop planting time should be adjusted to avoid peak adult activity periods during the early stage of growth.
- Trap crops like mustard, radish, and pak choi may be used to decrease the pest population.
- Biocontrol agents like *Beauvaria bassiana* (fungal pathogen) may be utilized.
- In severe infestation, insecticides, viz., carbaryl, bifenthrin/zeta-cypermethrin, and imidacloprid may be used as per recommended dose.

*Spotted Cucumber Beetle (***Diabrotica undecimpunctata***)*

Both adult and larvae are the damaging stages of this pest. Adult infestation occurs in roots, flowers, and foliage, whereas larval infestation mainly occurs on the underground roots and tunnel through stems, and it aggravates damages if an infestation occurs in the initial stage of seedling growth as it increases the infection of various root complex diseases, viz., *Fusarium* wilt (Webb 2005; Capinera 2008; Bessin 2010).

Control

Control measure of the destructive pest is expensively challenging due to its subterranean nature (Krysan and Miller 1986). Some of the measures found effective and recommended for integrated management are as follows:

- Use of the tolerance/resistant varieties.
- Follow various cultural practices to minimize the pest population, viz., early ploughing-disking delayed planting.

- Use biocontrol agents like tachinid flies (*Celatoria diabrotica* (Shimer)), fungus (*Beauveria*), and a nematode (*Howardula benigna* (Cobb)) (Capinera 2008) and entomopathogenic nematodes (Alston and Worwood 2012).
- The use of biopesticides, viz., pyrethrum and spinosad, and Kaolin clay were found effective against cucumber beetles to some extent (Snyder 2015).

Diseases

Damping-Off

Symptoms

It is a devastating disease that mostly occurs in a seedling stage in the nursery during the early season (preemergence infection), in which the growing apical portion is damaged before the seedling comes out from the soil, whereas collar rotting will appear in the young emerged seedling, which leads to sudden wilting of seedling due blockage of xylem column.

Control

Hot water treatment for 30 min at 50 °C of seeds followed by shade drying before sowing or seed treatment with captan/thiram/metalaxyl-mancozeb at the rate of 2.5–3 g/kg of seeds can also be done. Soil drenching of the nursery beds can be done with captan or thiram at 3 g/l of water, before 7 days of sowing. Soil solarization is also very effective to kill the fungus present in the soil. Adequate drainage is provided in beds. The plant population should be maintained in the seedbed. The affected seedlings shall be removed from beds once the symptoms become apparent. Flooding the beds should be avoided to check the spread of the disease.

Conclusion

Chenopodium album (Pigweed) and *Atriplex hortensis* (orache) possess immense nutraceuticals values with diverse medicinal uses. Because of their high nutraceuticals, values can be blended into various processed food products to fortify them with high nutritional value and consumer health-friendly. The commercial cultivation of these potential crops can be increased significantly by increasing the knowledge of their nutraceutical importance and suitability for the small and marginal farmer in a low-cost ecosystem in the rural farming community.

References

Abu-Tahon, M. A., Isaac, G. S., & Mogazy, A. M. (2021). Protective role of fat hen (*Chenopodium album* L.) extract and gamma irradiation treatments against fusarium root rot disease in sunflower plants. *Plant Biology*, *23*(3), 497–507.

Agrawal, M. Y., Agrawal, Y. P., & Shamkuwar, P. B. (2014). Phytochemical and biological activities of *Chenopodium album*. *International Journal of PharmTech Research*, *6*(1), 383–391.

Alston, D. G., & Worwood, D. R. (2012). Western striped cucumber beetle western spotted cucumber beetle (*Acalymma trivittatum* and *Diabrotica undecipunctata undecipunctata*). Utah State University Extension and Utah Plant Pest Diagnostic Laboratory.

Andrews, A. C. (1948). Orach as the Spinach of the Classical Period. *Isis*, *39*(3), 169–172.

Babb, M. F., & Kraus, J. E. (1939). *Orach: Its Culture and Use as a Greens Crop in the Great Plains* (No. 526). US Department of Agriculture.

Bassett, I. J., & Crompton, C. W. (1978). THE BIOLOGY OF CANADIAN WEEDS.: 32 *Chenopodium album* L. *Canadian Journal of Plant Science*, *58*(4), 1061–1072.

Bessin, R. (2010). Cucumber beetles. *ENTFACT-311., (University of Kentucky, 2003)*.

Bylka, W., Stobiecki, M., & Frański, R. (2001). Sulphated flavonoid glycosides from leaves of *Atriplex hortensis*. *Acta Physiologiae Plantarum*, *23*(3), 285–290.

Capinera, J. L. (2008). Spotted cucumber beetle or southern corn rootworm, *Diabrotica undecimpunctata* Mannerheim (Coleoptera: Chrysomelidae). *Encyclopedia of Entomology, Springer, Netherlands*.

Carlsson, R. (1975). *Selection of Centrospermae and other species for production of leaf protein concentrates* (Doctoral dissertation, Verlag nicht ermittelbar).

Duke, J. A. (1983). Handbook of energy crops. *Handbook of Energy Crops*.

El-Rokick Kowther, G., El-Masry, R. R., Ahmed, S. A. A., Mohan, S. A., & Messiha, N. K. (2018). Allelopathic effects of *Allium sativum* cloves on growth and yield of Helianthus annuus plants associating *Cyperus rotundus*. *Int. J. Environ*, *7*(3), 78–86.

Gale, J., & Poljakoff-Mayber, A. L. E. X. A. N. D. R. A. (1970). Interrelations between growth and photosynthesis of salt bush (*Atriplex halimus* L.) grown in saline media. *Australian Journal of Biological Sciences*, *23*(4), 937–946.

Greenway, H. (1968). Growth stimulation by high chloride concentrations in halophytes. *Israel J Bot*. 17: 169–177.

Handley, J. F., & Jennings, D. H. (1977). The effect of ions on growth and leaf succulence of *Atriplex hortensis* var. cupreata. *Annals of Botany*, *41*(6), 1109–1112.

Henson, I. E. (1970). The effects of light, potassium nitrate and temperature on the germination of *Chenopodium album* L. *Weed Research*, *10*(1), 27–39.

Holm, L. G., Plucknett, D. L., Pancho, J. V., & Herberger, J. P. (1977). *The world's worst weeds. Distribution and biology*. University press of Hawaii.

Hunt, S. P., Jarvis, D. E., Larsen, D. J., Mosyakin, S. L., Kolano, B. A., Jackson, E. W., Martin, S.L., Jellen, E.N. & Maughan, P. J. (2020). A Chromosome-Scale Assembly of the Garden Orach (*Atriplex hortensis* L.) Genome Using Oxford Nanopore Sequencing. *Frontiers in plant science*, *11*, 624.

Krysan, J. L., & Miller, T. A. (Eds.). (1986). Methods for the study of pest Diabrotica.

Lewis, J. (1973). Longevity of crop and weed seeds: survival after 20 years in soil. *Weed research*, *13*(2), 179–191.

Mabberley, D. J. (1997). *The plant-book: a portable dictionary of the vascular plants*. Cambridge university press.

Nicol, J. (1994). *Atriplex nummularia, Atriplex vessicaria* and other *Atriplex species*. *Aust J Med Herbalism*, *6*, 85–87.

Pagno, C. H., Costa, T. M., de Menezes, E. W., Benvenutti, E. V., Hertz, P. F., Matte, C. R., Tosati, J.V., Monteiro, A.R., Rios, A.O. & Flôres, S. H. (2015). Development of active biofilms of quinoa (W.) starch containing gold nanoparticles and evaluation of antimicrobial activity. *Food Chemistry*, *173*, 755–762.

Polunin, O. (1969). Flowers of Europe. A field guide. *Flowers of Europe. A field guide.*

Ramakrishnan, P. S., & Kapoor, P. (1974). Photoperiodic *Chenopodium quinoa* Requirements of Seasonal Populations of *Chenopodium Album* L. *The Journal of Ecology*, 67–73.

Reed, C. F. (1971). *Common weeds of the United States* (Vol. 366). Courier Corporation.

Rinchen, T., & Singh, N. (2015). Exploring nutritional potential of Atriplex hortensis. *Indian Hort*, *60*(2), 16–17.

Rinchen, T., Singh, N., & Maurya, S. B. (2018). Yield and Palatability of Potential Indigenous Leafy Vegetable of Cold Arid trans-Himalayan Ladakh, India. *Def. Life Sci. J.*, *3*(2), 136–139.

Rinchen, T., Singh, N., Maurya, S. B., Soni, V., Phour, M., & Kumar, B. (2017). Morphological characterization of indigenous vegetable (*'Atriplex hortensis'* L.) from trans-Himalayan region of Ladakh (Jammu and Kashmir), India. *Australian Journal of Crop Science*, *11*(3).

Ruas, C. F., Ruas, P. M., Stutz, H. C., & Fairbanks, D. J. (2001). Cytogenetic studies in the genus Atriplex (Chenopodiaceae). *Caryologia*, *54*(2), 129–145.

Rubatzky, V. E., & Yamaguchi, M. (2012). *World vegetables: principles, production, and nutritive values*. Springer Science & Business Media.

Saini, H. S., Bassi, P. K., & Spencer, M. S. (1985). Seed germination in *Chenopodium album* L.: further evidence for the dependence of the effects of growth regulators on nitrate availability. *Plant, Cell & Environment*, *8*(9), 707–711.

Siddiqui, B. S., Ahmed, S., & Khan, M. A. U. (1994). Triterpenoids of *Atriplex stocksii. Phytochemistry*, *37*(4), 1123–1125.

Snyder, W. E. (2015). Managing cucumber beetles in organic farming systems. *Department of Entomology, Washington State University Pullman, WA.*

Steinbach, G. (1996). Leksykon przyrodniczy, zioła i owoce leśne. *Świat Książki, Warszawa.*

Stevens, J.M. (1994). Orach-*Atriplex hortensis* L. Fact sheet HS-637. Gainesville, Fla.: Horticultural Sciences Department, Florida Cooperative Extension Service, Institute of Food and Agricultural Sciences, University of Florida.

Sukhorukov, A. P. (2014). The carpology of the family Chenopodiaceae in relations to problems of phylogeny, systematics and diagnostics of its representatives. *Grif i K, Tula.*

Webb, S. E. (2005). Insect Management for Cucurbits (Cucumber, Squash, Cantaloupe, and Watermelon): ENY-460/IN168, rev. 9/2005. *EDIS*, *2005*(14).

Welbaum, G. E. (2015). Vegetable seeds and crop establishment. *Vegetable production and practices*, 27–46.

Williams, J. T., & Harper, J. L. (1965). Seed polymorphism and germination: I. The influence of nitrates and low temperatures on the germination of *Chenopodium album*. *Weed Research*, *5*(2), 141–150.

Wilson, C., Lesch, S. M., & Grieve, C. M. (2000). Growth stage modulates salinity tolerance of New Zealand spinach (*Tetragonia tetragonioides*, Pall.) and red orach (*Atriplex hortensis* L.). *Annals of Botany*, *85*(4), 501–509.

Wright, K. H., Pike, O. A., Fairbanks, D. J., & Huber, C. S. (2002). Composition of *Atriplex hortensis*, sweet and bitter *Chenopodium quinoa* seeds. *Journal of food science*, *67*(4), 1383–1385.

Chapter 12
Production Technology of Underutilized Crops of Compositae Family

Rajkumari Asha Devi, Avinash Kumar, and Anil Kumar

Introduction

The Compositae, also popularly known as Asteraceae, is one of the most prominent families of the plant kingdom, with nearly 23,600 species with 1620 genera (Anonymous 2021). This group mainly consists of vines, herbaceous plants, shrubs, and trees. Apart from using its various species for ornamental purposes, this group also has a few members used as vegetables, including lettuce, endive, chive, dandelion, Spanish salsify, and black salsify tarragon, Jerusalem artichoke, globe artichoke, and chicory. However, most of these vegetables mentioned formerly are not popular in all parts of the world. They are still consuming in a few pockets of the world, and this lack of awareness confers them the tag of underexploited vegetables. Due to this, there is not much information available to date regarding the package of practice for these underutilized vegetables, nutritional values, and other beneficial uses. However, it is observed that the vegetables of this family have good healthy characters, viz., vitamin A, vitamin C, a good amount of fiber and calcium content. This book chapter will discuss the nutritional importance, climate, soil, sowing methods, planting season, spacing, field preparation techniques, manure and fertilizers, irrigation, intercultural operations, harvesting, diseases, etc., and will put the limelight on the production of these underutilized vegetable crops.

R. A. Devi (✉) · A. Kumar · A. Kumar
School of Agriculture, Lovely Professional University, Phagwara, India
e-mail: rajkumari.22847@lpu.co.in

Savita et al. (eds.), *Production Technology of Underutilized Vegetable Crops*,
https://doi.org/10.1007/978-3-031-15385-3_12

Origin and Distribution

This family includes around 32,000 species of plants, which are widely grown in different parts of the world either for vegetables or as flowering plants. The plants included in this family are either annual, biennial, or perennial, mostly with herbaceous growth habits. These also have a diverse distribution, which ranges from tropical to subpolar regions. Mostly they are found in tropical zones, viz., Brazil, South Africa, Central America, Central Asia, and southwest China. A larger chunk of the species is found mostly in different regions of temperate and subtropical latitudes. A list of the underutilized and cultivated crops of this family has enlisted in Table 12.1.

Nutritional Importance and Uses

Lettuce

The nutritional value of the lettuce is high as shown in Table 12.2. It is a good source of lutein. However, the nutritional value remains intact in raw lettuce since the vitamin C content is lost while cooking (Anonymous 2022).

Table 12.1 List of underutilized and cultivated crops

Common names	Botanical name	Area of cultivation
Lettuce	*Lactuca sativa* L.	Southwest Asia
Endive	*Cichorium endivia* L.	Egypt and Indonesia
Chicory	*Cichorium intybus* L.	Europe
Globe artichoke	*Cynara scolymus* L.	North Africa and other parts of the Mediterranean region
Jerusalem artichoke (sunchoke)	*Helianthus tuberosus* L.	North America
Dandelion	*Taraxacum officinale*	Eurasia
Tarragon	*Artemisia dracunculus*	Caspian Sea, Siberia
Salsify	*Scorzonera hispanica*	Southern Europe and the Near East

Table 12.2 Nutrient content per 100 g

Nutrient	Content	Nutrient	Content
Vitamin A (IU)	1650	Phosphorous (mg)	24
Vitamin K (mcg)	75.2	Calcium (mg)	27
Vitamin C (mg)	6	Protein	1.13

Source: nutritionvalue.org

Endive

This family includes around 32,000 species of plants, widely grown in different parts of the world either for vegetables or as flowering plants. These plants in this family are annual, biennial, or perennial, mostly with herbaceous growth habits. These also have a diverse distribution, ranging from tropical to subpolar regions. Mostly they are found in tropical zones, viz., Brazil, South Africa, Central America, Central Asia, and southwest China. A more significant chunk of the species is found mainly in different regions of temperate and subtropical latitudes. Endive may be annual or biennial but grown annually with a loose head of fine cut curly leaves. The dark green outer leaves have a slightly bitter taste, while the inner leaves are of a lighter color and milder in taste. The size, shape, and color of leaves may vary in different varieties. In some types, the leaves may be reddish or purplish. It is highly beneficial for treating several stomach ailments and act as a good laxative. This crop helps enhance appetite, favoring digestion, and increases the activity of the gall-bladder due to the presence of a compound "intibina," which imparts a bitter taste. It has low calories and fats, a good amount of vitamin B1, B2, calcium, folic acid, iron, and potassium, respectively, as shown in Table 12.3.

Chicory

Roots and tops are the edible part of it. The tops from a short stem may have an open rosette of leaves or a compact, leafy bud known as "Chicon." There are both annual and biennial varieties. It grows as a biennial, and its tender leaves are used as a salad vegetable. It contains 23 cal per 100 g of serving. Chicory raw contains 0 mg of cholesterol per serving as shown in Table 12.4.

Table 12.3 Nutrient content per 100 g

Nutrient	Content	Nutrient	Content
Vitamin-A (IU)	7405	Calcium (mg)	36
Vitamin-C (mg)	9.2	Potassium (mg)	194
Vitamin-K (mcg)	126.3	Protein (g)	1.36

Source: nutritionvalue.org

Table 12.4 Nutrient content per 100 g

Nutrition	Content	Nutrition	Content
Vitamin A (IU)	5717	Calcium (mg)	100
Vitamin K (mcg)	297.6	Potassium (mg)	420
Vitamin C (mg)	24	Protein(g)	1.70

Source: nutritionvalue.org

Globe Artichoke

It is a herbaceous perennial but grown annually with edible young flower heads or buds (capitulum). The enlarged receptacle, fleshy inner bracts, and choke (green flower buds, pappus, and bristles) are edible in the heads harvested early at the young immature stage. It contains 47 cal per100 g serving and nutrient content as shown in Table 12.5.

Jerusalem Artichoke

Its edible part is tubers and is fleshy in nature. The tubers are potato-like and oblong or elongated, weighing about 80–120 g and 7.5–10 cm long. The color of tubers ranges from white to yellow and from red to blue. The tubers contain insulin, a polymer of fructose, which diabetic patients consume. It includes 73 cal per 100 g serving and nutrient content as shown in Table 12.6.

Dandelion

A dandelion green, raw contains 45 cal per 100 g serving and nutrient content as shown in Table 12.7. It also has a small quantity of 3.10 mg of iron, as well.

Table 12.5 Nutrient content per 100 g

Nutrition	Content	Nutrition	Content
Vitamin A (IU)	13	Calcium (mg)	44
Vitamin K (mg)	14.8	Potassium (mg)	370
Vitamin C (mg)	11.7	Protein(g)	3.27

Source: nutritionvalue.org

Table 12.6 Nutrient content per 100 g

Nutrition	Content	Nutrition	Content
Vitamin A (IU)	20	Calcium (mg)	14
Vitamin K (mcg)	0.1	Potassium (mg)	429
Vitamin C (mg)	4.0	Protein(g)	2

Source: nutritionvalue.org

Table 12.7 Nutrient content per 100 g

Nutrition	Content	Nutrition	Content
Vitamin A (IU)	10,161	Calcium (mg)	187
Vitamin K (mcg)	778.4	Potassium (mg)	397
Vitamin C (mg)	35	Protein (g)	2.70

Source: nutritionvalue.org

Tarragon

Minerals like magnesium, copper, calcium, iron, and zinc, etc., are present abundantly in it. It is also rich in vitamins, viz., vitamin A, vitamin B complex, vitamin C, riboflavin, pyridoxine, niacin, etc., which act as cofactors as well as antioxidant for enzymes in the metabolism (Table 12.8).

Salsify

Salsify raw contains 82 calories per 100 g serving and nutrient content as shown in Table 12.9. It is very nutritious and helpful for diabetic patient as it contains the polysaccharide inulin.

Botany and Taxonomy

Lettuce

It is a glabrous annual herb containing latex, about 100 cm tall. The leaf is rosette, which afterward turns into a tall-branched flowering stem. It has shallow root with a strong fleshy taproot. The leaves are loose rosette generally pale to dark green or with red or brown anthocyanin pigment arranged in a more or less compact head, short petiole, undivided lamina to pinnatifid or saw-toothed, sometimes fringed and curly, stem leaves are arranged spirally, sessile, ovate to orbicular, entire, cordate, amplexicaul, progressively smaller. The heads are grouped in a dense corymbose, flat-topped panicle with three to four rows of lanceolate or ovate bracts and an involucre 10–15 mm long. Number of flowers per head are around 7–35, hermaphrodite;

Table 12.8 Nutrient content per 100 g

Nutrition	Content	Nutrition	Content
Vitamin A (IU)	4200	Calcium (mg)	1139
Carbohydrate (g)	50.22	Potassium (mg)	3020
Vitamin C (mg)	50.0	Protein (g)	22.77

Source: USDA National Nutrient database

Table 12.9 Nutrient content per 100 g

Nutrition	Content	Nutrition	Content
Carbohydrate (g)	19	Calcium (mg)	60
Fat (g)	0.2	Potassium (mg)	380
Vitamin C (mg)	8.0	Protein (g)	3.3

Source: USDA National Nutrient database

ligulate yellow corolla, five stamens with connate-shaped anthers; and inferior ovary, single-celled, bifid style. Fruit is a narrowly obovate achene 3–8 mm long, ribbed, compressed, white, yellowish, gray black, or brown, with a narrow beak, and a white pappus of two equal rows of soft hairs surmounted by a short beak.

Butter Head Lettuce It has originated in Western Europe. They are mostly growing in Europe. It has overlapping leaves with loose heads, pale green leaves, thin inner leaves, oily, buttery in texture, and slightly undulate. Heads weigh up to 350 g.

Chinese Lettuce It has generally long, non-head forming sword-shaped leaves, with a bitter and robust flavor unlike western types, appropriate for stews and stir-fried dishes. There are two types – "leaf use" and "stem use" types. It is cultivated for its fleshy, i.e., 30–50 cm long and 3–6 cm thick stem, which has a faint lettuce taste and a crispy texture. A rosette grows at the top of the stem, which has numerous leaves. Young leaves can also be eaten. Stem lettuce is originated in China and becomes most popular in East Asia and, to a lesser extent, Southeast Asia.

Crisp head (Iceberg) forms tight head as that of cabbage. They are popular more for their crunchy texture rather than for their flavor. Dark green thick, crispy leaves, with prominent flabellate veins and midribs; head weighed around 1 kg, interior white to creamy yellow. Non-heading to slightly heading types also occur. Iceberg lettuce are the most commonly grown in the USA.

Loose leaf are delicate, tender, and mildly flavored. Leaf lettuce, loose leaf lettuce, curled lettuce, bunched lettuce, and cutting lettuce; leaves broad, crinkled or curled, green or reddish, thin, smooth, arranged on a short stem or in a loose rosette; generally marketed in bunches of three to ten plants. It is mostly grown in African countries.

Romaine (cos): It is originated in Southern Europe and is tolerant to heat. It develops with a long head of sturdy leaves and a firm rib down the center. It forms an upright, cylindrical head with long, narrow leaves. It can be consumed both cooked as well as raw.

Endive It is a loose head lettuce and grows annually or biennially as herb up to 120 cm. The taproot contains a bitter milky juice. The leaves are alternate, broad, sessile, simple or pinnatifid, slightly crumpled, having margin entire or dentate (escarole type), or it can be very narrow as well, strongly curled (curly leaved type), and deeply pinnatifid, progressively smaller upward on the stem, slightly glabrous or pubescent, pale to dark green or yellowish, sometimes reddish along the midrib. Flowers are blue and sometimes white in color, 15–20 per head, all ligulate, 2 cm long corolla, five lobed at the apex; five stamens with anthers fused into a tube; and inferior ovary, single-celled, slender style, hairy, with two slender stigmatic lobes. Fruits are obovoid to cylindrical achene 2–3 mm × 1–1.5 mm, brown, with a pappus of one to three rows of small, persistent membranous scales.

Chicory

It is an herbaceous perennial that requires cold temperature and long days for bolting. It is grown as a biennial to use its tender leaves as a salad vegetable (Lucchin et al. 2008). It is cross-pollinating, mainly by insects, like honeybees and other bees, and self-incompatible. Stem hollow, 1.5 m in height, branching, herbaceous, glabrous to sparsely strigose and scabrous, and erect. Taproot are branched up to 75 cm long, deep, fleshy, with milky sap, forming a rosette near the ground, long petiole, basal leaves lyrate pinnatifid, dentate, pubescent above and below, and hirsute on midrib below.

Globe Artichoke

It is a herb which is perennial in nature. Stem generally erect, leafy branched, stout, glabrous to densely arachnoid – taproot, fleshy. Leaves are basal until bolting, pinnately lobed, oblong to lanceolate. On any given axis, the uppermost flower opens first followed by the lower or peripheral flowers. The involucre is hemispheric or ovoid, 3–6 cm in height. Phyllaries are surrounding flower head, with stout spines, overlapping in series, generally leathery, ovate, complete, receptacle flat, glabrous, fleshy, bristly (spiny guidelines on phyllaries) on immature flower head effortlessly differentiate this plant from the intently associated. Flowers are purple, with blue corollas, ±2 inches (5 cm), tube very slender, throat widened all at once, lobes linear; anther bases long-sagittate, guidelines oblong; fashion appendage lengthy, cylindrical, minutely papillate, and tip barely notched. Fruits are cylindric to obconic, 4-angled or compressed, glabrous, connected at base; pappus of many stiff bristles, 1–1.5 inches, in numerous series, white or brownish, plumose under, fused and falling together.

Jerusalem Artichoke

The tubers of those are fleshy and edible. The tubers are potato-like and oblong or elongated, weighing approximately 80–120 g and 7– 10 cm long. The coloration of tubers tiers from white to yellow and from red to blue. The tubers incorporate insulin, a polymer of fructose, beneficial for diabetic sufferers. The flora are like the ones of sunflowers. It is a perennial crop with tuber-bearing rhizomes. The tubers range from knobby to round clusters, variety from red to white, and are rougher in conformation than potato tubers. The fit for human consumption tubers is produced just below the floor on skinny white rhizomes. They may be segmented and knobby, 1–4 inches (2.5–10 cm) long, and have crisp, white flesh. Tubers are produced on the ends. The stem is erect, stout, and pubescent and grows 3–12 ft in top. Leaves are 1.14 to a few inches huge and 4–8 inches lengthy with the winged petiole. The thick, rough-textured leaves have coarse hairs at the upper floor and are finely pubescent to tomentose below. They're opposite on the lower part of the plant and exchange at the upper component. The leaves are company ovate to rectangular-lanceolate, step by step taper toward the end, and feature toothed margins. The flower heads (capitulum) are vibrant yellow in color and resemble to that of the cultivated sunflower however are smaller and produced at the ends of stems and axillary branches only. Each head is 2.5–3.5 inches across. Flowering heads have ten yellow ray flowers surrounding a significant cluster of immature, greenish disc plants and a slim band of mature, yellow disc vegetation. The bracts below the flower head identify characters of Jerusalem artichoke. These are sharply pointed and regularly fuzzy. There are 10–20 yellow ray florets with ligules 1–1.5 inches long and yellow to orange disk florets. The involucres of bracts are lanceolate to linear-lanceolate and feature pointed tips. An achene can reach 8 mm in duration. The seeds are flattened, mottled, wedge-shaped, smooth, and black in color. Only few flowers could develop into seeds. It can be propagated through fleshy rhizomes (underground stems) or seed, which endure small, potato-like tubers.

Dandelion

Dandelion is a broadleaved, herbaceous, perennial crop (Anonymous 2019).

The Black Salsify

The plant has heads of yellow ray flowers. The skinny black taproot grows up to 1 m in length and up to 2 cm (0.8 in) in diameter. It has black skin with white inner flesh. French tarragon is mostly a perennial regular plant, but it is mostly grown as an annual (A separate plant referred to as Russian tarragon is likewise a perennial). French tarragon is mainly a flowerless plant with aromatic leaves that evoke anise and mint. It grows from 12 to 2 inches tall; it spreads from tangled, underground rhizomes; its flowers produce sterile cloves (A special plant known as Russian tarragon can be grown from seed; however, its miles considered by means of maximum to be too bitter for culinary use). It can occasionally produce small greenish-white vegetation in branched clusters. Its leaves have long stems and are vivid, slender, darkish green, fragrant leaves to approximately 1-inch long and pointed at the end.

Agronomic Practices

Climate and Soil

Most of the plants beneath this own family require a cool climate and are pleasant thriving on rich friable and properly drained soil rich in organic count for his or her increase and improvement. Lettuce performs well below subtropical and temperate situations. Lower and higher temperatures above 30 °C affect seed germination and are great at 24–25 °C. The most useful temperature of 20 °C (day) and 10 °C (night) is needed for proper growth and development. Better temperature induces bolting, consequences bitter taste within the leaves, and accelerates the ailment "tipburn" and rot. Chilling remedy at 4–6 °C for 3–5 days facilitates in breaking dormancy. Sandy loam is preferred to get early yield. It's far susceptible to acidic soils. The pH of 6–8 is ideal for its boom. The superior temperature for both endive and chicory for his or her increase is 15–18 °C. Globe artichoke quality is adversely laid low with excessive temperatures. The most appropriate temperature requirement is 12–18 °C. Mostly loam as well as the light clay soils with a pH of 6–8 are suitable for cultivating the crop. Jerusalem artichoke may be grown on loam or sandy loam soils and even on bad soils. It does no longer require a whole lot nutrients and moisture. The top-rated temperature for developing is 18–26 °C. Dandelion can be grown under wide climatic condition (Giacomino et al. 2016). It indicates a wide variety of adaptability to light, developing vigorously in full sunlight or diffusing mild within the color of buildings or bushes (Longyear 1918). It can be raised under wide variety of soil (Giacomino et al. 2016); however, it thrives best in moist, precise first-class loam (Jackson 1982). It flourishes well in soils with a pH of 4.2–8.2 (Holm et al. 1997). Tarragon is proof against bloodless and warmth; it's far cold hardy to — 23–—33 °C, but freezing climate could prove fatal to the plant. Salsify is a hardy cool-climate root crop. It prefers soil wealthy in organic be counted, having a pH of 6.0–6.8. The soil must be well cultivated to the intensity of 20–30 cm; otherwise, it can motive roots to fork and breakup. It is planted before the temperature of the soil has reached 4.4 °C. As a winter crop, the salsify is cultivated in the early autumn. It calls for 120–150 days to get yield and is satisfactory with regard to adulthood in cool climate. It's far harvested after the first freeze in autumn. Postpone in harvesting, specifically all through excessive temperatures above 29 °C, makes the roots stringy and fibrous (Table 12.10).

Planting

To raise the lettuce seedlings for 1 hectare; 400–500 g seed is required to be sown in about 50–75 square meter area. Plant spacing is maintained at 30–45 cm × 15–25 cm within the rows. The direct-seeded crop usually needs thinning due to high plant density after 4–6 weeks of the emergence. It is recommended that the seedlings

Table 12.10 Propagation material, popular varieties, and edible parts of major underutilized Asteraceae crops

Crop name	Propagation	Popular varieties	Edible part
Lettuce	Seeds	Punjab lettuce no. 1, great lakes, slobolt, Chinese yellow, imperial 859, big boston, may king, white boston, and dark green	Leaves
Endive	Seeds	Majorly two types of varieties are cultivated, viz., escarole/broad-leaved endive (var. latifolia) and curly/frisee endive (var. crispum)	Rosette or curled leaves
Chicory	Seeds	Belgian endive, radicchio, and puntarella	Leaves and root
Globe artichoke	Axillary buds (ovoli) and seeds	Green globe, Magnifico	Flower buds
Jerusalem artichoke	Tuber	White French white, Columbia	Leaves and stem
Dandelion	Seeds		Petals and leaves
Tarragon	Divisions or cuttings	Russian or false tarragon, Mexican mint tarragon, and French tarragon	Young leaves
Salsify	Seed	Giant Russian, Sandwich Island mammoth, Scorzonera	Roots

should always be hardened before moving them to the main field for transplanting. For hardening, the withholding of the irrigation water for about 6–8 days before transplanting have been found helpful in acclimatization of the seedlings. Once the transplanting has been completed, the fields should be rinsed with sufficient amount of water.

Endive is sown in September–October and February–March in northern plains and March–April and August–September in the hills. Seedlings should be transplanted at closer spacing, 20–30 cm apart, to facilitate blanching of the leaves in the center, which is desirable to reduce the bitterness in the leaves. There are self-blanching varieties too.

Globe artichoke can be planted in February–March or September–October. It is propagated by suckers or rooted offshoots and by division from old crowns. It can also be grown from seeds. The planting is done in 2.5 m apart rows at a distance of 1.8 m between plants. Sow the salsify seed ½ inch (12 mm) deep and ½ inch apart. Thinning should be practiced for encouraging the successful seedlings to stand 3–4 inches (7–10 cm) apart. Build the rows 18–24 inches (45–61 cm) apart.

Tarragon seed is sterile. Divisions or cuttings can only propagate French tarragon. Most of the time when a buyer purchases the tarragon seed, he ended up receiving Russian tarragon which lacks the licorice flavor and smell of French tarragon. 6–8-inch stem cuttings should be rooted in damp sand. Allow for a 4-week period for the stems to root. Summer is the best time to take cuttings and root them. Because roots are so knotted, root division might be difficult. If required, cut through roots with a spade. Replant in only a moist planting mix after pruning roots to roughly 2 inches. In the spring, divide your plants. Plants should be divided every 3–4 years.

Jerusalem Artichoke

Planting can be done in February–March or September–October. It is grown from whole tubers or small tuber pieces (60–90 g). The tubers should be planted 5–7.5 cm deep on raised beds in rows that are 1–1.5 m apart and 60–90 cm between the plants.

Nutrient Management

The lettuce plant possesses a small and shallow root system. Therefore, the surface soil in which the lettuce is to be planted should be rich in nutrients. Application is of about 10–15 t/ha FYM with about NPK at 60:60:60 kg/ha. On sandy and sandy loam soil, apply NPK at 40–45:75–100:75–100 kg/ha in the absence of any farmyard manure application. On silt loam and the clay loam soil, apply NPK at 25:25:50–75 kg/ha. Globe artichoke requires about 10–15 tons of FYM, and 60–100 kg of N per hectare must be applied at the time of field preparation.

Irrigation

Because lettuce is a shallow-rooted crop, it requires regular but mild irrigation. Irrigation is administered first, right after transplanting. On light soils, irrigation should be done at every 4–5 days, and on heavy soils, every 8–10 days. In globe artichoke, the crop is irrigated after planting. Frequent irrigations are required during the initial stages of its growth. To keep the roots from becoming stringy, keep the salsify uniformly moist. Prior to planting, add well-decomposed compost to the planting beds. In the middle of the season, apply a side dressing of compost to the salsify. Roots can fork and split when manure is added too much or else the excess nitrogen is introduced into the growing medium before sowing. Until the plants are established, keep the French tarragon equally moist. Plants require only periodic watering once they have established themselves; the soil can become virtually dry between waterings.

Intercultural Operations

Light hoeing and 3–4 weedings are adequate during the cropping season to control the weeds, otherwise smother the young plants of lettuce. If there are rains during the cropping season, mulching in lettuce proves useful. It reduces soil erosion around the root zone and raises soil warmth in the winter, promoting plant growth. Mulching keeps the leaves from decomposing by preventing them from coming into contact with the soil. Rice straw can be utilized as a mulching material because of its easy availability.

Harvesting

Leafy varieties are harvested at their tender and immature stage. First harvesting is done 35 DAS. Head lettuce is harvested when the head is solid (crisp head: 80–85 days), butterhead and cos type (60–70 days), and loose leaf (40–50 days).

Endive is started harvesting from 70–100 days after transplanting. Chicory is harvested about 70–100 days after transplanting. The chicons and rosette of leaves are cut off at the base of the roots, and later the roots are removed.

Globe artichoke is ready for harvest about 6–7 months after planting. The crop can last for 6–7 years. However, the plants after harvest are cut just above the soil level. Jerusalem artichoke tubers are ready for harvesting in about 5 months after planting.

Salsify roots are harvested when the roots reach 12 inches (30 cm) long or longer. Lift the salsify whole with your hands or a garden fork keeping in mind not to damage or break the roots. It takes 120–150 days for roots to be ready for harvest. The salsify roots can endure freezing temperatures, so one can leave them in the fields/ground until they are needed. The longer salsify kept in the ground, the less oyster-like taste it will give. For the finest flavor from the harvest, the picking of the young top leaves in early summer shall be done. Cut back the plant's leafy top growth numerous times throughout the season so that the new development can occur. Early in the summer, and again at the conclusion of the season, prune the stems. Using a garden pruner or scissors, snip leaves and stems. Leaves should be handled with care since they bruise readily.

Yield

Lettuce: The average yield is about 10–14 tons per hectare. Globe artichoke: About 40–50 edible heads/buds are produced on each plant. The terminal buds are the largest (7.5–12.5 cm in diameter) and the earliest to mature followed by axillary buds (5–10 cm in diameter).

Diseases

Anthracnose (Shot-Hole)

Causal Agent: ***Microdochium panattonianum***

Symptoms

Small brown patches on the outer leaves that may enlarge and could change their color to straw-like-color, and appearance of the shot -hole.

Cause

The fungus usually survives in the debris of the preceding crop in soil, and the fungus spreads by splashing water in the filed during irrigation or in rain.

Management

Crop rotation, either removal of the debris or mixing it into the soil, controlling wild lettuce around the plantation, and also avoiding sprinkler irrigation.

Leaf Drop (Sclerotinia Drop)

Causal Agents: *Sclerotinia minor* and *Sclerotinia sclerotiorum*

Symptoms

The wilting of the exterior leaves is observed in the initial stages which later proceeds inwards until the entire plant gets infected. On the leaves, appearance of the soft watery lesions is observed, later on the infected leave(s) collapses to the ground, which finally turns into black fungal formations on the diseased leaf tissue as well as soil surface takes place.

Cause

Fungi can survive in the soil for a period of 8–10 years.

Management

Fungicides applied promptly after thinning the plants, ploughing the soil deeply, and rotating crops with nonhosts. These all practices can help to decrease the illness.

Powdery Mildew

Causal Agent: *Erysiphe cichoracearum*

Symptoms

The white and powdery fungal growth on the top as well as underside of the aged/older leaves, as well as yellowing or browning of the leaves is also observed, and little black fruiting bodies can also be seen.

Cause

The emergence of disease rages in warm humid weather, and sickness can be transferred over large areas by the wind movements.

Management

The sulfur application at the inception of the symptoms is recommended as long as the temperature is very high.

Septoria Leaf Spot

Causal Agent: ***Septoria lactucae***

Symptoms

The small, chlorotic spots with irregular appearance on the plant leaves, which turn brown, and then dries up; at later stages, the shot holes can also be observed; if the plant is seriously infected, the lesions may consolidate to produce massive necrotic patches and drooping leaves.

Cause

The illness spreads in humid or rainy environments because the fungus persists in contaminated seeds and crop waste. Splashing water can spread the fungus, and wild lettuce is an essential host of the fungus.

Management

Seed treatment with hot water treatment before planting could be helpful in reducing the level of disease infection. The planting areas should not have past infection, and the low rainfall regions are considered safe from attacks of Septoria.

Big-Vein Mirafiori Lettuce Big-Vein Virus (MiLBVV)

Causal Agent: ***Virus***

Symptoms

The veins are larger and clear, the leaves are puckered or ruffled, and the outer curled leaves are upright.

Cause

The virus is spread through the soil and transmitted to host by *Olpidium brassicae* which is a fungus. It appears most commonly in the cool weathers.

Management

The selection of resistant varieties supposed to be best method to control the MiLBVV.

Slugs and Snails (Gray Garden Slug, Spotted Garden Slug, Brown Garden Snail, European Garden Snail, etc.)

Causal Agents: *Decoratus reticulatum, Limax Maximusmaximus, Helix aspersa,* and *Cornu aspersum*

Symptoms

Irregular holes on stem and leaves; blooms and fruit can be affected too; leaves may be torn if the infestation is severe. On plant foliage, rocks, soil, etc., slime trails can be noticed; slugs and snails are typical garden pests; slugs are dark gray to black in color and can be grow up to 2.5–10 cm, i.e., 1–4 inch long; garden snails are comparatively smaller and have a round shell or a spiral shell.

Cause

Slugs as well as the snails love moist, shady environments and they hide in the weeds or in the organic waste; adults mostly lay their eggs in the soil all year.

Management

Sanitation should be done by eliminating wet habitat for slugs and snails and keeping your garden clean by removing garbage, weeds, and plant waste. To reduce the population of slugs, handpick them at night. Plants should be surrounded by wood ashes or eggshells. Leave out organic stuff, viz., grapefruit skins or lettuce, etc., to attract mollusks and remove any that are seen feasting on the lure. To drown the mollusks, place shallow dishes which are filled with beer in the dirt. In organic gardens, ferrous phosphate is used, while metaldehyde (e.g., Buggeta) and carbaryl (e.g., Sevin bait) are used in nonorganic gardens.

Bottom rot

Causal Agents: *Rhizoctonia solani*

Symptoms

Small red-brown spots leave near to ground (lower leaves), typically on the underside of the midrib, that can quickly spread and cause the leaves to decay; Leaf lesions may exude an amber-colored liquid; As the stems rot, the lettuce head becomes slimy, brown, and collapses. Infected tissue may have a brown mycelial growth evident.

Cause

The fungus survives in soil on the crop debris, and warm, wet weather favors disease onset.

Management

Ploughing soil prior to planting, rotating crops periodically, minimizing irrigation near to the harvesting of the crop, planting cultivars having erect growth habit to decrease the leaf contact with soil, and applying proper foliar fungicides are the most effective ways to treat the disease.

Downy Mildew

Causal Agents: *Bremia lactucae*

Symptoms

On the top side of leaves, light green or chlorotic angular lesions turn yellow; fluffy white growth on the underside of leaves is also observed.

Cause

The disease thrives in chilly, wet environments and can be spread by contaminated seed. The fungus can also be found in the debris of plant and the wild lettuce plants.

Management

Using resistant varieties and/or using proper fungicides are the most effective ways to combat the disease.

Physiological Disorders of Underutilized Asteraceae Crops

1. Tipburn

   Necrosis on the edges and margins of young leaves is a sign of tipburn. This condition arises primarily a few days prior to harvesting and is caused by a calcium deficit in a specific area. Tip burning is favored by environmental factors that encourage rapid plant development (high light intensity, warm temperatures, and extended photoperiod) or reduced transpiration (low air movement, high relative humidity, and water stress).

   Causes

   - High temperature
   - Excess N, deficiency of Ca and B
   - Control: raising dark period, RH, and Ca

2. *Russet spotting*

   Ethylene injury
3. *Rib Discoloration*

   When lettuce matures and becomes firmer, symptoms appear. Heat stress is closely linked to this physiological disorder. Small brown-colored streaks appear along the midribs of the leaves beneath those surrounding the head lettuce, signaling the start of symptoms. The lesions, which range in color from light brown to black, deepen over time and are frequently followed by rotting during storage, thereby lowering the quality of lettuce as well as marketing.
4. *Bolting*

   It is also known as "going to seed" and is the rapid extension of the main stem in the center of the plant before it has achieved ideal market weight. As a result, the lettuce's quality has deteriorated to the extent such that it is no longer suitable for marketing. This mostly occurs in romaine lettuce, where the heart should not be longer than one-third of the plant's overall length. It can be caused by a variety of reasons, including day lengths that surpass 13 h of sunlight, as in Quebec; the occurrence of high temperatures at specific periods of growth; and certain pressures such as a shortage of water.

Storage

Harvesting the crop on time is the most important step in maintaining the quality of the commodity as well as somewhat increasing the shelf life of the commodity. Harvesting time depends upon the market demand, and we have to harvest all the produce depending on that only. But the entire commodity is not going to be consumed by the consumers at the same time so, we have to save them for the future, and storing the commodity at appropriate environmental conditions can enhance the storage life of perishable commodities. Asteraceae crops are mostly leafy and have high perishable nature. Most of them can't be stored for a larger period. Like as lettuce, which is a leafy vegetable, can be stored up to 2–3 weeks at 1–2 °C along with 90–95% relative humidity, whereas, for globe artichoke, 0 °C in cold storage is considered as the best storage temperature (Haggag et al. 2017), but a range from 0 °C to 5 °C at 95% relative humidity is generally followed while storing globe artichoke. Endive can last for 15–50 days at 0 °C relative humidity of 95–98%. Endive's shelf life has also proved to be enhanced when the oxygen is maintained at 3–4% and carbon dioxide at 4–5%. Chicory is having a comparatively longer shelf life in proper storage conditions, i.e., 5–7 weeks at 0 °C. The longest shelf life is of Jerusalem artichoke and dandelion. The former can be stored for 4 months or even more when stored in peat or sand at 2 °C with a relative humidity of around 90–95% without much quality deterioration (Danilčenko et al. 2008) land former can also be stored for a long time at −17 °C.

In refrigerator, the salsify could be kept least for 3–4 weeks. The tops should be removed before keeping the salsify in refrigerator, and the roots can be kept in a cold and moist place for 2–4 months.

Fresh French tarragon is excellent. The wrapping of the leaves should be done in a paper towel, and place them in a plastic bag. This method can keep them fresh for 2 or 3 weeks.

Postharvest Management

This is an imperative step in enhancing the shelf life and quality of the harvested commodity. It involves a set of methods through which every crop should have to go to ensure quality maintenance. The leafy crops should be moved into an airy, shady as well as cool area directly after harvesting which should be free from strong winds, i.e., precooling, followed by sorting and grading. It is a basic rule that the parts of the commodity that are loose, discolored, damaged have soil on their surface, and diseased are removed.

Drying Hang tarragon stems upside down, after tying in small bunches, in a warm and dry spot away from the sun. Microwave could also be used for drying the tarragon by keeping it inside it for 1–3 min; check frequently, and remove leaves when they appear dry touch.

Freezing In a zippered plastic bag, French tarragon can also be frozen. Freeze leaf sprigs in a plastic bag, or make a paste with the salad oil, and freeze it in tiny containers.

Storing The storing of tarragon should be done in an airtight container.

Conclusion

The Asteraceae (Compositae) family includes nearly 13,000 species and comprises some highly nutritious vegetables, viz., dandelion, Spanish salsify, black salsify, tarragon, endive, lettuce, chicory, globe artichoke, Jerusalem artichoke; etc. The planting habit varies from herbaceous bushy to running vine. Generally, the members of the Asteraceae family possess taproots, but the fibrous roots have been also found in some members as well. Their stems are mostly aerial, branched, herbaceous, and cylindrical with glandular hairs present on them. The leaves can be alternate, opposite, or whorled. Asteraceae inflorescence contains flowers in dense flower heads called capitulum. The edible part of the vegetables belonging to this group includes flowers, fruits, leaves, roots, rhizomes, etc. The common pests are aphids, leafhoppers, spider mites, thrips, whiteflies, weevils, and scales, and diseases are stem rot, powdery mildew, downy mildew, etc.

References

Anonymous (2019). https://www.naturallifeenergy.com/dandelion-greens-for-your-salad dandelion-benefit/

Anonymous (2021). https://en.wikipedia.org/wiki/Asteraceae

Anonymous (2022). https://www.nutritionvalue.org/

Danilčenko, H., Jariene, E., & Aleknaviciene, P. (2008). Quality of Jerusalem artichoke (*Helianthus tuberosus* L.) tubers in relation to storage conditions. *Notulae Botanicae Horti Agrobotanici Cluj-Napoca, 36*(2), 23–27.

Giacomino, A., Malandrino, M., Colombo, M. L., Miaglia, S., Maimone, P., Blancato, S., Conca, E. & Abollino, O. (2016). Metal content in dandelion (*Taraxacum officinale*) leaves: influence of vehicular traffic and safety upon consumption as food. *Journal of Chemistry, 2016*.

Haggag, I. A. A., Shanan, S. A., Abo-Elhamd, A. S. A., Helaly, A. A., & El Bassiouny, R. E. I. (2017). Effect of Temperature and Modified Atmosphere Packaging on Globe Artichoke (*Cynara scolymus*, L.) Quality during Storage. *Adv Plants Agric Res, 6*(5), 00227.

Holm, L., Doll, J., Holm, E., Pancho, J. V., & Herberger, J. P. (1997). *World weeds: natural histories and distribution*. John Wiley & Sons.

Jackson, R. D. (1982). Canopy temperature and crop water stress. In *Advances in irrigation* (Vol. 1, pp. 43–85). Elsevier.

Longyear, B. O. (1918). *The dandelion in Colorado* (Vol. 236). Agricultural Experiment Station of the Agricultural College of Colorado.

Lucchin, M., Varotto, S., Barcaccia, G., & Parrini, P. (2008). Chicory and endive. In *Vegetables I* (pp. 3–48). Springer, New York, NY.

Chapter 13
Production Technology of Underutilized Vegetables of Cyperaceae Family

Radhika

Introduction

Chinese water chestnut is one of the most important crops belonging to the family Cyperaceae. It is introduced from southern China and is grown throughout the eastern part of China. Chinese water chestnut (*Eleocharis dulcis* (Burm. f.) Trin.) is an aquatic annual herbaceous plant. Therefore, it is cultivated under flooded conditions. In the warm season, this crop has long been cultivated for its corms with dormant sprouts or also called tubers. The size and shape of the corms are just like the gladiolus and have a crispy apple-like texture that is sweeter, and the outer skin of the corm is burgundy to blackish with white flesh. The corm is commonly used as a vegetable or also as a fruit. In India, dried chestnut converted into the flour called "singhare ka atta" which is used for making different kinds of products during fasts (Chandana et al. 2013). Apart from that, it is grown as a fodder crop which is used to feed the animals. Chinese water chestnut has a special taste and texture, and due to its unique flavor, it is very popular all over the nation (Song et al. 2019). Apart from that, it is also very popular for its medicinal properties which are used to cure cough, hypertension, type 1 or type 2 diabetes, or cancer (Klockeman et al. 1991; Liu et al. 2017). It is known by different names in different countries such as Aplid, Buslig (the Philippines); Cabezas de negrito (Spain); Chtaigne d'eau (France); Chikai, Dekang (Indonesia); Kalangub (the Philippines); Kohekohe (Hawaii); Mati (China); Nilaga (the Philippines); O-kuroguwai (Japan); O-yu, Peci (China), Pipi-wai (Hawaii); Pi't'si (China); Potok (the Philippines); Po-tsai (China); Sibosibolasan (the Philippines); Tek, Teki-tik (Indonesia); and Wu-yu (China). Due to the reason that it is grown underwater, the skin color of chestnut is brown with chestnutty flavor and texture with white flesh which gives rise to the English name, i.e., water chestnut.

Radhika (✉)
School of Agriculture, Lovely Professional University, Phagwara, Punjab, India

Savita et al. (eds.), *Production Technology of Underutilized Vegetable Crops*,
https://doi.org/10.1007/978-3-031-15385-3_13

Origin and Distribution

Chinese water chestnut is a common species that are cultivated in freshwater which is native to East Africa, southeastern Asia, the Philippines, Malaysia, Indonesia, Melanesia, and Fiji. It was originated or first cultivated in Indo-China or southeastern China. Apart from that, it is also commercially cultivated in Japan, Hong Kong, the Philippines, Hawaii, other Pacific islands, India, and the southern USA. It is the edible corm which is a common ingredient in food in China and has been commonly cultivated mainly in Anhui, Fujian, Guangdong, Guangxi, Hubei, Hunan, Jiangsu, Jiangxi, and Zhejiang provinces.

Nutritional Importance and Uses

The Chinese water chestnut is distributed all over the world for its nutritive values containing 74% water, carbohydrates (23.93 g), and 1% protein and contains low fat (0.1 g). It is also rich in vitamins such as thiamine (0.14 mg), riboflavin (0.2 mg), niacin (1.00 mg), vitamin C (4.00 mg), vitamin E (1.02 mg), and vitamin B6 (0.328 mg) and rich in minerals such as calcium (11.00 mg), iron (0.06 mg), magnesium (22.00 mg), phosphorous (63.00 mg), potassium (584.00 mg), and zinc (0.5 mg). The corm also contains 60% starch content. Six compounds were isolated during the extraction of Chinese water chestnut such as hexacosanoic acid, 5-alpha stigmastane-3, 6-dione, beta-sitosterol, stigmasterol, botulin, and tricin (You et al. 2007). Moreover, corms are rich in phenolic compounds, flavonoids, sterols, and polysaccharides which are good for health.

Polysaccharides play an important role in improving the immune system (Zheng et al. 2015). Mainly the corm consists of three tissues such as the epidermis, subepidermis, and parenchyma. The sugar content is high in the parenchyma cell walls (931.00 mg), and in epidermal cells, the phenolic compounds are high (21.07 mg). Moreover, corms contain some antibiotics, i.e., puchiin, that helps to suppress the growth of harmful bacteria which cause fungal infections. Apart from that, it is also well known for its medicinal values because it contains phenolics, tannins, and flavonoid bioactive compounds which help to inhibit cancerous cells (Baehaki et al. 2021). Moreover, it is used to treat hypertension, chronic nephritis, constipation, pharyngitis, laryngitis, and enteritis diseases. In China, it is used to cure diabetes, jaundice, urinary infection, and sore throat (Yang et al. 2020; Li et al. 2015). For hypertension, tea is prepared with chestnuts. It is also used to treat stomach pain, hernia, and a liver infection (You et al. 2007; Liu et al. 2017).

Mainly, Chinese water chestnuts are consumed as a vegetable, eaten as raw fruit, pickled, or used for canning purposes. The Chinese water chestnut is a very special and important crop because of its unique taste and flavor. During processing, different by-products are prepared and sold all over the world (Nie et al. 2021). Moreover, the Chinese water chestnut is starch-based food and is, therefore, well known for its gel-like as well as water-binder properties. Due to this property, starch flour is used as a thickening agent during food preparation such as ketchup, sauce, or soups. Apart from that, it is used in the pharmaceutical or paper textile industries (Syed et al. 2021). Different cultivars are used to prepare cookies, flour, or starches. Flour is used to prepare the cake, cookies, or pastries. In China, corms are used to make sweetmeat or different desserts and eaten with coconut milk or palm sugar. Apart from that, the dry stem of the plants as well as the peel of the corms are used as cattle or poultry feed, mulching, and also used for making baskets or mats for packing materials for horticultural produce (Zeng et al. 2020). Moreover, peels can be used as bioabsorbent which helps to remove the dyestuff in wastewater (Khan et al. 2013) and for fermentation to produce single-cell protein feed (Pen and Jiang 2003).

Botany and Taxonomy

Chinese water chestnut is perennial belonging to the family Cyperaceae and genus *Eleocharis*. The Chinese water chestnut is a developing, upright, aquatic grass-like plant cultivated under the water. Above the ground surface, the plants produce green cylindrical stems with fewer leaves or sometimes lack of leaves, whose photosynthetic systems have been transferred to the stems. Apart from that, underwater two types of modified stems are produced, i.e., corms or rhizomes. The rhizomes are elongated, depressed, and brown to black, and the diameters of the corm are up to 2–3 cm with 2–5 distinct rings and have white juicy flesh. The inflorescence is a single terminal spikelet. Flowers are bisexual, with three stamens, three stigmas, or three styles. The fruit is a nutlet, round in shape, and the size of nut is approximately 1.5–2.3 mm with shiny texture. The plant height is normally up to 2–6 feet under the aquatic system. Before vegetative growth, the plant produces insignificant flowers at the top of the stems. Male and female flowers are present on the same plant but not simultaneously. Firstly, female flowers appeared when the stem grows up to 15 cm above the water level. After that, when plants grow up to 24 inches, then male flowers (staminate) are produced in the same inflorescence. Tiny seeds are produced, but seeds are not having economic importance in cultivation because Chinese water chestnut is propagated vegetatively through rhizomes or corms. The plants of chestnut grow faster and produce two types of rhizomes, and they can be easily differentiated, i.e., rhizomes appear from the base of the plants and are produced earlier during the growth cycle, i.e., within 6–8 weeks after planting. Such types of rhizomes are developed horizontally under the surface of the soil from the parent plant.

Corms and corm producing rhizomes

Above-ground stems

In contrast, another type of rhizome is produced later in the growing season (in August to early September), i.e., just after the vegetative growth or the beginning of flowering. This type of rhizome is produced by the mother plants, and the daughter plants also produce tubers that develop at their tips. When corms are elongated, their size also increases. Young corms are white, but at maturity, the outer skin color changes to brown or black with white flesh. In the late spring or summer season, it requires long-day conditions. An increase in the aboveground foliage leads to an increase in the size of the rhizome up to 0.5 square meters per plant. While under short-day conditions, tubers are developed at the tip of the rhizomes. The stem reaches 1.5 m height and dies after the first frost, i.e., within 6–7 months after planting and after that, corms are ready to be harvested.

Climate and Soil

It requires long warm days for cultivation, with at least a 7-month frost-free period, and requires 30–35 °C temperature at the vegetative stage and 5 °C for tuber formation. Soil temperature should be 14–16 °C which is essential for germination of the rhizomes. The plants are cultivated under the water; therefore, areas with well-controlled irrigation facilities should be selected to overcome the irrigation problems throughout the growing season. If the temperature falls below 13.6 °C, the growth will also be affected (Kleinhenz et al. 2000). Planting should be done in leveled and fertile soil. Mostly muck soil, clay soil, or peat soil should be used because it is rich in organic matter. The pH should be between 6.9 and 7.3 because Chinese water chestnut does not tolerate high acidic soils. To overcome the acidic conditions, always add limestone to the soil.

Agronomic Practices

Seed Rate

A total of 500 kg of corms are used for planting a 1-hectare area. In China, seeds should be coated with mud which helps to prevent the seeds from dehydration or rotting and store overnight in pits. Apart from that farmers, store these in the refrigerator at 10–13 °C. The seeds should be planted before drying out or showing the first sign of sprouting. But propagation through seeds is not followed by the farmers because it takes more time to germinate, poor germination rate, and requires approximately 2 years for the development of corms.

Planting Method and Planting Time

Small corms are used for cultivation. Before planting, the land should be leveled properly. The corms may be planted directly into the field in rows and placed 10–12.5 cm deep in the surface. In a temperate climate, a nursery bed is prepared and when the plant reaches 20–30 cm in height then transplanted to the main field. During transplanting, spacing of 75 cm × 75 cm is maintained, but in China, triangular spacing is more common, i.e., 45–60 cm in between the plant (Midmore and Gerstelling 2004). The planting should be done manually, but in larger plantings, the furrows are opened with a plough or courter. After planting, the field is flooded for 24 h and then allowed to drain naturally, similar to paddy cultivation. When the plant height reaches 20–30 cm, the field should be flooded again with water, and it must be important to maintain the water level up to 10–12.5 cm throughout the growing season because lack of water gives poor germination or yield production.

The planting method is different in different countries, i.e., in the Philippines, planting should be done during the warm season at any time; at savannah or in Georgia, planting should be done in May, whereas in China, planting is done from March to late June. Chinese water chestnut does not require any weeding if the ground is clear and treated with herbicides before planting.

Nutrient Management

Chinese water chestnut being a heavy feeder of nutrients requires more fertilizers for better growth and yield. Approximately 1 ton per acre of organic manure should be applied and one-third should be applied during the soil preparation. Well rotten manure should be applied together with a half dose of lime which overcomes soil acidity and is always applied 2–3 weeks before planting. Organic manures (such as cow manure, poultry manure, peanut cake, etc.) should be mixed into the soil. Fourteen days after planting, night soil is used, when the plants start producing rhizomes with secondary plants. At that time, 50–60 kg peanut cake is applied to the main field and covered with the help of plant ash.

Another application of fertilizers is done 7–8 weeks after planting, and one-fourth dose of this total should be applied to the crop. When the crop produces new corms, a half dose of manure with lime is applied again. Apart from that 2.5 million tons per hectare, potassium and phosphorous should be applied at least 2 months after planting, and half-dose should be broadcasted during corm development.

Harvesting

Harvesting means picking the produce at the proper maturity stage or at the proper time with minimum damage or at minimum cost. About 20–30 days after the first frost killing or when the fruits attain desirable size, color, or texture, harvesting is done. Harvesting Chinese water chestnut is not easy. Harvesting is done manually or by using a mechanical harvester. Before harvesting, water should be drained out properly. The corms are harvested with the help of hand by using different tools such as spade because by using this tool soil is easily turned out and the corms are picked easily by using hands (Morton et al. 1988). Care should be taken during harvesting and protect the corms from digging tools. Another method of harvesting is to use the wire mesh screen. It is useful in harvesting small plots. Rubber paddles are also used which turn the soil carefully onto wire mesh and pick up the corms easily and drop them into the container or bucket which is filled with water. This method prevents bruising and injuries as well as removes dirt.

For large cultivation, harvesting should be done by using a small plough that helps to turn the furrow to a depth of approximately 5–6 inches. After that, corms should be picked and placed into the container and washed properly to remove the excess dirt or pathogens.

Wire mesh tray

Yield and Marketability

In southern China, the yield obtained is approximately 92,000 kg per hectare. Under good cultivation conditions, a single plant gives a significant yield of about 2.3 kg per season. For the marketable purpose, the corm size of 30–45 mm diameter is acceptable. Nearly about 60% of the harvested produce is marketable. Small corms are used for propagation, whereas damaged or bruised type corms are used for canning purposes.

Plant Protection

Insect/Pest

The main pest of Chinese water chestnut is grasshopper and snout moth larvae, and they can be controlled by insecticides. This pest causes severe damage and creates holes in corms. At present, billbug (*Calendra cariosa*) is the most serious pest. This pest feeds on the stem and also causes damage at the seedling stage. Apart from that, rodents and sometimes rabbits also attack chestnuts, especially rats or muskrats. Rodents and rabbits cause serious damage to corm yield. These animals feed on the stem, corms as well as on roots which cause damage by trampling, and due to this yield declines (Kleinhenz et al. 2000). The use of traps, poison baits, or predators is the best option to control the rodents. Another most serious pest is the aquatic Caddis fly (*Trichoptera* sp.). Their larvae feed on stems as well as on corms which cause severe damage. Various researchers found that there are currently no insecticides registered for use on Chinese water chestnut so control measure is limited, but with the help of special practices, we can kill the caddis fly larvae to maintain the sanitation and partly or completely dry out the water once in a week and use predators which feed on harmful larvae, but sometimes plant growth will be affected due to lack of water. Moreover, a weevil (*Lissorhoptrus oryzophilus*) damaged the crop, and it can be controlled by using Lorsban 500 EC.

Diseases

The primary disease is rust which is caused by *Uromyces* sp. Rust appears on the plant 5 months after planting. It causes brown spots on the stems, and after some time, if it is not treated properly, then brown spots change to yellowish-orange spots. It is controlled with the early application of sulfur dust. Apart from that stem, fungus also causes a serious problem in chestnuts. It is caused by *Cylindrosporium* sp., which attacks acidic soil, whereas *Fusarium* wilt caused by *Fusarium oxysporum* (Gongchen 1988) is also the most destructive disease in Chinese water chestnut. However, this disease is widely spread all over the area where Chinese water chestnut is grown. The main symptoms of this disease are yellow spots on the stem, chlorosis followed by necrosis, or stunted plants. If the disease is not cured at the time, then yellow spots are changed into brown to blackish, roots get infected, and at the end, plants get dried up and die (Zhu et al. 2018). In serious cases, pinkish liquid fluid is coming out from the base of the stem. This disease reduces the yield as well as the quality of the corms. The fungal disease can be managed by selecting the planting sites with good drainage facilities and by using different irrigation techniques (Zhu et al. 2014). Apart from that, Chinese chestnut yellow crinkle disease causes serious problems, and it is first reported in China. The causal organism is *Candidatus phytoplasma.* The main symptoms are wrinkled leaves, yellowing of leaves, short internode, and also reduced quality of fruit.

Postharvest Handling and Storage

Postharvest handling means maintaining the produce after the harvest or before the storage because it will help to reduce the losses after harvesting. Therefore, after harvest, corms should be graded or sorted properly to separate the broken, damaged, ungerminated corms. After grading, corms are washed in clean water to remove the dirt particles or soil-borne pathogens. The size of corms is very important. The chestnut diameter must be between 15 and 30 mm. Chestnut is collected from three different tanks from where they can be easily separated, and a cup elevator tank is used to transfer the corms from one tank to another.

Sorting table for chestnut

Perforated sizer for chestnut

Cup elevator belt

In some countries, water curing treatment is used which means that corms are kept in water for 8–9 days. The main aim of this method is to reduce fungal infection during storage. During the curing process, alcoholic fermentation takes place which helps to reduce the pH and allows diffusion of phenolic compounds from the epidermal cells into the flesh.

Water tank for curing chestnut

Storage

Storage is a major problem because chestnut is very susceptible to fungal infection. Chinese chestnut is perishable and to reduce deterioration reduce their temperature before storage. Different storage methods are now used to store the Chinese water chestnut which increases the shelf life of the product such as sand storage, ventilated wet air storage, cold storage, controlled atmosphere storage, etc. (Anese et al. 1997).

1. Sand storage: It is a traditional method in China. In this method, chestnut is placed inside the wet sand which helps to maintain the moisture or temperature around the chestnut. Before storage, corms are treated with $KMnO_4$ for 3 min. Chestnut is placed inside the wet sand or mud and maintains the ratio of sand/chestnut to 7:3. After 2 months, turn the sand and remove the damaged or infected corms. This procedure is done every 2 weeks.
2. Cold storage: Cold storage is very effective which helps to slow down the metabolic activity in chestnuts and increase the shelf life of the produce. Before storage, water cure treatment should be done and then packed in perforated plastic boxes or wooden boxes and then placed into the pallets, and the distance of 20 cm is maintained between the two pallets. The recommended storage temperature is 0 °C and 90% relative humidity, and corms stored in this manner survive successfully for 6 months. Storage temperature plays a very important role because if the temperature goes high, the sprouting of chestnut may occur. If sprouting is detected, then the crop should be dug into the soil or stored under a refrigerator to retard sprouting.
3. Controlled atmospheric storage: In CA storage, modify or control the atmospheric oxygen (3%) and carbon dioxide (15%) concentration. The main objective of this storage is to reduce or delay the sprouting in chestnuts. The recommended storage temperature for chestnut is 0 °C with 95% relative humidity, and corms are stored for 5 months (Kays and Sanchez 1984). Chestnut containers are placed inside the cold room, but before storage, wash the corms with water, and sometimes disinfectant should be added in water which removes the excess pathogens. After that, turn the oxygen burner which reduces the oxygen level to 2–3% and increases the carbon dioxide and humidity. It is very effective in inhibiting the sprouting and browning of fresh-cut chestnuts and also increases the shelf life of the chestnut (Gorny et al. 2002; Pan et al. 2015).

Processing

Canning Before canning, corms must be washed and peeled properly. Peeling is very difficult in Chinese water chestnuts. Corm peeling is usually done with the help of hands. After peeling, corms are cut into small pieces, and the processing method is similar to potatoes (LIUxin and Ai-mei 2006). The corms can be preserved in glass jars, but before storing the food, the can or corms must be processed. Firstly, corms are treated at 50 °C for 15–20 min and must be sterilized (heat treatment) at 115.6 °C in an autoclave or pressure cooker. Due to heat treatment, the food is sterilized by destroying the harmful bacteria and also increasing the shelf life of the products. Sometimes due to heat treatment, the flavor is also lost, but the texture remains the same. After processing, cooling is done at a temperature of 30–60 °C because if the cans are not cooled before filling the products, it may cause thermophilic spoilage (Morton et al. 1988). Different researchers found that peeled corms steam blanched for 4 min at 99–100 °C maintain the flavor and color as compared to an unblanched Chinese water chestnut.

Conclusion

Corms of Chinese water chestnut can be classified into two groups, i.e., one type contains long terminal buds with a flat bottom and is also rich in starch content. Whereas the other one has a short terminal bud with a curved bottom and low starch content, it is sweeter. Corms have uniform cell sizes with intercellular space. The cell wall of corms consists of ferulic acid, which is interlinked with the polysaccharides within the cell walls or between the cells, which maintain the texture and quality of Chinese water chestnuts. Chinese water chestnut is widely cultivated worldwide for its higher nutritional values like proteins, carbohydrates, fibers, and minerals. Its production is mainly low and produced in limited areas due to lack of awareness in farmers, lack of modern technology, or agronomic operations. Therefore, different researchers explored the recent cultivation methods, agronomic management, storage techniques, postharvest handling, or processing of Chinese water chestnut. Being a high-value cash crop, it can be sold as a fresh vegetable in the market, and farmers can gain more profit.

References

Anese, M., Manzano, M., & Nicoli, M. C. (1997). Quality of minimally processed apple slices using different modified atmosphere conditions. *Journal of Food Quality*, *20*(5), 359–370.

Baehaki, A., Widiastuti, I., Lestari, S., Masruro, M., & Putra, H. A. (2021). Antidiabetic and anticancer activity of Chinese water chestnut (*Eleocharis dulcis*) extract with multistage extraction. *Journal of Advanced Pharmaceutical Technology & Research*, *12*(1), 40.

Chandana, M., Rupa, M., & Chakraborthy, G. S. (2013). A review on potential of plants under Trapa species. *International Journal of Research in Pharmacy and Chemistry*, *3*(2), 502–508.

Gongchen, J. D. C. H. W. (1988). CHINESE WATER CHESTNUT WILT_A NEW DISEASE IN ZHEJIANG [J]. *Acta Phytopathologica Sinica*, *4*.

Gorny, J. R., Hess-Pierce, B., Cifuentes, R. A., & Kader, A. A. (2002). Quality changes in fresh-cut pear slices as affected by controlled atmospheres and chemical preservatives. *Postharvest biology and Technology*, *24*(3), 271–278.

Kays, S. J., & Sanchez, M. G. C. (1984). Storage of Chinese water chestnut [*Eleocharis dulcis* (Burm. F.) Trin.] corms. *Postharvest Handling of Vegetables 157*, 149–159.

Khan, T. A., Nazir, M., & Khan, E. A. (2013). Adsorptive removal of rhodamine B from textile wastewater using water chestnut (*Trapa natans* L.) peel: adsorption dynamics and kinetic studies. *Toxicological & Environmental Chemistry*, *95*(6), 919–931.

Kleinhenz, V., Midmore, D. J., & Lodge, G. (2000). A grower's guide to cultivating Chinese waterchestnut in Australia.

Klockeman, D. M., Pressey, R., & Jen, J. J. (1991). Characterization of cell wall polysaccharides of jicama (*Pachyrrhizus erosus*) and Chinese water chestnut (*Eleocharis dulicis*). *Journal of Food Biochemistry, 15(5), 317–329.*

Li, C., Huang, Q., Fu, X., Yue, X. J., Liu, R. H., & You, L. J. (2015). Characterization, antioxidant and immunomodulatory activities of polysaccharides from *Prunella vulgaris* Linn. *International Journal of Biological Macromolecules, 75*, 298–305.

Liu, X., Xie, J., Jia, S., Huang, L., Wang, Z., Li, C., & Xie, M. (2017). Immunomodulatory effects of an acetylated *Cyclocarya paliurus* polysaccharide on murine macrophages RAW264. 7. *International journal of biological macromolecules, 98*, 576–581.

LIUxin, Z. H. A. O., & Ai-mei, Z. H. O. U. (2006). Preliminary study on functional component and functional activities of waste slurry derived in processing water chestnut starch. *Food Science*, 02.

Midmore, D., & Gerstelling, P. (2004). Chinese waterchestnut. *The New Rural Industries Handbook: A Handbook for Farmers and Investors*, 181–187.

Morton, J. F., Sanchez, C. A., & Snyder, G. H. (1988). Chinese waterchestnuts in Florida: past, present, and future. In *Proceedings of the... annual meeting of the Florida State Horticulture Society (USA)*.

Nie, H., Chen, H., Li, G., Su, K., Song, M., Duan, Z., Li, X., Cao, X., Huang, J., Huang, S. & Luo, Y. (2021). Comparison of flavonoids and phenylpropanoids compounds in Chinese water chestnut processed with different methods. *Food Chemistry, 335*, 127662.

Pan, Y. G., Li, Y. X., & Yuan, M. Q. (2015). Isolation, purification and identification of etiolation substrate from fresh-cut Chinese water-chestnut (*Eleocharis tuberosa*). *Food chemistry, 186*, 119–122.

Pen, L. T., & Jiang, Y. M. (2003). Effects of chitosan coating on shelf life and quality of fresh-cut Chinese water chestnut. *LWT-Food Science and Technology, 36*(3), 359–364.

Song, M., Wu, S., Shuai, L., Duan, Z., Chen, Z., Shang, F., & Fang, F. (2019). Effects of exogenous ascorbic acid and ferulic acid on the yellowing of fresh-cut Chinese water chestnut. *Postharvest Biology and Technology, 148*, 15–21.

Syed, F. N., Zakaria, M. H., Bujang, J. S., & Christianus, A. (2021). Characterization, Functional Properties, and Resistant Starch of Freshwater Macrophytes. *International journal of food science, 2021.*

Yang, J., Tang, X., Shuai, L., Kwon, Y. S., & Kim, M. J. (2020). Chemical characterization, antioxidant properties and anti-inflammatory activity of Chinese water chestnut extracts. *SCIENCEASIA, 46*(2), 151–156.

You, Y., Duan, X., Wei, X., Su, X., Zhao, M., Sun, J., Ruenroengklin, N. & Jiang, Y. (2007). Identificationof major phenolic compounds of Chinese water chestnut and their antioxidant activity. *Molecules, 12*(4), 842–852.

Zeng, F., Chen, W., He, P., Zhan, Q., Wang, Q., Wu, H., & Zhang, M. (2020). Structural characterization of polysaccharides with potential antioxidant and immunomodulatory activities from Chinese water chestnut peels. *Carbohydrate Polymers, 246*, 116551.

Zheng, Y., Wang, W. D., & Li, Y. (2015). Antitumor and immunomodulatory activity of polysaccharide isolated from Trametes orientalis. *Carbohydrate polymers, 131*, 248–254.

Zhu, Z., Hu, X., Yu, C., Deng, W., & Huang, J. (2018). Occurrence and control of Fusarium wilt of Chinese water chestnut. *Journal of Plant Protection, 45*(2), 307–314.

Zhu, Z., Zheng, L., Pan, L., Hsiang, T., & Huang, J. (2014). Identification and characterization of Fusarium species associated with wilt of Eleocharis dulcis (Chinese water chestnut) in China. *Plant Disease, 98*(7), 977–987.

Chapter 14
Production Technology of Underutilized Vegetables of Cannaceae Family

Dipika Mal

Introduction

The starchy root crop edible canna (*Canna edulis*), often known as Queensland arrowroot, is cultivated seldom in the tropical highlands. The edible canna roots are mainly used for separating starch. Large granules and a high amylose concentration characterize edible canna starches. There are 25 species of Canna, which are found all across the tropics and subtropics. It is a monocotyledon plant and is perennial and herbaceous. The stems are generally purple, and the plant height ranges from 0.9 to 1.8 m. It has large, broad, pointed, whole leaves with a distinct, strong midrib. The fleshy rhizome is roughly 60 cm long, branched, and has thick fibrous roots. *C. edulis* dried rhizome contains 70–80% starches, which are said to be more digestible than other types of carbohydrates. The physiochemical characteristics and modification of *C. edulis* starch have been studied (Chuenkamol et al. 2007; Puncha-Arnon et al. 2007).

D. Mal (✉)
School of Agriculture, Lovely Professional University, Phagwara, Punjab, India
e-mail: dipika.21885@lpu.co.in

Savita et al. (eds.), *Production Technology of Underutilized Vegetable Crops*,
https://doi.org/10.1007/978-3-031-15385-3_14

Origin and Distribution

Canna edulis is said to have originated in South America's Andes region or along the Peruvian coast and spread from Venezuela to northern Chile (Andrade-Mahecha et al. 2012). It is used as a fodder crop in Japan. It is native to South America, Central America, the West Indies, and Mexico. It has also become naturalized in the southeastern United States and much of Europe, sub-Saharan Africa, Southeast Asia, and Oceania. For thousands of years, indigenous people of America have grown it as a minor food crop.

Botany and Taxonomy

Flowers are red, solitary, or in pairs, with 4–6 cm diameter and a bract of around 1.3 cm. The sepals are 1–1.5 cm long, and the color ranges from pale green to crimson or purple. The crimson or reddish colored 2.5–3 cm long corolla tube is around 1 cm. The staminodes have a vibrant crimson color. Hermaphrodite flowers are those that have both male and female reproductive organs. Bright green capsules with green to purple tubercles cover the fruits, green oblong or aid, gently echinate (spiny), and 2–2.5 cm long. Capsules are 4025 mm in diameter, with persistent outer tepals (sepals) near the tip. Leaves are lanceolate or ovate, measuring 10–30 cm long and 10–20 cm broad with huge laminae up to 60 cm long. A waxy glucose erect peduncle is approximately 30 cm long with two-flowered cincinnati inflorescence, with dark green leaves with white veins and edges. Carline, basic, alternating, and spiral are the four types. The petiole of the oblong leaves extends downward to form a sheathing base around the stem. The lamina is pinnately veined in a parallel pattern. The leaf edges are wavy and smooth, with a sharp apex. The leaves are enormous and foliaceous, reaching a length of 65–70 cm and a width of 30–35 cm. On the exterior, young rhizomes are yellowish-white or pinkish, with a yellowish-white color inside. They turn brownish on the outside as they mature due to a thick outer coating. It has sympodial axes, tuberous roots, and numerous roots that develop adaxially from the nodes. The stem is a pseudostem that may grow up to 1.5–2 m tall. It is glabrous, upright, herbaceous, robust, and cylindrical, with light green leaf bases enclosed by the sheathing. With a diameter of 2–5 mm and many root hairs, the roots are robust, tubular, and creamy white. There are also thinner primary and secondary lateral roots (Maas-Van de Kamer and Maas 2008; Thamburaj and Singh 2016).

Nutritional Composition and Uses

The nutritional composition of *C. edulis* rhizome contains 50.66% moisture, 4.17% carbohydrate, 4.81% protein, 2.85% ash, 4.35% lipid, and 33.16% fiber (Okonwu and Ariaga 2016).

Uses

S. no.	Plant parts used	Traditional uses	Nonpharmacological uses
1.	Rhizome	A decoction of the fresh rhizome is used to treat fevers, dropsy, dyspepsia, diuretic, antipyretic, gonorrhea, and irregular menses in women. Rhizomes that have been macerated are used to treat nosebleeds. Rhizome has been used to treat cancer alongside other medicinal herbs	It is utilized in the bread sector to make low-fat, high-fiber nutritional goods Rhizomes are used to propagate the plant The starch-rich cooked rhizomes are used to make cakes Pullaiah (2006)
2.	Roots	A root decoction that is used to cure gonorrhea and amenorrhea. The powdered root is used to treat diarrhea and dysentery, as well as being diaphoretic, diuretic, stimulant, and demulcent. It is also used to treat fevers and dropsy	The roots have a starchy texture Acid treatment and garlic treatment remove Cu^{2+}, Zn^{2+}, Ni^{2+}, Pb^{2+}, and Co^{2+} from aqueous solution Mahamadi and Chapeyama (2011)
3.	Leaves	Leaves are used to treat malaria, and a diuretic infusion is made from the leaves. Insecticides are reported to be present in the smoke produced by burning leaves. Baths made from freshly crushed leaves are used to treat rheumatic aches and arthritis, as well as ulcers	Leaves are used to wrap packages, and the fiber extracted from them is used to make paper Sucrose inversion and sugar transformation synthesis Putman and Hassid (1954)
4.	Flowers	The flowers were thought to be able to heal eye problems. A floral concoction is used to stop external wound bleeding	It works well as a mild steel corrosion inhibitor in an acidic environment Mathina and Rajalakshmi (2016)

Climate and Soil

Canna can be grown anywhere from sea level to 2700 m above sea level, but it thrives in temperate, tropical, or subtropical mountain climates between 1000 and 2000 m above sea level, with an average temperature of 14–27 °C and annual rainfall ranging from 500 to 1200 mm (Idárraga-Piedrahita et al. 2011). The substrate should be nutrient-dense, humectant-rich, and light. A deep, rich, and well-drained soil in a sunny location with a pH of 5.5–7.5 is the ideal substrate. After the last frost, plant the rhizomes at a depth of 10 cm. It may also be grown in subtropical and warm temperate climates. It thrives in environments with yearly daily temperatures ranging from 12 to 32 °C. The plant prefers a mean annual rainfall between 1000 and 4500 mm, but it can tolerate 500–5000 mm/year.

Agronomic Practices

Sowing

Rhizomes used for sowing are usually between 3000 and 4500 kg per hectare, with a planting density of not more than 22,500 plants per hectare. The sprouts should point upwards when the rhizomes are planted. Plants in a row should be spaced 60–70 cm apart, and rows should be spaced 70–80 cm apart. Because Canna grows fairly tall, it should be grown in areas with low wind speeds to avoid bending over. Germination occurs when the soil temperature rises above 16 °C, with a temperature range of 20–25 °C being ideal. Seedlings emerge 20–30 days after planting (Zang et al. 2010).

Harvesting

The roots can be harvested within 6 months from planting, though a higher yield is obtained after 8-10 months. Harvesting is done by hand, using a shovel or other digging instrument to dig out the crop, brushing off the soil, and then cutting the stems to separate the rhizomes (Caicedo Diaz et al. 2003).

Disease and Pest

Rust

The disease causes numerous tiny yellow, irregularly shaped powdery rust pustules on leaves, petioles, and flowers. These pustules are mostly seen on the bottom surface of leaves, with 1–2 mm expansive chlorotic lesions on the top surface. Spots on the upper leaf surface combine and turn dark brown as the illness progresses. Heavily diseased leaves eventually desiccate, collapse, and fall.

Management

Canna plants should be grown in well-drained, rich soil in well-ventilated, full sun locations. Apply a layer of compost or mulch around several inches deep plants. Canna should not be planted in shaded, poorly drained areas with high relative humidity. To promote quick plant development, use a modest amount of fertilizer (e.g., 5-10-5) once a month.

- It's best not to moisten the plant leaves. Sprinkler irrigation should not be used.
- Infected foliage should be removed and destroyed. Symptomatic leaves should not be composted (Nelson 2013).

Canna leaf roller moth (*Calpodes ethlius*) has been observed on Canna plants (Cook 2001). It is the plant's most significant pest and is mostly found in the South. By depositing its eggs in the buds of growing stalks, this insect causes harm.

Caterpillars utilize sticky webs to restrict the leaves from unfurling, protecting the eggs from predators and pesticides. The pupae then feed on the leaves, causing yield losses owing to decreased photosynthesis.

Another leaf ragging insect, the Japanese beetle (*Popillia japonica*) has minor repercussions for Canna plants. The portion of the leaf between the veins is where this beetle eats. It does not inflict significant harm in its own country of Japan. However, it has no natural predator in the United States and can cause substantial harm to Canna and other plants.

The bird cherry-oat aphid (*Rhopalosiphum padi*) has been shown to damage rhizomes that have been stored. Although this pest has not yet caused significant harm, it may affect greenhouse plants and can be combated by parasitic wasps (Reddy 2015).

Conclusion

Canna edulis can be a complementary starch source to cassava, rice for the food industry. It is more nutritional crop and has more properties which are good for human health. There is need of more attention in research and development product made from this crop.

References

Andrade-Mahecha, M. M., Tapia-Blacido, D. R., & Menegalli, F. C. (2012). Physical–chemical, thermal, and functional properties of achira (*Canna indica* L.) flour and starch from different geographical origin. *Starch-Stärke*, *64*(5), 348–358.

Caicedo Diaz, G. E., Rozo Wilches, L. S., & Rengifo Benítez, G. (2003). La achira: alternativa agroindustrial para áreas de economía campesina.

Chuenkamol, B., Puttanlek, C., Rungsardthong, V., & Uttapap, D. (2007). Characterization of low-substituted hydroxypropylated canna starch. *Food hydrocolloids*, *21*(7), 1123–1132.

Cook, I. (2001). "The gardener's guide to growing canna" Timber Press. ISBN 0-88192-513-6

Idárraga-Piedrahita, A., Ortiz, D.C., Callejas, R., & Merello, M. (Eds.). (2011). Flora de Antioquia: Catálogo, 2, 9-939. Universidad de Antioquia, Medellín.

Maas-Van de Kamer, H., & Maas, P. J. M. (2008). The Cannaceae of the world. *Blumea-Biodiversity, Evolution and Biogeography of Plants*, *53*(2), 247–318.

Mahamadi, C., & Chapeyama, R. (2011). Divalent metal ion removal from aqueous solution by acid-treated and garlic-treated Canna indica roots. *Journal of Applied Sciences and Environmental Management*, *15*(1).

Mathina, A., & Rajalakshmi, R. (2016). Corrosion inhibition of mild steel in acid medium using *Canna Indica* as green corrosion inhibitor. *Rasayan Journal of Chemistry*, *9*(1), 56–66.
Nelson, S. (2013). Rust of Canna lily. *Plant Disease*, *96*.
Okonwu, K., & Ariaga, C. A. (2016). Nutritional Evaluation of Various Parts of Canna indica L. *Annual Research & Review in Biology*, 1–5.
Pullaiah, T. (2006). *Encyclopaedia of world medicinal plants* (Vol. 1). Daya books.
Puncha-Arnon, S., Puttanlek, C., Rungsardthong, V., & Uttapap, D. (2007). Changes in physicochemical properties and morphology of canna starches during rhizomal development. Carbohydrate Polymers 70: 206217.
Putman, E. W., & Hassid, W. Z. (1954). Sugar transformation in leaves of Canna indica I. Synthesis and inversion of sucrose. *Journal of Biological Chemistry*, *207*(2), 885–902.
Reddy, P. P. (2015). Achira, *Canna edulis*. In *Plant protection in tropical root and tuber crops* (pp. 281–291). *Springer, New Delhi.*
Thamburaj, S. and Singh, N. (2016). Textbook of Vegetables, Tuber Crops and Spices. ICAR, New Delhi. Six edition. ISBN: 81-714-001-X.pp:447–448.
Zang, G. L., Sheng, G. P., Tong, Z. H., Liu, X. W., Teng, S. X., Li, W. W., & Yu, H. Q. (2010). Direct electricity recovery from *Canna indica* by an air-cathode microbial fuel cell inoculated with rumen microorganisms. *Environmental science & technology*, *44*(7), 2715–2720.

Chapter 15
Production Technology of Underutilized Vegetables of Marantaceae Family

Monisha Rawat and Khushboo Kathayat

Introduction

West Indian arrowroot scientifically known as *Maranta arundinacea* is a perennial plant which belongs to the Marantaceae family. The plant is known as "ararute" in the Caribbean language, which means a starchy root (Rukmana 2000). Mexico, Central, and South America are believed to be their center of origin. Its name, arrowroot, is because of the shape of its tubers and because the Indians in the Caribbean region used the pulp of its fresh rhizomes to treat wounds caused by poisonous arrows. The vegetable is usually grown for its starch-rich rhizomes and an ornamental plant owing to its medicinal values. It is a potential source of carbohydrates and functional foods (FAO 2007; Asha et al. 2015; Heuzé and Tran 2017). It is also known as common arrowroot, bamboo tuber, maranta, Bermuda arrowroot, and Saint Vincent arrowroot, while in India, it is commonly called desi arrowroot, murta, and sitalpati. It can be seen growing naturally in different habitats, from open areas with full sunshine to shady areas.

Origin and Distribution

West Indian arrowroot has its origin in tropical America (Florida). However, it has been grown for a long time in the Bahamas, Antilles, and especially in Jamaica and Saint Vincent Island of the West Indies which provides approximately 95% of the world's supply and also in Australia, South Africa, East Africa, and Southeast Asia

M. Rawat (✉) · K. Kathayat
Department of Horticulture, School of Agriculture, Lovely Professional University, Phagwara, Punjab, India
e-mail: monisha.22692@lpu.co.in

Savita et al. (eds.), *Production Technology of Underutilized Vegetable Crops*,
https://doi.org/10.1007/978-3-031-15385-3_15

(Govaerts 2018; Charles et al. 2016; Acevedo-Rodríguez and Strong 2012; Broome et al. 2007). Its cultivation has spread to other tropical countries, viz., Brazil, India, Sri Lanka, Indonesia, and the Philippines. Arrowroot was introduced from the Cape of Good Hope (South Africa) to India in 1831; afterward, it was cultivated in various regions of India (Peter 2007). Preferable conditions for this are at an altitude of 60–90 m from the sea level but can be cultivated at an altitude of up to 1000 m.

Nutritional Importance and Uses

It is mainly cultivated for its edible rhizomes that can be consumed either after boiling or roasting. Its rhizomes are a rich source of starch, i.e., dry white powder that is easily digestible (Villas-Boas and Franco 2016). Therefore, it is fed to infants as a substitute for breast milk and people with specific dietary requirements. It is one of the purest natural carbohydrates, highly viscous with good gelling ability (Charles et al. 2016; Hoover 2001). As it forms a gel upon cooling, it is widely used as a thickener in different foods and confectionaries like sauces, soups, dressings, candies, cookies, biscuits, ice creams, cakes, jellies, pastries, and puddings. It is also used as a primary constituent of cosmetics, glues, and soaps. The industries use the processed pulp of rhizomes for manufacturing cushions, paper, cardboards, and wallboards (Göhl 1982).

The pulp obtained from its fresh rhizomes has medicinal properties and is therefore used in poultices to heal wounds and ulcers. When used with water or milk, the starch extracted from its rhizomes helps relieve diarrhea and stomach pain. Its rhizomes are used as a substitute for maize in the ration for broilers, and after starch extraction, the remaining fibrous debris is used as fodder for animals and manure. It is used to manufacture tablets that require rapid disintegration and also carbonless paper for computers. The cultivars having brownish-purple and whitish-green colored leaves are grown as ornamental. Sometimes, the leaves are used as wrapping material and also as fibers (Flora of China 2018; Flora of North America 2018; FAO 2007; Kay and Gooding 1987; Erdman and Erdman 1984).

Arrowroot is mainly cultivated in the tropical regions but is of economic importance, especially in Saint Vincent Island of West Indies. It is also grown in various countries like the Philippines, India, Indonesia, Kenya, Sri Lanka, Rwanda, China, and Tanzania. In some Southeast Asian regions, it is generally grown in the home gardens, and its products are normally traded on a small scale in the local areas.

After harvesting, silage can be prepared from the aboveground portion of the arrowroot plant (Erdman and Erdman 1984). The residue left after extracting starch is known as bittie in Saint Vincent Island of West Indies, which has higher fiber content and can be used as a feed for livestock (Göhl 1982; Erdman and Erdman 1984; Valdés Restrepo et al. 2010).

Botany and Taxonomy

The Marantaceae family includes 31 genera and 550 species widely distributed worldwide, mainly in tropical regions. *Maranta arundinacea* has a chromosome number of $2n = 18, 48$ (Flora of China 2018). They are often robust plants having moderately sized pulvinate, petiolate, and paired leaves. They have asymmetric, mirror-image flowers (Stevens 2020). Thirty-four species of the genus Maranta are found in Central and South America. Still, the majority of its species are distributed in the Brazilian Atlantic and Amazonian forests and the Brazilian savannah (Vieira and Souza 2008). The plant parts of West Indian Arrowroot are shown in Fig. 15.1.

It is an erect, herbaceous plant that can be grown as an annual or perennial, and it reaches up to a height of 1 1/2 m and forms long, cylindrical, and pointed underground rhizomes, which, when boiled, tastes like corn. It has shallow roots, and the rhizomes grow deep into the soil. Rhizomes are fleshy, white, or red, fusiform, curved like arrows, 5–40 cm long with 2–5 cm diameter, and has small segments separated from each other with overlapping brownish-white persistent or deciduous scales. The stem is slender and is generally much forked toward the top. The leaves are large, lanceolate; petiole is often absent in upper leaves; the leaf blade is ovate to oblong, 10–30 cm in length with 3–10 cm width; and midrib is prominent with numerous fine and closely spaced parallel veins.

The inflorescence is paniculate; flowers are hermaphrodite, white (Djaafar and S. Rahayu 2012), and arranged in pairs. Fruits are 7 mm long berries, oblongoid in shape, leathery, glabrous to hairy, and brown. Seed three-sided, pinkish or red, and are rarely produced.

Varieties

Three important cultivars of Brazil are Common, Creole, and Banana (Leonel and Cereda 2002), while in St. Vincent, there are two main cultivars, Creole (having long thin rhizomes that spread more widely and penetrate deep into the soil) and

a b c

Fig. 15.1 West Indian arrowroot, *Maranta arundinacea* (**a**) plant, (**b**) flowers, and (**c**) rhizomes

Banana (having short, thick rhizomes which are less fibrous and are formed near the soil surface, thus facilitating mechanical harvesting).

Climate and Soil

M. arundinacea requires warm and humid climatic conditions for its optimum growth and development. It prefers an optimum temperature of 17–34 °C with annual average precipitation ranging from 700 to 4000 mm. It requires sufficient water in the soil during its growth as even rainfall distribution helps achieve optimum yield. When dormant, the rhizomes can tolerate minimum temperature of approximately 5 °C. It can be grown in areas having abundant sunshine as well as partial shade but can also tolerate up to 50% shade without loss in yield (Useful Tropical Plants 2017; FAO 2007; Kay and Gooding 1987).

Arrowroot can grow in soils like sandy, loam, and clay; achieving better yield requires well-drained, deep, loose, and slightly acidic loam soils. Although it can survive under saturated soil and waterlogging, it does not produce storage rhizomes under such conditions. The plants can be grown even in poor light conditions, infertile land, and shady places (Deswina et al. 2015). When it is cultivated on soft soils, providing shade to the plants has been proven beneficial. Sandy loam soils of Saint Vincent containing minerals of volcanic origin have been found most suitable for its cultivation. Soils having a pH of 5.5–6.5 are suitable, but they can also tolerate soil pH of 5–8.

Agronomic Practices

West Indian arrowroot can be grown as a sole crop for 5–6 years (FAO 2007). It is generally vegetatively propagated by rhizomes and suckers, but seeds can also propagate it. Approximately 3000–3500 kg rhizome bits are required for planting a 1-hectare area. In certain Asian regions, these pieces are often treated with smoke to achieve early germination. Sometimes suckers are also used for its propagation. It is usually planted during the onset of the rainy season, after thorough plowing and harrowing of the soil. 8 to 15 cm deep holes are prepared, maintaining a spacing of 75 cm × 37.5 cm, within which 4–7 cm long rhizome bits with buds are placed and then covered with soil. It forms adventitious roots within 6–7 days, and the shoots arise within 1–3 weeks. Farmers must remove weed during the first 3–4 months, and for its control, preemergence application of suitable weedicides can be made. Sometimes, rhizome bits left in the field after harvesting resprout easily, posing a weed problem and are challenging to eradicate. Under optimal growing conditions, if the rhizomes are not harvested timely, the plant continues to grow producing suckers and rhizomes, leading to withering of the older leaves and the formation of new plants from the rhizomes. The plant usually flowers after 3–6 months of planting, and the flowers are removed as and when these appear.

Plant Protection

Generally, arrowroot is not attacked by any serious pests or diseases. A larger canna leaf roller, also known as Brazilian skipper (*Calopodes ethlius*) is an important pest in Saint Vincent. Great southern white (*Ascia monuste orseis*), northern mole cricket (*Neocurtilla hexedactyla*), and tawny mole cricket (*Scapteriscus vicinus*) attack the crop occasionally in Brazil and Venezuela. In thc wet districts of the Caribbean, it is often affected by a disease, i.e., black root rot caused by *Rosellinia bunodes*. The plant is also affected by leaf blight diseases in India that are caused by *Rhizoctonia solani* and *Pellicularia filamentosa*. Cigar roots, a disorder has been reported on the plants grown in the Caribbean in which the rhizomes are elongated and become very fibrous, and nutritional deficiencies are believed to be its cause.

Harvesting and Yield

The rhizomes are ready to harvest after 8–12 months of planting when the leaves start wilting and dying. The rhizomes are manually removed at this stage from the soil and are separated from the leafy stem. Maximum starch content is found in the rhizomes after 12 months of planting, but at this stage, more fibers are developed in the rhizomes, thus making starch extraction more difficult. Leaving the storage rhizomes in the field for more than 12 months leads to the gradual conversion of starch into sugar. Generally, the rhizomes are left in the area till these are required for processing. After harvesting, rapid deterioration occurs as the rhizomes are more prone to rotting; therefore, these must be processed soon after harvest. The cultivar Banana must be processed within 2 days and Creole within 7 days of harvesting.

Generally, 12.5 tons rhizomes can be obtained from 1-hectare areas; however, higher yields up to 31 t/ha can be achieved under favorable conditions. The starch content in the rhizomes is approximately 20%, of which 17–18% starch can be extracted (Cecil 1992; Kay and Gooding 1987). In Saint Vincent, 8–16% of commercial starch yield can be obtained. The average yield of starch is 2500 kg per hectare, but by adopting the improved methods of the Saint Vincent Ministry of Agriculture, the average yield of 3700 kg–5600 kg starch can be achieved.

Conclusion

West Indian arrowroot is a promising underexploited vegetable crop with higher yield potential and superior quality starch that can replace conventional starch. It offers several health benefits for people suffering from different health issues like autism, diabetes, and digestive disorders. Starch flour extracted from its rhizomes can be processed to prepare various value-added products. Due to the enormous potential of this crop, it is required to develop wide yielding varieties that can be suitably cultivated in different regions of the world.

References

Acevedo-Rodríguez, P., & Strong, M.T. (2012). Catalogue of seed plants of the West Indies. 1192 pp.

Asha, K. I., Krishna Radhika, N., Vineetha, B., Asha Devi, A., Sheela, M. N., & Sreekumar, J. (2015). Diversity analysis of arrowroot germplasm using ISSR markers.

Broome, R., Sabir, K., & Carrington, S. (2007). Plants of the Eastern Caribbean. Online database. *Plants of the Eastern Caribbean. Online database.*

Cecil, J. E. (1992). *Small-, medium-, and large-scale starch processing* (No. 98). Food & Agriculture Org.

Charles, A. L., Cato, K., Huang, T. C., Chang, Y. H., Ciou, J. Y., Chang, J. S., & Lin, H. H. (2016). Functional properties of arrowroot starch in cassava and sweet potato composite starches. *Food Hydrocolloids*, *53*, 187–191.

Deswina, P., Indrayani, S., Paradisa, Y. B., & Mulyaningsih, E. S. (2015). Analisis keragaman genetic tanaman garut (*Maranta arundinacea* L.) koleksi kebun plasma nutfah cibinong science center. In *Prosiding Seminar Nasional Perhimpunan Biologi Indoneia Cabang Jakarta. Fakultas Biologi Universitas Nasional. Jakarta* (Vol. 11, pp. 338–452).

Djaafar, T. F., & S. Rahayu, S. (2012). Karakteristik rimpang garut (*Marantha arundinacea*) pada berbagai umur panen dan produk olahannya.

Erdman, M. D., & Erdman, B. A. (1984). Arrowroot (*Maranta arundinacea*), food, feed, fuel, and fiber resource. *Economic botany*, *38*(3), 332–341.

FAO, (2007). Crop Ecological Requirements Database (ECOCROP).

Flora of China Editorial Committee. (2018). Flora of China. *Flora of China.*

Flora of North America Editorial Committee. (2018). Flora of North America North of Mexico. *In*: Flora of North America North of Mexico St. Louis, Missouri and Cambridge, Massachusetts, USA: Missouri Botanical Garden and Harvard University Herbaria

Göhl, B. (1982). Les aliments du bétail sous les tropiques. *FAO, Division de Production et Santé Animale, Roma, Italy*, *12*, 543p.

Govaerts, R. H. (2018). 101 Nomenclatural corrections in preparation for the Plants of the World Online (POWO). *Skvortsovia*, *4*(3), 74–99.

Heuzé, V., & Tran, G. (2017). Arrowroot (*Maranta arundinacea*). Feedipedia, a program by INRA, CIRAD, AFZ, and FAO.

Hoover, R. (2001). Composition, molecular structure, and physicochemical properties of tuber and root starches: a review. *Carbohydrate polymers*, *45*(3), 253–267.

Kay, D. E., & Gooding, E. G. B. (1987). TDRI crop and Product digest No. 2. *TDRI, London*, 166–173.

Leonel, M., & Cereda, M.P. (2002). Physicochemical characterization of some starchy tubers. *Food Science and Technology*, *22*(1), 65–69.

Peter, K.V. (2007). Underutilized and Underexploited Horticultural Crops: Vol 1 New India Publishing Agency. *New Delhi*, 378.

Rukmana, R. (2000). Garut. *Yogyakarta: Kanisius*.

Stevens, P. F. (2001 onwards). Angiosperm Phylogeny Website. Version 14, July 2017 [and more or less continuously updated since]. http://www.mobot.org/MOBOT/research/APweb/.

Useful Tropical Plants. (2017). Useful tropical plants database. In: Useful tropical plants database: K Fern.

Valdés Restrepo, M. P., Ortiz Grisales, S., & Sánchez, T. (2010). Plant morphology, yield and quality of sagú starch. *Acta Agronómica, 59*(3), 372–380.

Vieira, S., & Souza, V.C. (2008). Four new species of *Maranta* L. (Marantaceae) from Brazil. *Botanical Journal of the Linnean Society, 158*(1), 131–139.

Villas-Boas, F., & Franco, C. M. (2016). Effect of bacterial β-amylase and fungal α-amylase on the digestibility and structural characteristics of potato and arrowroot starches. *Food hydrocolloids, 52*, 795–803.

Chapter 16
Protected Cultivation of Underutilized Vegetables

Nikhil Ambish Mehta and Savita

Introduction

Horticultural crops have the potential to improve the region's economy and food security both directly and indirectly. The Agribusiness industry is yet to realize its full potential. Specific scientific-technical interventions with the potential for diversification and long-term production must be used to attain the abovementioned. Protected cultivation is a way of cultivating crops that involve plastic components to address adverse growth conditions. The polyethene polymer was first discovered and developed in the late 1930s, further introduced in the early 1950s and gave rise to new products like plastic films, polysheets, mulching sheets and drip irrigation pipes, which transformed the commercial horticulture crop production practices under protected cultivation technologies. Discovery and development of polyethylene polymer in the late 1930s, followed by its introduction in the early 1950s in plastic films, mulches, and irrigation tubing, transformed commercial horticultural crop production and gave rise to new protected product cultivation technologies.

Protected cultivation techniques are those cropping techniques in which the microclimate surrounding the plant canopy is partially or completely managed to meet the needs of the plant species. The microclimate is modified using a variety of techniques to bring it closer to natural growth periods. Protected farming is a comparatively new and novel method of cultivating high-value crops to maintain agricultural production and productivity. The pressure from an ever-increasing population on the planet leads to the development of precision farming technology to meet current food demands within the same accessible cultivable area on

N. A. Mehta (✉)
School of Agriculture, Lovely Professional University, Phagwara, Punjab, India

Savita
Graphic Era Hill University, Dehradun, Uttarakhand, India

Savita et al. (eds.), *Production Technology of Underutilized Vegetable Crops*,
https://doi.org/10.1007/978-3-031-15385-3_16

scientifically based lines in a sustainable manner. It can solve the problem by offering attractive returns and a multifold increase in cropping intensity on the same piece of land, as many technological innovations around the world have shown. Greenhouse technology has been around for almost 200 years, and Europeans are considered pioneers in this industry. Approximately 90% of new greenhouses are being developed with ultraviolet (UV) stabilized polythene sheets as the glazing material, rather than the more expensive non-flexible glazing material. The protected cultivation technology has been standardized according to different agroclimatic zones is still in its developmental stage in India.

Around 115 countries in the world are growing vegetables in a greenhouse. As per the world scenario, the protected cultivation area is approximately 623,030 hectares in world, whereas the overall total greenhouse vegetable production area under protected cultivation is 402,098 hectares. Of the aggregate world, the greenhouse vegetable area accounts for 95,000 ha (Hickman 2011), whereas in India, the area under-protected cultivation for different horticulture crops was around 40,000 ha (Singh 2014).

Food resources that are currently underutilized, such as minor grains, millets and pulses, root and tuber crops, fruits and vegetables, leafy herbs, and non-timber forest products, have the great potential to contribute significantly to nutritional and food security. Also, these crops can protect against internal and external market disruptions and climate uncertainties, resulting in improved ecosystem functions and services, and sustenance (Keatinge et al. 2010). Higher use of underutilized crops and species, either intercropped with main staple crop in cereal-based systems or as stand-alone crops, would provide a variety of options for incorporating temporal and spatial heterogeneity into uniform cropping systems, improving resilience to biotic and abiotic stress factors and, ultimately, leading to a more sustainable supply of diverse and nutritious food.

Advantages of Protected Cultivation

- Compared to open-field cultivation, the production of vegetables under protected conditions can minimize the quantity of water and pesticides utilized in the production of high-value underutilized vegetables.
- Year-round production of vegetables.
- Multiple cropping is possible on the same piece of land.
- Shielded growing systems can be effective to counter adverse climate for vegetable production.
- Vegetables are grown in the off-season to give growers a higher return.
- Protected cultivation of vegetables can boost productivity while reducing prices per land, water, energy, and labor. It encourages the farmers to produce high-quality, clean products.
- It enables vegetable growing in areas where it is impossible in open conditions, such as at high altitudes.

- Under a protective structure, disease-free seed production of expensive and underutilized vegetables becomes easier.
- The potential of protected conditions vegetation to meet the demand for good nutritional healthy food and pesticide-free crops can be explored.
- Early production of nursery plants, off-season production of vegetables, seed production, and the protection of valuable germplasm all require controlled environmental conditions.
- Vegetable crops can be grown in all types of weather and at any time of year.

Disadvantages of Protected Cultivation

- High input and high initial cost
- Need more technical know-how
- Labor-intensive farming
- Prone to damage by storm

What Exactly Is a Greenhouse?

A greenhouse is a framed structure with a transparent plastic/net covering large enough to grow crops under wholly or partially controlled climatic conditions for maximum growth and productivity. A greenhouse/protected structures are covered structures that protect plants from adverse climatic conditions, i.e., wind, rainfall, hailstorm, excessive sun radiation, temperature extremes, and significant insect – pest and disease attacks. Further, it also provides optimum light, temperature, humidity, CO_2, and air circulation for plant growth to achieve maximum yield and quality. Greenhouse technology is a technique for providing plants with favorable environmental conditions. The greenhouse is primarily used for the production of seasonal and nonseasonal crops, as well as the creation of high-quality flowers and vegetables, and the preparation of tissue culture nurseries (Singh 2014).

Principle of Greenhouse

A translucent material, such as plastic, PVC sheet, or glass covers the greenhouse. The greenhouse cover transmits the majority of the sunlight due to its transparency. The crop, floor, and other items within the greenhouse absorb the light that is allowed inside. These objects then generate long-wave thermal radiations, which cause the greenhouse-covering material to become less transparent, trapping solar energy and increasing the temperature inside the greenhouse which is referred to as the greenhouse effect.

Difference Between Greenhouse and Polyhouse

A greenhouse is a broad term among the different types of greenhouses; the polyhouse is a structure made of polyethylene sheets as the primary building material. When compared to other types of greenhouses, the polyhouse is more durable, and it all depends upon the quality of the polyethylene sheet used.

Types of Greenhouse/Polyhouse

Greenhouses are classified according to their shape, construction, material, and ventilation. Each conservatory is best suited to a specific set of circumstances. Different types of greenhouses are intended to meet particular requirements. Saw-toothed teeth are common in India. Vegetables and cut flowers are grown in a natural ventilated greenhouse.

Based on Shape

The peculiarity of the cross-section of the greenhouses might be considered criteria in classification. The following are the most popular types of greenhouses based on their shape:

- Lean type Greenhouse
- Greenhouse with an even span
- Greenhouse with an uneven span
- Ridge & furrow type
- Saw-tooth type Greenhouse
- Quonset Greenhouse
- Interlocking ridges and furrows type Greenhouse
- Ground to ground greenhouse

Based on Construction

- The structure is constructed of wood.
- Structure made of pipes.
- The structure is framed by trusses.

Based upon Covering Material

- Glasshouse – Covering material is glass.
- Plastic film greenhouse – Polyethylene sheet used.
- Rigid panel greenhouses – Polycarbonate sheet used.

Based on Designs and Ventilation

1. Hi-Tech Polyhouse / Fan & Pad Polyhouse
2. Naturally Ventilated Polyhouse (NVPH)
3. Net House
4. Shade Net House
5. Walk-in Tunnel

Types of Protected Structures/Greenhouses (Figs. 16.1, 16.2, 16.3, 16.4, 16.5, 16.6, 16.7, 16.8, 16.9 and 16.10)

Fig. 16.1 Hi-Tech PolyhouseN

Fig. 16.2 Hi-Tech Nursery Unit

Fig. 16.3 Dome Shape Net House

Fig. 16.4 Flat Net House

Fig. 16.5 NVPH

Fig. 16.6 NVPH

Fig. 16.7 Shade Net House

Fig. 16.8 Shade Net House

Fig. 16.9 Walk in Tunnel

Fig. 16.10 Walk in Tunnel

Importance of Underutilized Vegetables

Malnutrition, poverty, and economic success are being addressed more widely and efficiently with underutilized vegetables. They are essential biological resources of the rural poor people and can improve the lives of millions of native populations. Vitamins, minerals, and other health-promoting components, such as high antioxidant activity, are abundant in underutilized vegetables. They have an essential role in dietary diversity, resulting in a more balanced source of micronutrients. Furthermore, underutilized vegetables are resistant to various biotic and abiotic stresses, and they can also give sustenance to poor people by satisfying the nutrient needs of vulnerable groups.

A variety of underutilized vegetables are natural sources of vitamins, minerals, and antioxidants. Underutilized vegetables play an essential role in the lives of people living in rural areas. They are an integral part of the local population's food and nutrition. Many of them have been valued for their medicinal, therapeutic, and nutritional values since time immemorial. They are consumed as raw or cooked vegetables as traditional delicacies, with sales from the underutilized vegetables. Malnutrition and related food scarcity are shared among the poor rural population. They have the potential to help reduce poverty by providing job opportunities and money and improving the efficiency and profitability of farm family labor in both rural and urban areas. There is a technique to minimize the risk of overreliance on a small number of essential crops by using underutilized vegetable crops. They can help to long-term livelihoods by improving household food security by increasing food edibility options. They supplement the diet with nutrients and are sometimes a convenient dish for low-income urban residents. They are adaptable to vulnerable conditions and can help maintain agroecosystem stability, especially in dry and semiarid plains, mountains, steppes, and tropical forests. They offer a diverse range of crops to boost production and global food security as well as meet new market demands (Hughes and Ebert 2011).

Underutilized Vegetables as Their Potential Role in Food Security and Nutrition

Many neglected and underutilized vegetables are nutritionally and well-suited to low-input cultivation. Whether wild, managed, or cultivated, the extinction of these species has immediate implications for the poor's food security and well-being. Their enhanced use may result in improved nutrition in the community area. Many underutilized vegetables, for example, have higher levels of vitamin C and provitamin A than commercially accessible species and variations. Focusing emphasis on underutilized and neglected veggies is a valuable method to maintain a diverse and balanced diet and prevents micronutrients, minerals shortages, so-called hidden hunger, and other nutritional deficits, especially among the poor people (Sharma 2003).

Increased Income for the Rural Areas

Consumer demand for variety and innovation in foods in emerging countries creates new market niches for underutilized and neglected veggies. If these crops have a comparative return over major staple crops or commercial crops in harsh environmental conditions, these market prospects might produce additional cash for needy farmers. Furthermore, modern advanced technologies such as hydroponics, aeroponics, and aquaponics have created new opportunities to develop new production technology, allowing these crops and their products to be sold in adverse climate conditions while also extending their shelf-life and improving their quality.

Stability in Ecosystem

Climate change and the degradation of land and water resources have prompted a spike in interest in these underutilized crops and species that are suited to rugged habitats, including desert boundaries, those with deficient soil or degraded vegetation, and those that are drought-prone.

Biodiversity in Culture

Plants have traditionally played an essential role in local cultures and rituals. Many underutilized vegetables have a crucial role in preserving cultural diversity related to cooking habits, health practices, religious rites, and social interactions.

Nursery Production Under Protected Cultivation with Underutilized Vegetables

To grow underutilized vegetables in the off-season under protected conditions, we have to produce their seedlings 1 month earlier under suitably protected structures. We have to grow the nursery first, in case of the small seeded vegetable crops like broccoli, red cabbage, amarath and some herbs like kale, oregano, thyme, sage, basil etc., further transplant under suitable field conditions and protected structures (Fig. 16.11).

Fig. 16.11 Hi-tech nursery unit and nursery of broccoli

Production of Underutilized Vegetables Suitable for Protected Cultivation

Compared to worldwide common vegetables like tomato and cabbage, many traditional or indigenous vegetables have high nutritional content (Keatinge et al. 2011). In conventional and underutilized vegetables, legume crops can play a vital role in achieving nutritional security as sources of essential vitamins, minerals, protein, and other plant-based nutrients. Traditional underutilized vegetables like Amaranth (*Amaranthus* spp.), Malabar spinach (*Basella alba*), winged bean (*Psophocarpus tetragonolobus*), and several gourd species are considered significant for sustainable food production since they decrease the impact of climate change and soil fertility and can be grown with minimum inputs like fertilizer-pesticides.

Amaranth (Amaranthus *spp.)*

Amaranth is an annual summer plant that grows in warm and humid climates all over the world. Amaranth leaves contain 17.5–18.3% protein (dry weight basis), including 5% lysine, an essential amino acid deficient in most cereal and tuber-based diets. The leaves are high in vitamin C, K, and folate, as well as antioxidants (Masarirambi et al. 2012).

Gherkin (Cucumis sativus *var.* anguria*)*

Gherkin is a cucumber variety that has nearly all of the nutrients, which are also found in cucumber. It has an average number of calories and fats, which aids in weight maintenance and lowers the risk of serious illnesses like heart disease and cancer. It is generally used in the form of pickle as part of food habits, which can be further stored as pickle for longer commercial storage. For the production of gherkins, pollinators are used for cross-pollination, or hand pollination is required under protected cultivation (Doltu et al. 2020).

Basella (Basella alba, Basella rubra*)*

Basella, often known as vine-type spinach, is a popular tropical leafy green vegetable commonly grown as a home garden herb. *Basella alba*, which is recognized with its green stem and deep green foliage, and *Basella rubra*, which is recognized by the purplish stem and dark green leaves with pink veins, are the two main cultivars and also called Malabar spinach (Cook 2010). *Basella alba* is high in vitamins (A, C, B), minerals (Ca, Mg, Fe), and other vital antioxidants. It can be grown in protected conditions year-round (Gayathree et al. 2020) (Fig. 16.12).

Winged Beans (Psophocarpus tetragonolobus *L.)*

It is a tropical legume that grows in large quantities in hot, humid equatorial nations such as India, Burma, Sri Lanka, Thailand, and the Philippines. It's also known as the "wonder bean" because of the high protein content in the seeds, making it a multipurpose legume (Peyachoknagul et al. 1989). The winged bean is a plant that is underutilized but can become a significant multipurpose food crop. The entire plant of the winged bean is edible. The leaves, blossoms, roots, and bean pods can all be consumed fresh or cooked; the pods are even edible when unripe. After cooking, the seeds are edible. They can be grown under protected structures year-round (Kabalci and Sari 2018).

Fig. 16.12 Malabar spinach

Fig. 16.13 Broccoli, red cabbage and colored cauliflower. (Apahidean and Apahidean 2009)

*Broccoli (***Brassica oleracea** *var.* **italica***)*

Broccoli is rich in protein, phenols, and antioxidants; it can easily be grown in open conditions. Under protected conditions, it's grown in winter seasons only; the quality of broccoli is higher inside protected conditions (Demir and Polat 2011) (Fig. 16.13).

*Red Cabbage (***Brassica oleracea** *var.* **capitata f. rubra***)*

The varieties of colored cabbage are rich in antioxidants, dietry fiber and minerals. In the extreme winters during snowfall, It can be easily grown under protected conditions. The quality of red cabbage is better under protected structures with no attack of caterpillars. It is suitable for crop rotation with cucurbits under nethouse cultivation in plains (Strauch et al. 2019).

Herbs (Kale, Oregano, Thyme, Sage, and Basil)

Some herbs which are high-value crops like kale, oregano, thyme, sage, and basil can quickly be grown under protected conditions under shade net house year-round. These herbs also have export value and are also used for making essential oils. Year-round, the production of these leafy herbs can be an excellent viable option to grow under protected structures and get a pesticide-free harvest (Shiwakoti et al. 2016).

*Snow Pea, Snap Pea, and Sugar Pea (***Pisum sativum** *spp.)*

Some of the indeterminate varieties of legume crops like snow pea, snap peas, and sugar pea can be grown under net house to get good yield and better quality; by utilizing vertical space in protected structures and with a support system, we can

Fig. 16.14 Pea, snow pea, snap pea, and sugar pea

grow these types of pea varieties in the winter season and can get a good return for a longer time (Kong et al. 2018) (Fig. 16.14).

Developmental Barriers to Underutilized Vegetable Crops

(a) There aren't enough studies on these crops that are grown in protected environments.
(b) There is a scarcity of desirable seeds and planting materials.
(c) Advance on-farm agro-methods can only be used up to an extent.
(d) Insufficient marketing support, also transportation, storage, and insufficient processing infrastructure.
(e) There is a lack of understanding among farmers regarding the nutritional and therapeutic significance of underutilized vegetable crops.
(f) There is a lack of knowledge about postharvest management practices.
(g) Protected cultivation for underutilized vegetables with high crop productivity needs various development programs and selection due to a lack of implementation of creative and revolutionary technologies such as biotechnology.

Conclusion

Underutilized vegetables with high nutritious potential and the ability to withstand harsh weather conditions could be a blessing to growers, consumers, and environmentalists if adequately tamed. Lack of availability of quality planting material, lack of understanding of nutritional and medicinal importance, and lack of information on growing practices are all possible reasons for the low availability of underutilized vegetables, despite their acknowledged importance. In this scenario, to assure future food and nutritional security, there is a need to explore the programs like genetic resource identification, germplasm management, and exploitation for the development of underutilized vegetable crops. Underutilized vegetables play a significant impact on the national economy. The climate and soil in India are ideal for growing a variety of underutilized vegetables. Underutilized vegetable crops require special attention and promotion to take advantage of their potential to treat various lifestyle disorders. In terms of nutritional value, research in the area of domestication and usage is fundamental. Increased area and production of these vegetable crops will ensure nutritional security and strengthen the region's economy through the export of fresh rare vegetable products and seeds, in addition to agro-based industries, packing, storage, preservation, canning, and transportation. Vegetable production will address the deficit of per capita consumption availability, address nutritional issues while creating jobs and increasing rural people's income, and ultimately contribute to the national economy. Although these neglected/underutilized vegetables can be cultivated in an open field during a certain season, in harsh climate conditions throughout various agroclimatic zones, these vegetables can be produced under protected structures for year-round production to get reasonable market price and better quality production.

References

Apahidean, A. I., & Apahidean, S. A. (2009). Research Regarding Technological Particularities of Cauliflower Grown in Polyethylene Film Greenhouse. *Bulletin UASVM Horticulture, 66*, 1.

Cook, A. (2010). Linnaeus and Chinese plants: a test of the linguistic imperialism thesis. *Notes and records of the Royal Society, 64*(2), 121–138.

Demir, H., & Polat, E. (2011). Effects of broccoli-crispy salad intercropping on yield and quality under greenhouse conditions. *African Journal of Agricultural Research, 6*(17), 4116–4121.

Doltu, M., Jerca, I. O., Sora, D., Tănasă, V., & Drăghici, E. M. (2020). Influence of some rootstocks on two *Cucumis sativus* genotypes cultivated in greenhouse. *Biotechnol Lett, 25*(3), 1526–1531.

Gayathree, T. H. I., Karunarathne, S. I., Ranaweera, L. T., Jayarathne, H. S. M., Kannangara, S. K., Ranathunga, A. P. D. T., Weebadde, C. & Sooriyapathirana, S. S. (2020). Green-spinach, red-spinach, and tree-spinach ('three-fold spinach'in Sri Lanka): An insight into phylogenetics and consumer preference. *Emirates Journal of Food and Agriculture*, 82–91.

Hickman, G. W. (2011). A review of current data on international production of vegetables in greenhouses. *Statistics Cuesta Roble Consulting, Mariposa, CA., Available from www.cuestaroble.com.*

Hughes, J. D. A., & Ebert, A. W. (2011, June). Research and development of underutilized plant species: the role of vegetables in assuring food and nutritional security. In *II International Symposium on Underutilized Plant Species: Crops for the Future-Beyond Food Security 979* (pp. 79–92).

Kabalci, B. B., & Sari, N. (2018, August). Morphological characterization, plant growth, yield and pod properties of some winged bean (*Psophocarpus tetragonolobus*) genotypes. In *XXX International Horticultural Congress IHC2018: II International Symposium on Plant Breeding in Horticulture 1282* (pp. 363–370).

Keatinge, J. D. H., Yang, R. Y., Hughes, J. D. A., Easdown, W. J., & Holmer, R. (2011). The importance of vegetables in ensuring both food and nutritional security in attainment of the Millennium Development Goals. *Food Security, 3*(4), 491–501.

Keatinge, J. D., Waliyar, F., Jamnadas, R. H., Moustafa, A., Andrade, M., Drechsel, P. Hughes, J.D.A., Kadirvel, P. & Luther, K. (2010). Relearning old lessons for the future of food—by bread alone no longer: diversifying diets with fruit and vegetables. *Crop Science*, *50*, S-51.

Kong, Y., Llewellyn, D., & Zheng, Y. (2018). Response of growth, yield, and quality of pea shoots to supplemental light-emitting diode lighting during winter greenhouse production. *Canadian Journal of Plant Science*, *98*(3), 732–740.

Masarirambi, M. T., Dlamini, Z., Manyatsi, A. M., Wahome, P. K., Oseni, T. O., & Shongwe, V. D. (2012). Soil water requirements of amaranth (*Amaranthus hybridus*) grown in a greenhouse in a semi-arid, sub-tropical environment. *American-Eurasian Journal of Agricultural and Environmental Sciences*, *12*, 932–936.

Peyachoknagul, S., Matsui, T., Shibata, H., Hara, S., Ikenaka, T., Okada, Y., & Ohno, T. (1989). Sequence and expression of the mRNA encoding the chymotrypsin inhibitor in winged bean (*Psophocarpus tetragonolobus* (L.) DC.). *Plant molecular biology*, *12*(1), 51–58.

Sharma, D.V. (2003). Transfer of technology for increasing the scope of underexploited horticultural crops. Winter School on "Exploitation of Underutilized Horticultural Crops, 5–25th November, Department of Horticulture, College of Agriculture, MPUAT, Rajasthan, 313–320.

Shiwakoti, S., Zheljazkov, V. D., Schlegel, V., & Cantrell, C. L. (2016). Growing spearmint, thyme, oregano, and rosemary in Northern Wyoming using plastic tunnels. *Industrial Crops and Products*, *94*, 251–258.

Singh, B. (2014). Protected cultivation of horticultural crops in India: Challenges and opportunities. *Agrotechnol*, *2*(4), 51.

Strauch, R. C., Mengist, M. F., Pan, K., Yousef, G. G., Iorizzo, M., Brown, A. F., & Lila, M. A. (2019). Variation in anthocyanin profiles of 27 genotypes of red cabbage over two growing seasons. *Food chemistry*, *301*, 125289.

Chapter 17
Seed Production of Underutilized Vegetables

Sandeep Bagri and Harmeet Singh Janeja

Introduction

In the last decade, excellent progress in the field of seed production of main dietary crops, including vegetables as the most common ingredient in the development bucket, has been seen in India. Since the human population is increasing, the demand for a surplus supply of vegetables and other crops has also been increasing by 4% or more/annum. Synergistic assessment of seed production technologies, institutions, and farmers is needed in the present scenario to uplift the overall production of various crops. Farmers' primary focus is on major vegetables and cash crops, which are crucial to their daily demands and help achieve nutritional security. Vegetables account for approximately 58% of annual production, which consists of nearly 163 MT of cultivated crops spread across ten million hectares of cultivable land (Arora et al. 1980; NHB 2015). However, the proportion of underutilized vegetables in these major products is extremely low. Therefore, the idea of producing underutilized vegetables needs to be taken into consideration. Underutilized crops are those that have not been exploited for trade purposes and are not taken as part of food security and environmental issues (Jaenicke and Hoeschle-Zeledon 2006). Vegetables grown in rural areas are underutilized because they are grown not only for local consumption but also for national trade. These underutilized vegetables have good nutritional value but somehow are not part of a balanced diet due to community sectors, as these crops are believed to be produced and consumed by the poor (Jaenicke and Hoeschle-Zeledon 2006). In this chapter, the basic elements of seed production along with agro-techniques of some underutilized vegetables are discussed. Classification of these vegetables based on the mode of pollination and the mechanism followed by the plant to ensure seed production is also considered.

S. Bagri (✉) · H. S. Janeja
Department of Genetics and Plant Breeding, School of Agriculture, Lovely Professional University, Phagwara, Punjab, India

Savita et al. (eds.), *Production Technology of Underutilized Vegetable Crops*,
https://doi.org/10.1007/978-3-031-15385-3_17

Seed

A seed is a matured ovule (a result of fertilization) in flowering plants (Merriam-Webster 2021). It consists of an intact embryo, its nourishing part endosperm, and cotyledon covered with a protective layer, i.e., seed coat. Seeds can refer to vegetatively growing crop production material such as bulbs, roots, tubers, rhizomes, stem cuttings, setts, and so on.

Characteristics

- *Genetic purity* – It refers to the presence of crop seeds free from the seeds of other varieties of the same crop and to other crop seeds. However, there is a permitted range of contamination of 0–0.1%.
- *Freedom from other plant seeds* – Some crops produce similar seeds to crop seeds from physical admixtures at the same time of maturity, which is extremely difficult to separate mechanically.
- *Objectionable plant seeds* – Weed plant seeds are not acceptable in physical purity as they are harmful to crop plants and are toxic to both humans and animals. These are similar to crop seeds in terms of size and shape and are difficult to separate mechanically, for example, cucurbits and wild cucurbits species, lettuce, and wild lettuce.

Genetically pure seeds are crucial for the production of true-to-type seeds. It varies depending on the level of seed production and the type of seed used.

Type of seed	Genetic purity
Nucleus seed	100%
Breeder seed	100%
Foundation seed	99.5%
Certified seed	98% (variety); 95% (hybrid)

Agarwal (2018)

1. *Physical purity* – Seed should also be of acceptable physical purity (98% in general). It should be free from inert matter, defective and diseased plants.
    - Inert matter refers to nonliving soil particulates, stones, pebbles, straw, etc.
    - Defective seeds are pest-infested, diseased, and broken ones that are not fit for germination and are not even suitable for crop production.
    - Diseased plant seeds – To keep seed lots free of contamination, the seed certification agency has provided some specifications for plant diseases, for example, *Alternaria* blight of tomato and virus attack in brinjal.

2. *Germination rate (GR) and seed vigor* – The high germination rate and high seed vigor reduce the amount of input and provide a good yield. The minimum standards for seed GR are high for cereals (80–90%) and low for vegetables (60–70%).
3. *Uniform size and shape* – For quality seed production, seeds with uniform size and shape are considered separately. Separation is accomplished through the sorting and grading process.
4. *Moisture content* – Seed moisture content is very crucial for the processing operations and safe storage of seeds in godowns. Drying of seeds results in optimum moisture level of seeds for storage and germination for the next sowing. For example, pulses require 8–10% moisture content, while most vegetables require 5–6% moisture content.

Importance

Seeds are essential for growing a variety of crops. High-quality seeds provide a good yield. It has even been mentioned in one of the ancient scripts as "Subeejam Sukshetre Jayate Sampadyathe." It is a cheap source for crop production and food supply. The seed is a carrier for improved traits. Quality seed production is very useful in terms of yield, being more tolerant to biotic and abiotic constraints, and producing true-to-type pure seed.

Classification of Underutilized Vegetables

The underutilized vegetables are classified on basis of mode of pollination.

- *Self-pollinated crop* – The crops promoting self-pollination to more than 95% of degrees and cross-pollination less than 5% are self-pollinated crops.
- *Cross-pollinated crop* – The crops promoting cross-pollination to more than 95% of degrees and self-pollination nearly 5% are cross-pollinated crops.
- *Often cross-pollinated crop* – When self-pollinated crops have more than 5% cross-pollination, they are often cross-pollinated crops.

Crop	Mechanism
A. Self-pollinated crops	
Basella	Cleistogamy
Lettuce	–
Cherry tomato	Cleistogamy
Shallot	–
Pigeon pea	–

(continued)

Crop	Mechanism
Winged bean	–
Lima bean	–
Water spinach	–
B. Cross-pollinated crops	
Colocasia	Protogyny
Mint	Entomophily (bee as a pollinator)
Coriander	Entomophily (honeybee as a pollinator)
Asparagus	Sporophytic self-incompatibility
Celery	–
Brussels sprout	Entomophily (bee as a pollinator), sporophytic self-incompatibility
Chinese cabbage	Slight protogyny, sporophytic self-incompatibility
Kale	Self-incompatibility
Parsley	Entomophily (insects as a pollinator)
Agathi	Entomophily (bees as a pollinator)
Parsnip	Entomophily (insects as a pollinator)
Leek	Entomophily (bees, flies as a pollinator)
Artichoke	Bulb propagation, recessive male sterility
Rhubarb	Anemophily (wind pollination)
Oriental pickling melon	Entomophily (insects, bees as pollinators)
Orach	Anemophily (wind pollination)
Elephant foot yam	Dichogamy, entomophily (beetle as a pollinator)
Amaranth	–
Broccoli	–
Gourds	–
C. Often cross-pollinated crops	
Rutabaga	Entomophily (bee as a pollinator)
Bathua	–

Agro-Techniques for Seed Production of Underutilized Vegetables

Vegetable name	Botanical name	Method of propagation	Spacing	Chromo-some no.	Family	Origin
Asparagus	*Asparagus officinalis* var. *altilis*	Seeds and crowns	30 × 30 cm	20	Liliaceae	Asia minor
Basella	*Basella alba*	Stem cuttings and seeds	2 × 2 m in pits	44	Basellaceae	South Asia
Colocasia	*Colocasia esculenta* L. Scott	Root cuttings and layering	12 × 12 m	42	Araceae	Southeast Asia
Brussels sprout	*Brassica oleracea* var. *gemmifera*	Seeds 500 g/ha	60 × 50 cm	18	Brassicaceae	North America
Butter bean	*Phaseolus lunatus*	Seeds 25 to 35 kg/ha	35 × 25 cm	22	Fabaceae	Peru
Celery	*Apium graveolens*	Seeds (125 g/ha)	60 × 15 cm	22	Apiaceae	Mediterranean region
Bathua	*Chenopodium album*	Seeds (50 kg/ha)	30 × 15 cm	36	Amaranthaceae	South Asia
Snake gourd	*Trichosanthes cucumerina* L.	Stem cuttings	45 × 60 cm	22	Cucurbitaceae	South East Asia and Australia
Chinese cabbage	Heading type – *Brassica campestris* var. *pekinensis* Non-heading type – *Brassica campestris* var. *chinensis*	Seeds (375 g/ha)	45 × 45 cm	18	Cruciferae	China
Coccinia	*Coccinia indica*	Stem cuttings from pistillate plants	2 × 2 m	24	Cucurbitaceae	Africa
Knol-khol	*Brassica caulorapa*	Seed (1.5 kg/ha)	30 × 25 cm	18	Cruciferaceae	Northern Europe
Lettuce	*Lactuca sativa*	Seed (500 g/ha)	30 × 15 cm	18	Compositae	Asia minor, Mediterranean region
Mint	*Mentha viridis*	Cuttings	15 × 15 cm	72	Lamiaceae	Northern America
Spinach (Jayaweera 1982; Malalavidhane et al. 2001)	*Beta vulgaris* var. *Bengalensis*	Seeds 20–25 kg/ha	20 × 10 cm	12	Amaranthaceae	Iran
Winged bean	*Psophocarpus tetragonolobus*	Seeds	1 × 0.5 m	18	Fabaceae	New Guinea

Isolation Distance Required for the Production of Underutilized Vegetables

Isolation is carried by time and spacing. Keeping an isolation distance is a very important step for quality seed production. It is the separation distance to be maintained between the main crop and other crops or another variety of the same crop to prevent contamination of seeds and avoid unfavorable pollination. The isolation distance varies depending on the class of seed produced. It is the maximum for breeder seeds, then decreases for subsequent seed classes, and is the minimum for certified seed production. The table shows the recommended isolation distance for underutilized vegetables (Singh et al. 2010):

	Isolation distance (m)	
Crop	Foundation seed	Certified seed
Basella	1600	1000
Colocasia	1000	500
Brussels sprout	1500	1000
Bean (Jana 2007; Duke 1981)	10	5
Celery	500	300
Bathua	1500	1000
Snake gourd	1000	500
Chinese cabbage	1500	1000
Knol-khol	1600	1000
Lettuce	50	25
Spinach	1600	1000
Winged bean	10	5
Amaranth	400	200
Pointed gourd	1000	500
Parsley	500	300

In above all crops, the maximum permission of off-types produced during seed production – 0.10% (foundation seed) and 0.20% (certified seed).

Seed Production of Some Underutilized Vegetable Crops

Amaranth

Amaranth is an annually growing plant that provides an important source of protein, vitamins, fiber, phytochemicals, and phenolic compounds, as reported by Shukla et al. (2006). It is also helpful in the prevention of cancer, aging, and arteriosclerosis. It is a nutritious crop as it contains protein, minerals, and vitamin A and C along with oxalates and nitrates (Devadas and Mallika 1991).

Planting It is a tropical summer crop that is grown in a humid environment with temperatures ranging from 20 to 30 °C. Amaranth thrives in fertile sandy loam and loamy soil having a pH of 5.5–7. In the plains, sowing takes place from February to March (summer), June to July (Kharif), and May to July in the hilly regions. The seed is sown 1.0 cm deep in the soil at a rate of 2 kg per hectare (Sirohi and Sarkar 2001).

Nutrient requirement FYM is the most common source of nutrition for plants. At the time of field preparation, 20–30 tons of FYM need to be applied per hectare of land. NPK at 50:50:25 kg/ha is the standard recommended dose.

Rouging Frequent weeding is necessary during the early stages of plant growth to reduce the competition between the main crop and weeds and off-types. All the diseased, damaged, and off-types should be ripped out of the main field to prevent genetic contamination. Their identification and removal are based on morphological traits like the height of the plant, leaf color, shape, flower color, fruit, etc. The maximum permission of off-types produced during seed production is 0.10% (foundation seed) and 0.20% (certified seed).

Harvesting and yield After 25–40 days of sowing, the crop is ready to harvest (Shree et al. 2020). The average yield ranges from 10 to 15 tons per hectare.

Elephant Foot Yam

It is a perennial growing plant also known as Suran or Zaminkand. Ramalingam et al. (2010) reported its medicinal properties as it contains betulinic acid, stigma sterols, triacotane, and lupeol.

Planting It grows best in tropical and subtropical regions with a temperature range of 25–35 °C. Well-drained fertile loamy or sandy loam soil having a pH value of 5.5–7.0 is suitable for the crop. This crop favors vegetative propagation by corms. Considerable propagating material is 500 g of corm or tuber. Planting is carried out in pits of 60 × 60 × 60 cm that is 90 cm spaced apart in February–March (Singh 1989).

Nutrient requirement Approximately 25–40 tons per hectare of FYM as well as the total required quantity of phosphorus and half of the required quantity of nitrogen and potassium are applicable as a basal dose at the time of field preparation. A standard nutrient dose of 150 kg of nitrogen, 60 kg of phosphorus, and 100 kg of potassium is required.

Rouging Weeding and hoeing are necessary at the initial stages of crop development. Further weeding along with earthing up is essential at the time of fertilizer application to control weeds. All the diseased, damaged, and off-types should be ripped out of the main field to prevent genetic contamination. Their identification and removal are based on morphological traits like the height of the plant, leaf color, shape, flower color, and fruit. The maximum permission of off-types produced during seed production is 0.10% (foundation seed) and 0.20% (certified seed).

Harvesting and yield After 7–8 months, the leaves turn yellow and begin to droop, indicating that the crop is ready for harvest. However, the crop is generally harvested earlier in November because of higher prices in the market. The average yield obtained is nearly 8–10 tons/ha.

Asparagus

Asparagus or Satawar (Madhvi et al. 2020) is a perennial plant with herbaceous nature having succulent type shoot "spear." It continues to give yield for more than 10 years. It contains the enzyme asparagine that is in medicinal use for its diuretic properties. Asparagus is highly nutritious with 2.9 g carbohydrate, 2 g protein, 0.7 mg iron, 15.8 mg calcium, 25 mg vitamin C, and 762 (IU) vitamin A.

Planting Asparagus is generally cultivated in moderate temperate to subtropical regions with a day-night temperature difference of 25 to 18 °C. However, 24 °C is favorable for germination. At the time of seed sowing, it requires heavily ploughed sandy loam soil having a pH value of 6.5–7.5 in March to May (for hilly regions) and July to Nov. (in the plains). The varieties in cultivation are classified into two groups, i.e., with green spears for local market use or with white, light green spears for processing purposes (Singh 1989). Some cultivated varieties are perfection (80–100 q/ha), selection 841 (90–100 q/ha), UC 66, selection 831, Mary Washington1, Mary Washington 500 W (Toledo 1989), etc.

Planting Methods

There are three types of planting methods:

Seed material	Crown planting	Direct seedling	Transplanting
Sowing	15 cm deep in the soil	2 cm deep in the soil a single row (W shape bed); double row (V shape bed)	Less than 1.25 cm deep; single row (W shape bed); double row (V shape bed)
Spacing	30 × 150 cm	5 cm in a row, 5 foot spacing in beds	5 × 5 cm
Seed rate	21,750 crowns/ha	2.5–3.5 kg/ha	–

Nutrient requirement Essentially, the required nutrients are FYM (Farm Yard Manure) at 150–200 q/ha (basal dose) and N-P-K dose at 80-80-60 kg/ha. For healthy plants, timely irrigation should be given at the seedling stage and continuously maintain moisture for first 2 months (Rahman 1996).

Rouging All the diseased, damaged, and off-types should be ripped out of the main field to prevent genetic contamination. Their identification and removal are based on morphological traits like the height of the plant, leaf color, shape, flower color, fruit, etc. The maximum permission of off-types produced during seed production is 0.10% (foundation seed) and 0.20% (certified seed).

Harvesting and yield (1 harvestable spear/row foot) The maximum spear length is 22.5 cm. Harvesting is limited in the first and second years to 2–3 weeks and 6–8 weeks in the third year. If the temperature is more than 28 °C, daily harvesting is necessary. The yield ranges from 25 to 40 spears/ha.

Storage Spears is stored in wet tissue paper at 95% moisture content and 0–2 °C temperature (2–3 weeks).

Lettuce

Lettuce is an annually grown herbaceous plant. Its vegetable parts are included in the diet as a salad. However, it is known for its oil, vitamins, and minerals like iron and calcium (Madhvi et al. 2020). It also contains 1.2 g of protein, 8 mg of ascorbic acid, 0.3 mg of niacin, and 900 (IU) of vitamin A. It is helpful in the prevention of cancer and lung diseases. Kirtikar and Basu (1989) reported the blood purifying property of lettuce.

Planting Lettuce grows in moderate tropical regions with a temperature range of 20–22 °C. Temperatures below and above this are unsuitable for growth. For proper germination, a temperature of 22–24 °C is suitable. For planting, it is necessary to have well-drained soil having a pH value of 6.0 to 7.0.

Varieties

Head type	Non-heading type
Crisphead (ex-New York 515, Great lakes)	Cos/Romaine – Eiffel tower
Butterhead (ex – Cobham green)	Leaf/bunching – salad bowel
	Asparagus/stem – celery lettuce

Some cultivated cultivars are PL no. 1 (88 q/ha), Alamo 1(230 q/ha), etc.

Planting time	March–June (hilly areas) Sept.–Oct. (in plains)
Planting method	Direct seed or raising seedlings on nursery beds
Seed rate	400–500 g/ha
Spacing	45 × 45, 45 × 30, 30 × 30 cm

Nutrient requirement About 100–150 q/ha of farmyard manure and an N-P-K dose of at 50-50-90 kg/ha is necessary for the plant development. A full dose of FYM followed by half of the required quantity of nitrogen and the total required quantity of phosphorus and potassium are applicable as basal at the time of field preparation. The remaining nitrogen application is done in split doses. Four to five irrigations are required for proper vegetative growth at the time of pre-sowing, after transplanting, and then after an 8- to 10-day interval.

Intercultural operations Three to four weeding, application of herbicides for weed control.

Rouging All the diseased, damaged, and off-types should be ripped out of the main field to prevent genetic contamination. Their identification and removal are based on morphological traits like the height of the plant, leaf color, shape, flower color, and fruit. The maximum permission of off-types produced during seed production is 0.10% (foundation seed) and 0.20% (certified seed).

Harvesting and yield The crop is ready to harvest after 50–60 days of sowing (leaf type) and 60–70 days of sowing (head type) with an average yield of 100–160 q/ha. The harvest is stored in conditions with 90–95% relative humidity and a 0 °C temperature for 3–4 weeks.

Chinese Cabbage

These are of two types; one is the heading type, and the other is the non-heading type. Chinese cabbage is a good nutritional crop, having 3 g of carbohydrates, 1.2 g of protein, 0.6 mg of iron, 25 mg of vitamin C, and 625 (IU) of vitamin A (Kalloo and Rana 1993).

Planting It is grown on a limited scale in temperate regions with a temperature range of 15–21 °C. Sowing is suitable in well-drained loamy soil having a pH value of 6.8–7.8 (Rahman 1996). The general seed rate for sowing is 375 g/ha with the required spacing of 45 × 45 cm. The months of June to August is the best time for planting.

Nutrient requirement 10 t/ha of farmyard manure and N-P-K dose at 90-125-100 kg/ha are essential nutrients for the crop. At the time of field preparation, half of the required quantity of nitrogen and the total required quantity of potassium and phosphorus are applied as the basal dose. The remaining N dose application can be top-dressed after 4 weeks (Singh et al. 1959).

Rouging During the early stages of plant growth, the crop requires three to four times hoeing and weeding. All the diseased, damaged, and off-types should be ripped out of the main field to prevent genetic contamination. Their identification and removal are based on morphological traits like the height of the plant, leaf color, shape, flower color, and fruit. The maximum permission of off-types produced during seed production is 0.10% (foundation seed) and 0.20% (certified seed).

Harvesting and yield Crops are harvested when the heads are fully developed and compact, which take between 80 and 100 days depending on the variety. The yield is 25–35 t/ha on average, depending on the climate and the variety.

Celery

Celery is a minor vegetable crop grown in Central India for seeds during the winter months (Browers and Orton 1986). It contains 0.8–1 mg of protein, 2–3 mg of carbohydrate, 1 mg of vitamin C, and 10% vitamin A per 100 g (Malhotra 2006).

Planting Well-drained loamy soil having a pH value of 6–6.5 is suitable for the sowing of celery in areas with a temperature range of 16–21 °C. The standard seed rate is 1.5–2.0 kg/ha with a spacing of 10–20 cm × 60 cm. The usual planting time of the crop is February to August in different seasons, i.e., Neerbogam season (Feb. to April), Karbogam season (April to June), and Kadaipokam season (June to August). Some of the cultivated varieties are Fordhook Emperor, Giant Pascal, and Wrights Groove Giant (Kalloo and Bergh 2012).

Nutrition requirement 20 t/ha of farmyard manure and NPK at 200:100:150 kg/ha are the required dose of fertilizers to be applied. Nitrogen application is in two to three split doses at different stages of plant vegetative growth. Irrigation is provided once a week in summer and at a 10–14-day interval in winter (Singh 1989).

Rouging Regular weeding and hoeing are required to control the weeds and off-type plants. All the diseased, damaged, and off-types should be ripped out of the main field to prevent genetic contamination. Their identification and removal are based on morphological traits like the height of the plant, leaf color, shape, flower color, fruit, etc. The maximum permission of off-types produced during seed production is 0.10% (foundation seed) and 0.20% (certified seed).

Harvesting and yield After 115 days of planting, the crop is ready for harvest with 30–40 tons per hectare of yield on average.

Conclusion

Seed production of vegetable crops is of prime importance because the seed is the basic unit of agriculture. Production of hybrids/HYVs needs to be promoted on high priority to improve productivity and increase production of underutilized vegetables, etc. The production of underutilized vegetables is limited because of the unavailability of quality seeds. There is a need for promoting joint venture collaborations between industry and national and international institutions related to seed. If the government also extends the benefit of subsidy to truthfully labeled seeds of promising hybrids produced by the private sector, it may be beneficial for the seed sector. Sharing of germplasm is imperative for crop improvement; while the national repository makes available the germplasm to the researchers and national seed companies, the private sector must also come forward and share their valuable germplasm with the public sector institutions for research purposes. Also, these need to be stored in the Gene Bank for posterity. The need for the creation of awareness among farmers and the general public on the benefits of quality seed production of vegetable crops is also very important.

References

Agarwal, R.L. (2018). Seed Technology (second edition), Published by Oxford and IBH Publishing; 216.

Arora, D., Chandel, K.P.S., Joshi, B.S., & Pent, K.C. (1980). Rice bean: Tribal pulse of eastern India. Economic Botany; *34*:260–263.

Browers, M. A., & Orton, T. J. (1986). Celery (*Apium graveolens* L.). In Crops I (pp. 405–420). Springer, Berlin, Heidelberg.

Devadas, V. S., & Mallika, V. K. (1991). Review of research on vegetables and tuber crops-Amaranthus. Kerala Agricultural University, Thrissur.

Duke, J.A. (1981). Handbook of legumes of world economic importance. Plenum Press, New York and London, 288–291.

Jaenicke, H., & Hoeschle-Zeledon, I. (2006). Strategic Framework for Underutilized Plant Species Research and Development, with Special Reference to Asia and the Pacific, and to Sub Saharan Africa. ICUC, Colombo and GFU, Rome, 33p.

Jana, J.C. (2007). Use of traditional and underutilized leafy vegetables of Sub-Himalayan Terai region of West Bengal. Acta Horticulture; *752*:571–575.

Jayaweera, D.M.A. (1982). In: Medicinal Plants (indigenous and exotic) used in Ceylon. Part 11.

Kalloo, G., & Bergh, B. O. (Eds.). (2012). Genetic improvement of vegetable crops. Newnes.

Kalloo, G., & Rana, M. (1993). Chinese cabbage: Brassica pekinensis, B. chinensis. In Genetic Improvement of Vegetable Crops (pp. 179–186). Pergamon.

Kirtikar, K.R., & Basu, B.D. (1989). Indian Medicinal Plants, Published by Lalit Mohan Basu, Allahabad, India, 4, 2609–2610.

Madhvi, Sharma, D., Chauhan, A., Banyal, S.K., & Divya. (2020). Underutilized Vegetable Crops: A Lost Treasure. *Int.J.Curr.Microbiol.App.Sci.* *9*(04): 2172–2178. https://doi.org/10.20546/ijcmas.2020.904.259

Malalavidhane, S., Wickramasinghe, S. M. D. N., & Jansz, E. R. (2001). An aqueous extract of the green leafy vegetable Ipomoea aquatica is as effective as the oral hypoglycaemic drug tolbutamide in reducing the blood sugar levels of Wistar rats. *Phytotherapy Research*, *15*(7), 635–637.

Malhotra, S. K. (2006). Celery. In Handbook of herbs and spices (pp. 317–336). Woodhead Publishing.

Merriam-Webster. (2021). Seed. In Merriam-Webster.com dictionary. Retrieved June 29, 2021, from https://www.merriam-webster.com/dictionary/seed.

NHB. Indian Horticulture Database. National Horticulture Board, Gurgaon, Haryana, India, 2015.

Rahman, A. F. M. H. (1996). Vegetable seed production (No. RESEARCH). AVRDC.

Ramalingam, R., Bindu, K. H., Madhavi, B. B., Nath, A. R., & Banji, D. (2010). Phytochemical and anthelmintic evaluation of corm of Amorphophallus campanulatus. *International Journal of Pharma and Bio Sciences*, *1*(2).

Shree, S., Dei, S., Saha, R., Bamaniya, B. S., & Kumari, R. (2020). Potential of Underexploited Vegetable Crops and Vegetable Legumes to Nutritional Security and Sustainable Production System. TTPP, 523.

Shukla, S., Bhargava, A., Chatterjee, A., Srivastava, A., & Singh, S. P. (2006). Genotypic variability in vegetable amaranth (*Amaranthus tricolor* L) for foliage yield and its contributing traits over successive cuttings and years. *Euphytica*, *151*(1), 103–110.

Singh, H. B., Thakur, M. R., & Bhagchandani, P. M. (1959). Vegetable Seed Production in Kulu Valley-II Seed Production of Cabbage, Cauliflower, Knol Khol and Chinese Cabbage. Indian Journal of Horticulture, 16(3), 153–161.

Singh, P.M., Singh, B., Pandey, A.K., & Singh, R. (2010). Vegetable Seed Production - A Ready Reckoner. Technical Bulletin No.37, IIVR, Varanasi.

Singh, S. P. (1989). Production technology of vegetable crops. Agricultural Research Communication Centre, Karnal, India.

Sirohi, P. S., & Sarkar, M. (2001). For the Farmers… Pre-harvest management of vegetables: a key of success. Indian Horticulture, *46*(2), 26–31.

Toledo, J. (1989, June). Asparagus production in Peru. In VII International Asparagus Symposium 271 (pp. 203–210).

Chapter 18
Integrated Pest Management of Underutilized Vegetables

Johnson Wahengbam, S. Sanathoi Devi, A. M. Raut, and A. Najitha Banu

Abbreviation

IPM integrated pest management

Introduction

Globalization and urbanization accelerate the food demands and reduce agricultural farmlands. Deforestation and turning undisturbed natural ecosystems into agricultural farmlands and urban areas led to disturbances in ecosystem and biodiversity. Extensive exploitation of nonrenewable resources results in difficulty meeting future demands; advanced technology has been evolving, though many are not eco-friendly and harms the environment. The conventional pest management method used many toxic pesticides; the residues remained for a longer period, leading to environmental degradation. Many of the traditional insecticides are nonselective that harm other beneficial invertebrates and vertebrates. Contamination of soil, water, and air; toxic effects on nontarget organisms; and food contamination have led the environmentalists and agriculturalists to raise a voice against their injudiciousness. Before discovering insecticides, pest control was mostly achieved by manipulating cultural practices and utilizing natural enemies, mechanical and physical means. Later during the 1970s of the last century, the concept of integrated pest management was floated, which brought wide publicity and acceptance among the plant protection experts. Integrated pest management (IPM) is a suitable

J. Wahengbam (✉)
Department of Entomology, Szent Istvan University, Budapest, Hungary

S. S. Devi
Department of Horticultural Engineering, Szent Istvan University, Budapest, Hungary

A. M. Raut
Department of Entomology, School of Agriculture, Lovely Professional University, Phagwara, Punjab, India

A. N. Banu
Department of Zoology, School of Bioscience and Bioengineering, Lovely Lovely Professional University, Phagwara, Punjab, India

Savita et al. (eds.), *Production Technology of Underutilized Vegetable Crops*,
https://doi.org/10.1007/978-3-031-15385-3_18

combination of various nonchemical methods, and pesticides are employed when all other methods fail.

Groups of insects belonging to different orders, families, genera, and species attacked the underutilized vegetables causing considerable damages, either reducing the yield or deteriorating the product's quality and reducing market value. The major groups of insects attacking underutilized vegetables are lepidopteron, hemipteran, coleopteran, and dipterans. Despite losses caused by insect pests, proper management practices have not been followed as the crops are neither grown on a large scale for commercial production nor trading. Underutilized vegetables possess high nutritional values, and medicinal properties contain several vitamins and minerals that are not available in major vegetables. Besides, the vast zones of agroclimatic conditions of India are suited for growing varieties of underutilized vegetables. Proper cultivation practices could provide the demand for nutrition up to some extent in the increasing population.

Important Insect Pest of Underutilized Vegetables

Abelmoschus moschata

During the vegetative stage, defoliators are leaf roller, cotton semi looper, and flea beetle. Sucking pests complexes are aphid, mealybug, and mite.

Leaf Roller (*Sylepta derogata*) Fully grown larva is 20–25 mm long, greenish with a dark brown head. In the early stage, caterpillars feed on the epidermis of the ventral surface of leaves for 2 days, then roll the leaves, live within it, and are characterized as the hanging of rolled leaves on a plant and sometimes the presence of more than one caterpillar in the rolled leaf.

Cotton Semi Looper (*Anomis flava*) Caterpillar is 20–25 mm long, pale yellowish-green with five longitudinal dorsal lines. Initially, small holes on the leaves later feed voraciously, and only mid dribs and veins are left.

Flea Beetle (*Podagrica* spp.) The beetle causes small holes on the leaf present mainly on the leaves' upper surface, and the damage is severe.

Sucking Pest Aphid (*Aphis gossypii*) on upper leaves and mite (*Tetranychus urticae*) are found on lower leaves sucking sap and devitalized the plants. Mealybug (*Phenacoccus solenopsis*) attacks tender parts of plants.

In the reproductive stage, blister beetle (*Mylabris pastulata*) voraciously feeds on the flower and fruit borer (*Earias vitella*) bore on the fruits, thus limiting fruit productions and red cotton bug (*Dysdercus cingullatus*) suck sap from developing fruit caused by shriveling and premature fall off (Pandit and Das 2012).

*Amaranth (*Amaranthus *spp.)*

Hypolixus truncatulus (Curculionidae; Coleoptera): Major pest of cultivated *Amaranthus* spp. Both larva and adult caused damage, larva tunnel the stem and adult feed on leaves, under severe infestation 33–35 larvae recorded from a single plant. The active period of weevil is from April to October and is overwintered in cracks and crevices. Adult feeding on leaves caused irregular scratches on the leaves, and larvae zigzag tunneling on stem causes the loss of plants' vigor and vitality (Tara et al. 2009).

Spodoptera litura: Polyphagous pest with a wide host range, early stages feed gregariously by scratching the leaf epidermis and later stage feed voraciously and defoliate the plants (Ebert et al. 2011).

Helicoverpa armigera: The larvae bore into tender shoots, and inflorescences cause wilting, dropping, and drying of inflorescences (Levin et al. 2004).

Grasshopper, aphids, stem weevils, leaf miner, mole crickets, stink bugs, and mites also attack amaranths and cause considerable damage under favorable conditions (Ebert et al. 2011).

Sesbania grandiflora pers *(Agathi)*

Ceroplastodes spp. (scales) nymphs and adults colonize on tender parts of the plant and devitalize it (Shah et al. 1989). Other insect pests species attacking sesbania are *Adelphocoris lineolatus* (lucerne bug), *Bactrocera cucurbitae* (melon fly), and *Spodoptera litura* (taro caterpillar) (CABI).

Sechium edule *(Chow-Chow)*

Cotton aphid (*Aphis gossypii* Glov.): Both adults and nymphs suck the cell sap from leaves and tender parts of the plants, and the infested plants show stunted and sickly appearance.

Pumpkin fruit fly [*Bactrocera* (Dacus) *cucurbitae* Coquillett]: Adult puncture on the tender fruits and the numbers of punctures per fruit likely occur more in spineless fruits than spine fruits. The infested fruit drops off prematurely.

The minor pests attacking Chow-Chow as root feeders are as follows:

- White grubs (*Phyllophaga* spp.)
- Mole cricket (*Gryllotalpa fussor* Fab.)
- Field cricket (*Brachytrypes portentosus* Licht.)

Aboveground, feeders are pumpkin beetle (*Aulacophora foveicollis* Lucas.), blue pumpkin beetle (*Aulacophora lewisii* Baly), flea beetle (*Monolepta signata* Oliv.), tobacco caterpillar (*Spodoptera litura* Fab.), brown weevil (unidentified), stink bug [*Coridius* (Aspongopus) *janus* Fab.], spotted beetle (*Epilachna pusilanima* Mulsant.), banded blister beetle (*Mylabris phalerata* Pallas), hairy caterpillar, brown bug (*Agonoscelis nubila* Fab.), and stem-boring beetle (Neupane et al. 2006).

Parkia roxburghii *(Tree Bean)*

Cadra cautella Walker: The larvae are voracious feeders of green kernel, highly mobile tunnel through the kernel with the formation of a web with yellowish excreta near the entrance. Under heavy infestation, the whole kernel is eaten up, leaving only the cover with a tiny entrance hole (Thangjam et al. 2003). Other pest attacks are stem borer and bark caterpillars.

Psophocarpus tetragonolobus *(Winged Bean)*

Many insect species visited the winged bean as the fruits of the winged bean are tender and succulent. Though the pest incidence's severity depends on the climatic condition and other environmental factors prevail in the area. In the northeast part of India winged, the bean is severely attacked by a sucking pest complex consisting of aphids and spiraling whiteflies. The attack of spiraling whiteflies is very severe almost occurring at all the Manipur regions where winged beans are grown. The other pest complex attacking winged bean includes *Myllocerus curvicornis*, *Myllocerus undatus*, *Hypolixus truncatulus*, *Luperomorpha discoidea*, *Pagria signata*, *Parasa lepida*, *Syntomis passalis*, *Euproctis scintillans*, *Dysdercus olivaceus*, *Riptortus pedestris*, and species of *Helopeltis*, *Leptocentrus*, *Zoriada*, *Laius*, *Euops*, *Brachyplatys*, and *Cletus* (Shanthichandra et al. 1990).

Coccinia grandis

The insect species defoliating on *C. grandis* belong to coleopteran, *Epilachna vigintioctopunctata*, *Aulacophora* sp., and the lepidopteran defoliator *Diaphania indica* Saunder. The leaf's defoliator feeds by scratching on leaf epidermis results in lace-like skeletonized appearance and later stage holes on the leaves and completely skeletonized the infested leaves (Vijayasree et al. 2011).

*Sweet Gourd (*Momordica cochinchinensis*)*

Insect pest complex occurred in cucurbits attacks on sweet gourds. Among all species of insect pest attacking sweet gourd, red pumpkin beetle and fruit fly are more severe, causing losses to a great extent (Mukherjee et al. 2017).

*Basella (*Basella alba*)*

A group of sucking pest and defoliator attack on *B. alba*; among the sucking pest, 3 species of thrips attack *B. alba*: namely, *Haplothrips gowdeyi* and *Megalothrips ventralis* where both attack on the flower (Oparaocha and Okigbo 2003; Deshmukh and Gaikwad 2013) and *Frankliniella occidentalis* attacks on leaf, flower, and stem and white flies (*Bemesia tabaci*). The infestation by sucking pests results in yellowing of leaves, curling, and crumbling, and the whole plant looks sick. In addition to their direct injury, they are also a potential vector of plant pathogens. Their honeydew production led to the development of sooty mold that interferes with plants' photosynthetic activity and invites ants, limiting the exposure to natural enemies. Major defoliators infesting on *B. alba* are *Helicoverpa armigera*, *Spodoptera exigua*, *Phylotreta* sp., and leaf miner *Liriomyza* sp. (Deshmukh and Gaikwad 2013).

Nelumbo nucifera

Sucking pest: A group of sucking pests, namely, winged aphids, whitefly, mites (*Oligonychus* spp. and *Tetranychus* spp.) (Anonymous 2020), and flower thrips (*Scirtothrips dorsalis*) (Zhang et al. 2004) fed on the above-water parts of lotus. Whitefly, mites, and aphids feed on tender stems and leaves; due to the sucking up of plant cells, the infested parts become yellow and curl. Production of honeydew invites sooty mold development that interferes with the photosynthetic activity of the plant. Thrips infestation led to necrosis of infested tissue turning silvery to brown or black color, presence of feeding scars, distortion, and discoloration of flowers and buds (Jha et al. 2010).

Leaf-eating pest: Water lotus is vulnerable to attacks by several leaf roller species, leaf miner, and water lily beetle (*Galerucella nymphaeae*) (Anonymous 2020). Leaf roller cut the edges of leaves, rolled the leaves, and fed within. Lily beetles feed on the leaves and overwinter in plant debris, and many species of leaf miner mines on lotus leaves.

Root and stem pests: Species of caddisfly larvae feed on stems, roots, and flower buds. Fungus gnats (*Sciaridae* spp.) larvae feed on roots and stems, leading to wilt, discolor, and unhealthy appearance. They also act as a vector of Pythium, Fusarium, and Verticillium fungal infections (Anonymous 2020).

Integrated Pest Management

Before discovering insecticides, pest control was mostly achieved by manipulating cultural practices and utilizing natural enemies, mechanical and physical means. Later during the 1970s of the last century, the concept of integrated pest management was floated, which brought wide publicity and acceptance among the plant protection experts. In principle, integrated pest management (IPM) is a judicious combination of various nonchemical methods, and pesticides are employed when all other methods fail.

Biological agents are the primary tool in the IPM program. However, the toxic effect of insecticides often limits their effectiveness. Pesticides may be detrimental to natural enemies directly in them or by adversely affecting their longevity, fecundity, development of offspring and sex ratio, searching ability, aggressiveness, etc. (Jepson 1989; Croft 1990).

Cultural Method of Pest Management

It involves pest management through the manipulation of agronomic practices detrimental to pest growth and development such as tillage, crop rotation, intercropping, resistant or tolerant variety, irrigation, weeding, fertilizer application, etc. (Ehler 2006). Some of the cultural management practices are:

Resistant Variety The variety or cultivars that have resistance against insect attack may govern the plant's phonological or genotypic characteristics. Phenotypic or morphological characters are thorns, hairy leaves, unattractive color, absence of required nutrients, stem thickness, etc., which make it difficult for insects to feed and oviposit on the plant (Panda and Khush 1995; Yates and Michel 2018).

Tillage Ploughing timely disturb the normal life cycle of insects as the insect may undergo dormancy in the soil after the crop is harvested or shift to weed hosts that serve as carrying over pest for the next season. Ploughing exposes the hibernating or aestivating larva or pupa residing in soil and crop stubbles to extreme environments (Päts 1996). Furthermore, weeds are killed, and thus it eliminates the alternate source of host for insect pests. Tillage before planting exposes the white grub (Oliveira et al. 2000), and tillage exposed hairy caterpillar pupa to natural enemies and harsh environment (Bagenia and Meena 2017).

Date of Planting/Sowing Insects are cold-blooded animals, and their activity depends on environmental conditions, and different species are different in their activity period. Therefore change in planting date unsynchronized the crop susceptibility stage and activity period of insects.

Crop Sanitation and Other Practices Alternate hosts or weeds can serve as breeding sites in the absence of host plants that shift to host during the cropping season, causing tremendous losses. Cleaning plant residue and weeds or other alternate hosts will help in reducing the carryover pest population.

Cropping Pattern Seasonal activity and host preferences differ for the different pests. In monophagous and oligophagous pests, crop rotation with nonhost crops would eliminate the population carryover. Intercropping of marigold with tomato can reduce the attack by tomato fruit borer as the adults of tomato fruit borer prefer more on marigold for oviposition which can be collected and destroyed. Furthermore, root exudation from marigold also repelled nematodes. Strip cropping of mustard between rows of cabbage attracts diamondback moths that can be eliminated and save the sole crop (Sarkar et al. 2018). Flowering intercrop plants such as cowpea, buckwheat, coriander, etc., attract natural enemics and pollinators, increasing yield and reducing the pest population.

Nutrient Management Crop vigor highly depends on nutrient supply which may attract more insect pests. Balance fertilizer is always required, and also an increase in phosphorus, potassium, and other micronutrients harms insect growth and development.

Physical Control

The use of physical forces for the elimination of pests, such as the use of traps during the peak emergence of insects. The traps are based on several factors: physical factors like light and sticky trap, chemical factors include pheromone and bait trap, and mechanical factors like suction pitfall and emergence trap. The operation of light traps attracts both pests and nontarget organisms like natural enemies. To avoid such losses of natural enemies, it must be operated at a specific timing.

Sticky traps trap soft-bodied aerial insects such as leafhopper whiteflies, aphids, thrips, etc. The attraction toward trap highly depends on the trap's colors. The sticky substances are mostly vegetable oils or Vaseline, etc. Pigeon pea flies are attracted more toward white colors, whitefly and aphids on a yellow color, and thrips on blue color (Gobber et al. 2018).

Activated Clay Kaolinite clay mixed with grains killed the insect due to desiccation as it absorbed the lipids and led to the loss of the integrity of the insect cuticle.

Mechanical Control

This method involves killing insects through mechanically hitting or manually.

Collection and Destruction Handpicking of egg masses, larva, or nymphs and destruction. The first instar of hairy caterpillar and armyworm are gregarious feeding habits that could be easily collected and destroyed. Moringa hairy caterpillar on the bark of trees can be burnt and killed; grasshopper, leafhopper, and ear head bug are accessible to the net and killed.

Biological Control

It is the method of controlling pests like insects, mites, pathogens, and weeds using living organisms like predators, parasitoids, and parasites with human beings' intervention. Three basic principles of biological control are the introduction, augmentation, and conservation. The introduction includes the importation of bioagents that have the potential to destroy target pests. The introduced bioagents must be imported from the similar climatic condition, high adaptability rate, and species-specific: specific to particular species or host range, easy to multiply in laboratory and field condition, should not become a pest, free from hyper-parasitoids, high searching capability, and high fecundity and short life span. The active period of bioagents must synchronize with the activity period of pests.

The augmentation involves the steps toward the establishment of pests. It consists of two types of release: inundative release and inoculative release. In inundative release, a large number of bioagents are released to bring immediate control of the pest, while in an inoculative release, the individuals are released periodically, and management is expected from the progeny of released individuals.

Once the bioagents are established in the ecosystem, conservation practices must be taken to maintain the pest population below the threshold level and minimize pesticide use. Some of the conservation practices are as follows:

(a) *Selective pesticide*

If the pesticide application is needed in combating pests, the selected pesticide must be very specific only to target organisms and safe for nontarget organisms.

(b) *Avoidance of harmful agronomic practices*

Certain agronomic practices disturb the ecological balance, such as the burning of crop residues, high N fertilizer, weeding, and other operations. Many of the flowering weeds provide shelter and food sources of adult parasitoids; therefore, leaving some strips of weed during intercultural operation may help establish the natural enemies' population. The application of organic fertilizers is considered beneficial due to arthropods' conservation, natural enemies, and microbial bioagents.

(c) *Provision for food and shelter*

Intercropping with pollen and nectar-bearing plant along with sole crops provide the source of food for adult bioagents. The creation of reservoirs such as some areas in the field is left out from agronomic practices that could encourage natural enemies.

(d) *Beneficial management practices*

Irrigation and manipulation of relative humidity increase the activity of entomopathogens, chisel ploughing encourage soil-inhabiting bioagents, retention of crop stubbles and weed helps in establishment of natural enemies, and inactive stages of bioagents can be collected for laboratory multiplication, etc.

(e) *Suppression of ants*

Ants and hemipteran pests have a symbiotic relationship; ants sometimes protect the predator or parasitoid attack on prey; therefore, ants suppression would increase the efficiency of natural enemies.

(f) *Influence of plant types*

The morphological features of plants such as leaf structure, types of flower, colors, and other volatile components have a direct influence on predator and parasitoids efficiency. The list of bioagents against the insect pests of underutilized vegetables is mentioned in Table 18.1.

Botanicals Pesticides

A wide variety of plant species produce various secondary metabolites, namely, peptides, amines, antibiotics, alkaloids, cyanogenic glucosides, glucosinolates, organic acids, nonprotein amino acids, phenolics, polyphenols, polyacetylenes, terpenoids, lipophilic terpenes, sesquiterpene lactones, quinones, and defensive proteins, viz., chitinases, lectins, beta-1,3-glucanases, vicilins, systemins, arcelins, and enzyme inhibitors (Wahengbam et al. 2021). These compounds act as either allelochemical or antibiosis, and secondary metabolites are produced under stress conditions like insect infestation and other environmental stress, light drought, injury, etc. Research on botanical compounds is an emerging trend in today's agriculture as they are sustainable and nontoxic to the environment. Nawaz et al. (2016) reported that approximately 6000 plant species possessing insecticidal properties had been identified. Some of the commonly used essential oil and their active compounds are listed in Table 18.2.

Conclusion

There is neither a proper survey nor management tactics on the incidence of insect pests against underutilized vegetables as these vegetables are not cultivated on a large scale for commercialization. However, these plants' nutritional and medicinal value is far more, and some vitamins, minerals, and pharmaceuticals values present in underutilized vegetables cannot be compensated from major crops. Several pests and diseases occurred in underutilized vegetables, but there are no IPM tactics developed. Moreover, either the crops are left untreated or treated with very rare pesticides. The best possible way to manage pest incidences in underutilized

Table 18.1 List of parasitoids against some important pest of underutilized vegetables

Pest	Crop	Bioagents	References
Sylepta derogata	Polyphagous	*Apanteles opacus*	Kang et al. (2006)
Anomis flava	Polyphagous	*Elasmus flabellatus* (Fonscolombe), *Elasmus nudus* (Nees), and *Elasmus viridiceps*	Kravchenko et al. (2014)
Podagrica sp.	Polyphagous	*Brachymeria podagrica* (Fabricius) *Sarcodexia lambens* (Wiedemann)	Marchiori et al. (2002)
Aphis gossypii	Polyphagous	*Aphidius colemani* Viereck *Aphidius matricariae* (Haliday)	Zamani et al. (2007)
Tetranychus urticae	Polyphagous	*Cryptolaemus montrouzieri* Mulsant *Neoseiulus californicus* McGregor *Aphidius colemani* Viereck)	Urbaneja et al. (2008)
Phenacoccus solenopsis	Polyphagous	*Aenasius bambawalei*	Sahito et al. (2017)
Mylabris pustulata	Polyphagous	*Rhynocoris marginatus*	Ambrose and Claver (1999)
Earias vitella	Polyphagous	*Bracon lefroyi* *Bassus fabiae*	Prasanth and Shetgar (2018)
Dysdercus cingulatus	Polyphagous	*Rhynocoris marginatus*	Ambrose and Claver (1999)
Hypolixus truncatulus	Polyphagous	*Entedon* sp.	Louw et al. (1995)
Spodoptera litura	Polyphagous	*Campoletis chlorideae Uchida Eriborus argenteopilosus* (*Cameron*)	Bajpai et al. (2006)
Helicoverpa armigera	Polyphagous	*Habrobracon hebetor* Say (*Hh*)	Allahyari et al. (2020)
Ceroplastodes spp.	Polyphagous	*Lakshaphagus fusiscapus* (Agarwal)	Gupta and Joshi (2013)
Adelphocoris lineolatus	Polyphagous	*Peristenus digoneutis* Loan *Peristenus relictus* (Ruthe)	Pansa et al. (2012)
Bactrocera cucurbitae	Polyphagous	*Pachycrepoideus vindemmiae* Rondará	Zhao et al. (2013)
Bactrocera (Dacus)	Polyphagous	*Fopius arisanus* (Sonan), F. vandenboschi (Fullaway), *Diachasmimorpha longicaudata* (Ashmead), and *Psyttalia incisi* (Silvestri)	Gu et al. (2018)
Phyllophaga spp.	Polyphagous	*Tiphia vernalis* Rohwer *Tiphia popilliavora* Rohwer *Istocheta aldrichi* (Mesnil)	Redmond and Potter (2010)
Aulacophora foveicollis Lucas	Polyphagous	*Trichogramma* spp. *Brachymeria tachardiae* *Trichospilus pupivora*	Lal et al. (2014)
Bemesia tabaci	Polyphagous	*Delphastus catalinae* *Delphastus catalinae* *Macrolophus caliginosus* *Chrysoperla carnea*	Gerling et al. (2001)
Phylotreta sp.	Polyphagous	*Bracon* spp.	Mureithi et al. (2015)

Table 18.2 List of some commonly used essential oils in pest control

Essential oils	Insecticidal compounds	Insects	References
Wintergreen oil	Methyl salicylate	Various insects	Zhang et al. (2014)
Geranium oil	Trans/cis-rose oxides, p-menthone, l-menthone, b-bourberene, b-citronellol, and geraniol	Various insects	
Pennyroyal oil	p-menthone, l-menthone, and pulegone	Various insects	
Rosemary oil	1,8-cineole and camphor	Various insects	
Clove oil	Eugenol carvacrol thymol cinnamaldehyde	BMSB and various insects	
Spearmint oil	l-carvone,	BMSB	
Ylang Ylang oil	4-methylanisole, methyl benzoate, and benzyl acetate	BMSB	
Lemongrass oil	Citronellal, citral α-pinene	BMSB and various insects	
Pongamia glabra, syn. P. pinnata (Karanj)	Karanjin and pongamol	*Dysdercus koenigii* (IGR), *S. oryzae* mites	Walia et al. (2017)
Curcuma longa (turmeric)	Curcumin – I, II, III	Red flour beetle, brown plant hopper, diamond back moth	
Neem	Azadirachtin saponins	Various insects	
Neem oil+ turmeric oil		Spilosoma obliqua, D koenigii	
Tagetes species (Marigold)	Ocimenone, β-ocimene, tagetone, tagetenone, dihydrotagetone, limonene, and β-caryophyllene	*Manduca sexta, Pieris rapae, Simulium verecundum* (blackfly), and *A. aegypti* larvae, Mexican bean weevil (insecticidal), *Triatoma infestans* (repellants), *Spodoptera frugiperda* (antifeedants)	
Anethum sowa (Indian dill)	Dillapiole (5-allyl-6,7-dimethoxy-1,3-benzodioxole)	Yellow stem borer *Scirpophaga incertulas* (ovicidal activity), grain beetle, *C. maculatus* (oviposition deterrents)	
Eupatorium adenophorum (Crofton weed)	Stigmasterol	Mites and insecticidal properties	Nong et al. (2017)
Eucalyptus species	1,8-cineole citronellal Z- and α- citral α-pinene	Insecticidal and repellant activity	Batish et al. (2008)
Cinnamomum zeylanicum	Cinnamaldehyde, linalool	Stored grain pest	Brari and Thakur (2015)

(continued)

Table 18.2 (continued)

Essential oils	Insecticidal compounds	Insects	References
Ocimum spp.	p-cymene estragosl linalool linoleic acid eucalyptol eugenol camphor citral thujone limonene ocimene and others	Various insects	Umerie et al. (1998)
Cedrus wood oil	Pentane, acetonitrile, atlantone, and himachalene	*Callosobruchus chinensis*, *Plutella xylostella* (larvicidal activity)	Chaudhary et al. (2011)
Acorus calamus	β- asarone	Insecticidal and antigonadal action	Koul et al. (1990)

vegetables is to promote biological control agents and other eco-friendly pest management tactics. Neglecting chemical pesticides in management tactics provides chemical residue-free vegetables and is safer for the environment and nontarget organisms. Pest surveys and monitoring on their seasonal activity, eco-biology, and other environmental resistance against the pest need to be discussed to develop effective management strategies.

References

Allahyari, R., Aramideh, S., Safaralizadeh, M. H., Rezapanah, M., & Michaud, J. P. (2020). Synergy between parasitoids and pathogens for biological control of Helicoverpa armigera in chickpea. *Entomologia Experimentalis et Applicata, 168*(1), 70–75.

Ambrose, D. P., & Claver, M. A. (1999). Suppression of cotton leafworm Spodoptera litura, flower beetle *Mylabris pustulata* and red cotton bug *Dysdercus cingulatus* by *Rhynocoris marginatus* (Fabr.)(Het., Reduviidae) in cotton field cages. *Journal of Applied Entomology, 123*(4), 225–229.

Anonymous (2020). Pests on Water Lotus Plants. https://homeguides.sfgate.com/pests-water-lotus-plants-66760.html Date of accessed: 1/07/2020.

Bagenia, P. S., & Meena, K. A. (2017). A study on adoption of integrated pest management practices for red hairy caterpillar, *Amsacta moorei* Butler in groundnut in Rajasthan. Agriculture Update, 12, 714–719.

Bajpai, N. K., Ballal, C. R., Rao, N. S., Singh, S. P., & Bhaskaran, T. V. (2006). Competitive interaction between two ichneumonid parasitoids of Spodoptera litura. BioControl, 51, 419–438.

Batish, D. R., Singh, H. P., Kohli, R. K., & Kaur, S. (2008). Eucalyptus essential oil as a natural pesticide. For. Ecol. Manag, 256, 2166–2174

Brari, J., & Thakur, D. R. (2015). Insecticidal efficacy of essential oil from *Cinnamomum zeylanicum* Blume and its two major constituents against *Callosobruchus maculatus* (F.) and *Sitophilus oryzae* (L.). J. Agril. Tech., 11, 1323–36.

CABI. Sesbania grandiflora (sesbania). Invasive Species Compendium. https://www.cabi.org/isc/datasheet/49455#26D60D78-FD70-4E4C-B30E-45AA0AB484C5

Chaudhary, A., Sharma, P., Nadda, G., Dhananjay Kumar, T., & Bikram, S. (2011). Chemical composition and larvicidal activities of the Himalayan cedar, *Cedrus deodara* essential oil and its fractions against the diamondback moth, *Plutella xylostella*. J. Insect Sci. 11.

Croft, B. A. (1990). Arthropod biological control agents and pesticides. John Wiley and Sons Inc..

Deshmukh, S. A., & Gaikwad, D. K. (2013). A Review of the package of practices, insects and non-insect pests of *Basella alba* L. J Crop Sci Technol, 3, 1–3.

Ebert, A. W., Wu, T. H., & Wang, S. T. (2011). Vegetable amaranth (Amaranthus L.). AVRDC publication, (11–754), 9.

Ehler, L. E. (2006). Integrated pest management (IPM): definition, historical development and implementation, and the other IPM. Pest Manag. Sci., 62, 787–789.

Gerling, D., Alomar, Ò., & Arnò, J. (2001). Biological control of *Bemisia tabaci* using predators and parasitoids. Crop Prot., 20, 779–799.

Gobber, C., Alonso, J. L., & Jonsson, Å. (2018). U.S. Patent Application No. 15/780,065.

Gu, X., Cai, P., Yang, Y., Yang, Q., Yao, M., Idrees, A., Ji, Q., Yang, J. & Chen, J. (2018). The response of four braconid parasitoid species to methyl eugenol: optimization of a biocontrol tactic to suppress *Bactrocera dorsalis*. Biol. Control, 122, 101–108.

Gupta, A., & Joshi, S. (2013). Additions to the fauna of parasitic wasps (Hymenoptera: Chalcidoidea) and coccoids (Hemiptera: Coccoidea) from the Andaman and Nicobar Islands, India, with illustrations and diagnosis. J. Threat. Taxa, 5, 4542–4555.

Jepson, P. C. (1989). Pesticides and non-target Invertibrates.(ed.) Intercept. Wimborne. Dorset, UK.

Jha, V. K., Seal, D. R., & Kakkar, G. (2010). Chilli thrips *Scirtothrips dorsalis* Hood (Insecta: Thysanoptera: Thripidae). EDIS, 2010(1).

Kang, X. X., Zhao, G. M., Gong, Y. F., U, I., & Yang, Y. Z. (2006). Biological Characteristics of Adult Apanteles opacus, a Parasitoid of *Sylepta derogata*. Chinese J. Biol. Control, 22, 275.

Koul, O., Smirle, M. J., & Isman, M. B. (1990). Asarones fromAcorus calamus L. Oil. J. Chem. Ecol., 16, 1911–1920.

Kravchenko, V. D., Müller, G. C., Allan, S. A., & Yefremova, Z. A. (2014). Seven invasive owlet moths (Lepidoptera: Noctuidae) in Israel and their potential parasitoids (Hymenoptera: Chalcidoidea). Phytoparasitica, 42, 333–339.

Lal, J., Sharma, K. C., Nama, C. P., & Panwar, D. (2014). Pests of Cucurbitaceous Vegetables and Their Management. Popular Kheti, 2, 78–82.

Levin, L., Ranjith, A. M., & Mathew, M. P. (2004). Record of *Helicoverpa armigera* (Hübner) on amaranthus in Kerala. Insect Environ., 10, 108–109.

Louw, S., Van Eeden, C. F., & Weekds, W. J. (1995). Curculionidae (Coleoptera) associated with wild and cultivated *Amaranthus spp*.(Amaranthaceae) in South Africa.

Marchiori, C. H., Pereira, L. A., Silva Filho, O. M., Ribeiro, L. C. S., & Borges, V. R. (2002). *Brachymeria podagrica* (Fabricius)(Hymenoptera: Chalcididae) as parasitoids of *Sarcodexia lambens* (Wiedemann)(Diptera: Sarcophagidae) in Brazil. Arq. Bras. Med. Vet. Zoo. 69, 121–122.

Mukherjee, A., Karmakar, A., & Barik, A. (2017, June). Bionomics of *Momordica cochinchinensis* fed *Aulacophora foveicollis* (Coleoptera: Chrysomelidae). In Proceedings of the Zoological Society (Vol. 70, No. 1, pp. 81–87). Springer India.

Mureithi, D. M., Mworia, J. K., Meyhoefer, R., Murungi, L. K., Losenge, T., Akutse, K. S., Ekesi, S. & Fiaboe, K. K. (2015). Survey for pest and natural enemies of amaranth and African nightshades in Kenya and Tanzania.

Nawaz, M., Mabubu, J. I., & Hua, H. (2016). Current status and advancement of biopesticides: microbial and botanical pesticides. *J Entomol Zool Stud*, *4*(2), 241–246.

Neupane, F. P., Sharma, M. D., & Neupane, K. R. (2006). Incidence of Insect Pests on Chayote, *Sechium edule* (Swartz.) in Nepal. J. Inst. Agri. and Animal Sci., 27, 161–164.

Nong, X., Yang, Y., Yang, G., Chen, F., Tang, M., & Wang, G. (2017). Toxicity of stigmasterol isolated from crofton weed, Eupatorium adenophorum Spreng. against a rabbit ear mite, psoroptes cuniculi. Pak. J. Zoo., 49, 1197–1200.

Oliveira, L. J., Hoffmann-Campo, C. B., & Garcia, M. A. (2000). Effect of soil management on the white grub population and damage in soybeanPesqui. Agropecu. Bras., 35, 887–894

Oparaocha, E. T., & Okigbo, R. N. (2003). Thrips (Thysanoptera) of vegetable crops (okro, spinach, garden egg and pumpkin) grown in Southeastern Nigeria. Plant Prot. Sci.-UZPI (Czech Republic).

Panda, N., & Khush, G. A. (1995). Host plant resistance to insects. CAB international.

Pandit, M. K., & Das, B. K. (2012). Nutritional cum medicinal properties and insect-pest complex of Abelmoschus moschatus-an underutilized vegetable. In Proceedings of the International Symposium on Minor Fruits and Medicinal Plants for Health and Ecological Security (ISMF & MP), West Bengal, India, 19–22 December, 2011 (pp. 47–50). Bidhan Chandra Krishi Viswavidyalaya.

Pansa, M. G., Guidone, L., & Tavella, L. (2012). Distribution and abundance of nymphal parasitoids of Lygus rugulipennis and Adelphocoris lineolatus in northwestern Italy. Bull. Insectology, 65, 81–87.

Päts, P. (1996). Management of crop residues to reduce the aestivating population of stemborers in maize. International J. of Pest Manag., 42, 151–156.

Prasanth, Y., & Shetgar, S. S. (2018). Life table of Earias vitella (F.) on okra cultivars. Indian J. Entomol., 80, 1045–1052.

Redmond, C. T., & Potter, D. A. (2010). Incidence of turf-damaging white grubs (Coleoptera: Scarabaeidae) and associated pathogens and parasitoids on Kentucky golf courses. Environ. Entomol., 39, 1838–1847.

Sahito, H. A., Kousar, T., Mangrio, W. M., Mallah, N. A., Jatoi, F. A., Shah, Z. H., & Kubar, W. A. (2017). Stage Specific Life Table of Invasive Pest Mealybug, Phenacoccus Solenopsis (Tinsley) Under Cotton Field Conditions. Current Res. Agril. Sci., 4, 43–50.

Sarkar, S. C., Wang, E., Wu, S., & Lei, Z. (2018). Application of trap cropping as companion plants for the management of agricultural pests: a review. Insects, 9, 128.

Shah, N. K., Belavadi, V. V., & Pal, R. N. (1989). Occurrence of the scale insect Ceroplastodes sp.(Homoptera: Coccidae) on Sesbania. J. Andaman Sci. Assoc., 5.

Shanthichandra, W. K. N., Gunasekera, S. A., & Price, T. V. (1990). Diseases and pests of the winged bean (*Psophocarpus tetragonolobus* [L.] DC.) in Sri Lanka. Int. J. Pest Manag., 36, 375–379.

Tara, J. S., Azam, M., Ayri, S., Feroz, M., & Ramamurthy, V. V. (2009). Bionomics of Hypolixus truncatulus (F)(Coleoptera: Curculionidae. Lixinae: Lixini) a major pest of *Amaranthus caudatus* L. Munis Entomol. Zoo., 4, 510–518.

Thangjam, R., Damayanti, M., & Sharma, G. J. (2003). Cadra cautella Walker (Lepidoptera: Crambidae: Phycitinae)–a pest on *Parkia timoriana* (DC.) Merr. in Manipur. Current Sci., 85, 725.

Umerie, S. C., Anaso, H. U., & Anyasoro, L. J. C. (1998). Insecticidal potentials of Ocimum basilicum leaf-extract. Bioresource Tech., 64, 237–239.

Urbaneja, A., Pascual-Ruiz, S., Pina, T., Abad-Moyano, R., Vanaclocha, P., Montón, H., ... Jacas, J. A. (2008). Efficacy of five selected acaricides against Tetranychus urticae (Acari: Tetranychidae) and their side effects on relevant natural enemies occurring in citrus orchards. Pest Manag. Sci.: formerly Pesticide Sci., 64, 834–842.

Vijayasree, V., Nalinakumari, T., & Xavier, G. (2011). Seasonal abundance and damage potential of major defoliators infesting Coccinia, *Coccinia grandis* (L.) Voigt. in Kerala. Pest Manag. Horti. Ecos., 17, 113–120.

Wahengbam, J., Bhushan, L. S., Patil, J. B., & Pathma, J. (2021). Insecticides Derived from Natural Products: Diversity and Potential Applications. In *Current Trends in Microbial Biotechnology for Sustainable Agriculture* (pp. 403–437). Springer, Singapore.

Walia, S., Saha, S., Tripathi, V., & Sharma, K. K. (2017). Phytochemical biopesticides: some recent developments. Phytochem. Reviews, 16, 989–1007

Yates, A. D., & Michel, A. (2018). Mechanisms of aphid adaptation to host plant resistance. Current opinion in insect science, 26, 41–49.

Zamani, A. A., Talebi, A., Fathipour, Y., & Baniameri, V. (2007). Effect of temperature on life history of *Aphidius colemani* and *Aphidius matricariae* (Hymenoptera: Braconidae), two parasitoids of Aphis gossypii and *Myzus persicae* (Homoptera: Aphididae). Environ. Entomol., 36, 263–271.

Zhang, Q. H., Schneidmiller, R. G., Hoover, D. R., Zhou, G., Margaryan, A., & Bryant, P. (2014). Essential oils as spatial repellents for the brown marmorated stink bug, H alyomorpha halys (S tål)(H emiptera: P entatomidae). J. Appl. Entomol., 138, 490–499.
Zhang, W. Q., LU, Y. F., & Tong, X. L. (2004). A New Insect Pest of Lotus Flower——*Scirtothrips dorsalis* Hood and its Control Strategies [J]. J. Chinese Landscape Architecture, 4.
Zhao, H. Y., Zeng, L., Xu, Y. J., Lu, Y. Y., & Liang, G. W. (2013). Effects of host age on the parasitism of *Pachycrepoideus vindemmiae* (Hymenoptera: Pteromalidae), an ectoparasitic pupal parasitoid of *Bactrocera cucurbitae* (Diptera: Tephritidae). Florida Entomologist, 451–457.

Chapter 19
Cultivation of Underutilized Vegetables in a Hydroponic and Aeroponic System

Sophiya Bhatta and Savita

Introduction

Among all the countries in the world, India is one of the countries with a vast population. Industrialization and urbanization have also increased rapidly, decreasing cultivable land by a considerable percentage. Indian population is approx. 1.34 billion and continuously growing. Feeding such a huge population is one of the most challenging tasks, so to meet the growing population's food requirement, different vegetable production systems are developed and are in use such as hydroponic systems and aeroponic systems. Hydroponics is a method of growing plants in water without the use of soil, utilizing mineral nutrient solutions. In place of soil, sometimes artificial growing media are used. Perlite, vermiculite, expanded clay, coir, brick shards, polystyrene packing peanuts, and wood fiber are also common materials used in the hydroponic system. In the current situation, soilless farming can be successfully launched and considered an alternative option for growing nutritious food plants, crops, or vegetables (Butler and Oebker 1962).

The method of growing plants without the use of soil in an air or mist environment is called aeroponics, whereas cultivation with soil is named geoponics. In hydroponic and aeroponic systems, vertical space is used to produce vegetables that are unutilized in conventional farming. The utilization of vertical space enables farmers to produce more food in a small land area. A hydroponics system can be established outdoors as well as inside a protected structure. Establishing hydroponics and aeroponics inside a protected structure comes with numerous advantages. Horticultural crops grown inside a secure facility give high yield and high quality of

S. Bhatta (✉)
School of Agriculture, Lovely Professional University, Phagwara, Punjab, India

Savita
Graphic Era Hill University, Dehradun, Uttarakhand, India

Savita et al. (eds.), *Production Technology of Underutilized Vegetable Crops*,
https://doi.org/10.1007/978-3-031-15385-3_19

resources that are used efficiently. Horticultural crops typically acquire four to eight times higher profits than other crops. This type of agricultural production system can help generate a reliable source of income and employment among rural people and the youth. Therefore, underutilized vegetable crops like lettuce, celery, parsley, Chinese cabbage, broccoli, and rhubarb can be grown under a hydroponics system of cultivation, whereas Tania, giant taro, and swamp taro crops can be grown under an aeroponic system of cultivation.

Overview and Concept of Hydroponic and Aeroponic System

The land is a foremost basic requirement for any agriculture production. Due to urbanization, there is significantly less or no agricultural land left in urban areas, which has become one of the strong limitations for any type of crop production in urban areas. Therefore, scientists developed hydroponic and aeroponic systems to overcome this constraint and make farming possible in any part of the earth. The term "hydroponics" is derived from the Greek terms "hydros," which means water, and "ponos," which means to work. Hydroponics is a soil-free form of plant cultivation in which plants are grown in mineral nutrient solutions (Beibel 1960). In a hydroponic system, artificial growing media/inert media are used instead of soil to improve the roots of the plants. A nutrition solution is placed over the roots and supplied with oxygen and water (Maharana and Koul 2011). Aeroponic comes from the Greek word "aer" which means "air" and ponos which means "labor" (Beibel 1960; Reyes et al. 2011). Aeroponics is a method of growing plants in the air with only water and air and artificial support. In aeroponics, plants are grown without soil or an aggregate medium (Osvald et al. 2001). The roots are suspended in the air inside a closed container in a dark room in an aeroponic system. Roots of the plant will get nutrients to form nutrient-rich spray through an atomizer. Researchers have identified plants that are grown aeroponically and hydroponically attain faster growth, higher yield, and are superior in quality than conventionally grown plants (Table 19.1) (Fig. 19.1).

Table 19.1 Comparison between the yield obtained from hydroponics and normal soil (Singh and Singh 2012)

	Yield	
Name of crop	The agricultural equivalent per acre	The hydroponic equivalent per acre
Potato	8 tons lb	70 tons
Beet root	9000 lb	20,000 lb
Cabbage	13,000 lb	18,000 lb
Peas	2000 lb	14,000 lb
Tomato	5–10 tons	180 tons
Cauliflower	10–15,000 lb	30,000 lb
French bean	–	42,000 lb
Lettuce	9000 lb	21,000 lb
Lady's finger	5–8000 lb	19,000 lb

Fig. 19.1 Lettuce is grown under a hydroponic system

Types of hydroponics system The amount of water, nutrients, and photoperiod in most hydroponic systems is automatically adjusted to meet the needs of individual plants (Resh 2012). Hydroponic systems can be set up in a variety of ways. However, all the factors such as availability of space availability of suitable growing media, availability of technical and human resources, etc must be taken into account when choosing a system. Hydroponics system is subdivided into two approaches which are as follows:

(i) Circulating methods: It is also known as "Active or Closed System" or a "Continuous-Flow Solution Culture," and it includes the following techniques: (a) Ebb and flow method, (b) nutrient film technique (NFT), and (c) deep flow method (DFT).
(ii) Noncirculating methods: It is also known as "passive or open systems" or "static solution culture," and it includes the following techniques: (a) Root dipping technique, (b) floating technique, (c) wick technique.

Circulating Methods

The excess solution containing nutrients is stored as it runs off and reused in a circulating technique in the reservoir. Because the extra fluid is reused, it makes better use of nutritional solutions; also, even though a circulating technique does not require accurate monitoring of the watering cycles, it typically allows for a less expensive timer. This system can experience large fluctuations in nutrient intensity and pH, necessitating periodic checks and modifications.

(a) *Ebb and flow*: The growing tray in the ebb and flow system is temporarily filled with the nutrient solution before being drained back into the reservoir. A pump is submerged in nutrient solution and connected to a timer to carry out this action. A water pump floods nutrient solution and water from a tank into a grow bed until it reaches a given level and stays there for a certain period, delivering nutrients and moisture to plants. Furthermore, a wide range of vegetable crops can be grown, but the problem of algae, mold, and root rot is very common

(Nielsen et al. 2006). As a result, a redesigned system equipped with a filter unit is necessary.

(b) *Nutrient film technique (NFT)*: This technique was invented in England in the mid-1960s by Dr. Alen Cooper to mitigate the limitations of the ebb and flow system. The nutrient film method is one of the most widely used and efficient hydroponics techniques used worldwide. In the growing tray, there is a continuous flow of a solution containing nutrients, and hence, the timer is not required in this technique (Domingues et al. 2012). A pump delivers the solution containing nutrients into the growing tray. Growing tray in nutrient film technique does not contain any growing medium. The system is slightly inclined to flow the nutrient solution through the roots and back into the tank/reservoir. Plant roots are placed in a channel with a continuous flow of nutrient solution. The main drawback of this technique is that the roots become more susceptible to fungal diseases due to their continuous exposure to water-containing nutrients. This technique can be used to grow several leafy vegetables, with lettuce being the most popular.

(c) *Water culture*: This is an active hydroponics system. In this system, plant roots are completely immersed into the nutrient solution. Styrofoam is used as a platform to hold plants. To supply oxygen, an air pump is used. The air pump produces bubbles in the nutrient solution, which provides sufficient oxygen to the plant roots to help them attain healthy and productive growth. Monitoring the oxygen and nutrient concentrations, salinity, and pH (Domingues et al. 2012) as algae and molds can increase in the reservoir. This technique is best for taller plants that produce fruits, such as cucumbers and tomatoes, which thrive well in this environment.

Noncirculating Method

The runoff is not collected using the noncirculating approach. It also requires a more reliable timer so that watering cycles may be modified to ensure that plants receive adequate nutrients (Awad et al. 2017). The nutrient concentration and pH will not alter because the excess solution containing nutrients is not drained back into the tank/reservoir.

(a) *Root Dipping Technique*

Planting is done in tiny pots using tiny growing materials. The pots are positioned so that the bottom 2–3 cm are buried in the solution containing nutrients. Only a few roots are immersed in the solution, whereas some are exposed to the air, which is critical for air and nutrient absorption. This approach is straightforward and may be implemented with readily available materials. This growing "low-tech" approach is economical to build and requires minimal care. This strategy does not necessitate the purchase of expensive items such as power, a water pump, channels, and so on. Root crops, on the other hand, must be grown in an inert medium (beet, radish, etc.).

(b) *Floating Technique*

In this method, small and shallow containers (10 cm deep) are used. Plants in these containers are affixed to a Styrofoam sheet or other light plate and allow their roots to be immersed in the artificially aerated solution containing nutrients (Singh and Singh 2012).

(c) *Wick System*

This is the simplest and most passive type of hydroponic system (Shrestha and Dunn 2013). In this method, a wick is used to draw nutrient solution from the tank/reservoir into the growing medium. Wick uses capillary action for removing the nutrient solution from the reservoir. Perlite and cocopeat are considered the growing media for the wick system (Figs. 19.2, 19.3, 19.4 and 19.5).

pH and Electrical Conductivity (EC) Management in a Hydroponic System

Water-soluble plant nutrients-based fertilizers used in hydroponics are generally inorganic and ionic types. Different chemical combinations are used to supply all 17 nutrients/elements required for plant growth. Hoagland's solution is the most often used nutrient solution in hydroponic systems (Table 19.2).

Due to osmotic pressure, higher EC prevents nutrient absorption, while lower EC harms plant health and yield. As a result, proper EC management in the hydroponics technique can significantly increase crop yield and quality (Gruda 2009). The

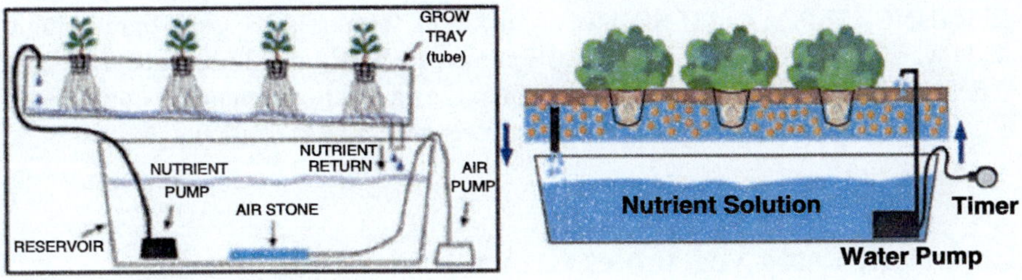

Fig. 19.2 Illustration of ebb and flow

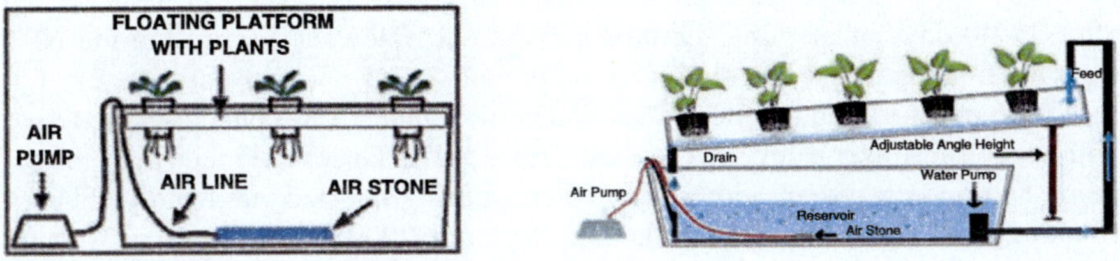

Fig. 19.3 Illustration of nutrient film technique

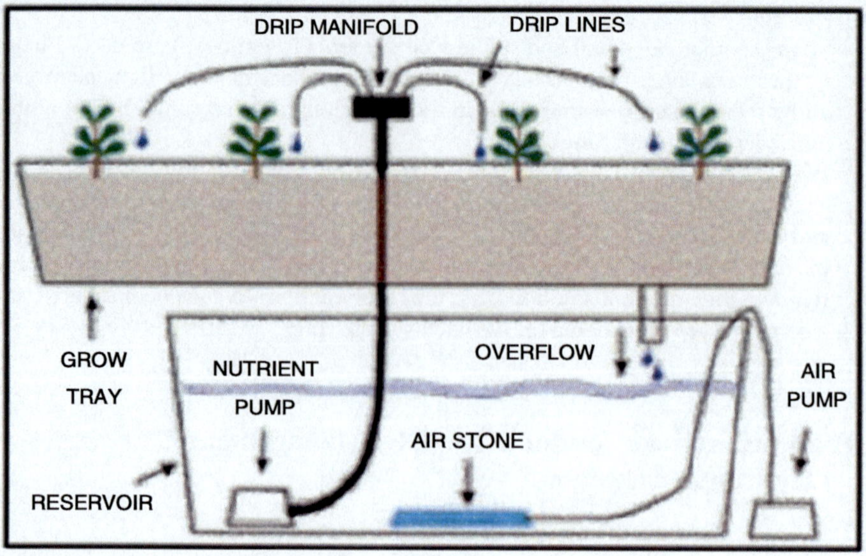

Fig. 19.4 Illustration of water culture

availability of critical plant components in a nutrition solution is determined by pH. For most species, 5.5–6.5 is the ideal pH requirement of nutrient solutions for plant growth and development (Trejo-Téllez and Gómez-Merino 2012), but certain species may fall outside of this pH range. As the plants mature, they will alter the amount of the nutrients in the solution by reducing some nutrients more quickly than others. To maintain the 5.5–6.5 pH of an ideal solution, a combination of three acids (HNO_3, H_3PO_4, and H_2SO_4) was found to be far more successful than a single acid (Wang et al. 2017). A change in pH can create nutritional imbalance, resulting in deficiency or toxicity symptoms in the plant. As a result, maintaining optimal pH, EC, and nutrient levels in hydroponic solutions necessitates caution.

Important Leafy Vegetables Grown Under Hydroponics

The life cycle of lettuce cultivated in a hydroponic system is shorter than lettuce grown by the traditional method. Hydroponically grown lettuce can be harvested after 35–40 days of planting. Lettuce may be cultivated successfully in the NFT (nutrient film technique) method, with up to eight crops per year possible. For lettuce yield optimization, the horizontal and vertical hydroponic systems were tested with various fertilizer solutions (Touliatos et al. 2016). Growing lettuce in a recirculating hydroponic system with a 10 × 20 cm spacing boosted yield and yield contributing traits significantly (Maboko and Du Plooy 2009). There was a substantial difference in nitrate content of lettuce leaves in both soilless and soil culture; however, other features such as ascorbic acid content, dry weight, and leaf area were

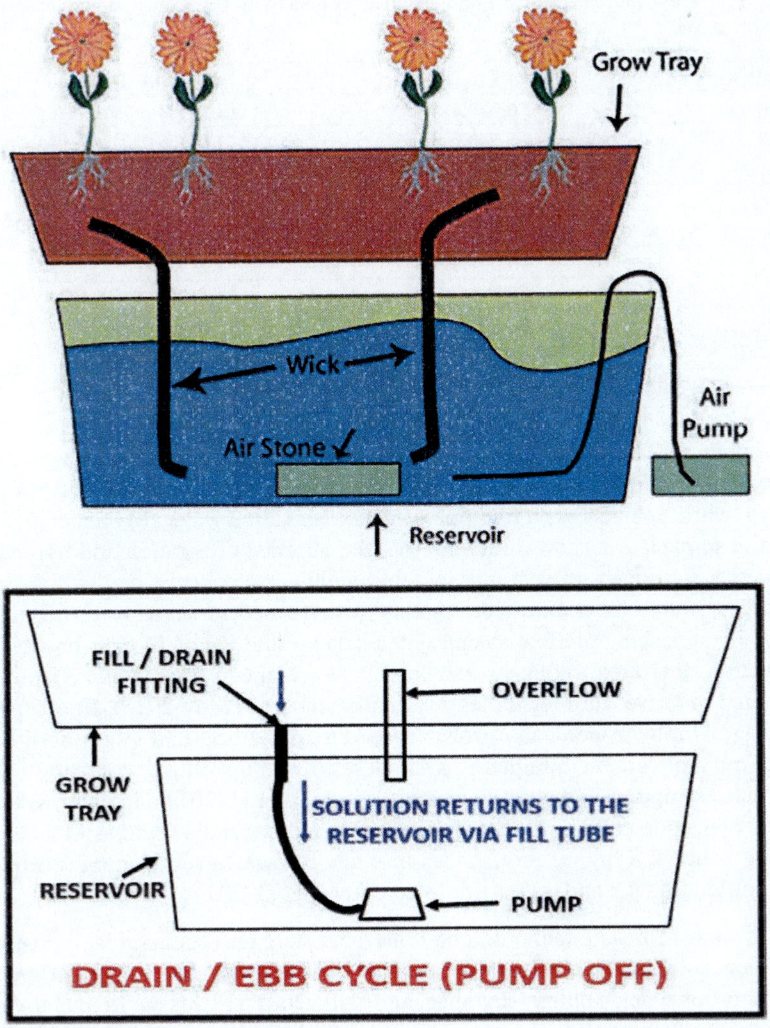

Fig. 19.5 Illustration of a wick system

unaffected. The air space between the nutrient solution and the tank cover also influences the best lettuce output in a noncirculated and non-aerated system (Frezza et al. 2005). Fallovo et al. (2009) found that in a floating system, the composition of nutrients in the solution did not affect leaf area index and marketable yield of lettuce cultivated.

Delaying harvesting, on the other hand, improved production, reduced nitrate levels, and reduced health risks. Apart from lettuce, various hydroponic experiments employing spinach as a model crop have lately been done. The yield of spinach in hydroponic, aquaponic, and standard systems was compared (Ranawade et al. 2017). The plants were supported by perlite (aquaponics) and sphagnum moss

Table 19.2 The requirement of pH and electrical conductivity (EC) for hydroponically grown vegetable crops

Crop	pH	EC (dSm^{-1})
Asparagus	6.0–6.8	1.4–1.8
Basil	5.5–6.0	1.0–1.6
Broccoli	6.0–6.8	2.8–3.5
Celery	6.5	1.8–2.4
Leek	6.5–7.0	1.4–1.8
Lettuce	6.0–7.0	1.2–1.8
Pak choi	7.0	1.5–2.0
Parsley	6.0–6.5	1.8–2.2
Rhubarb	5.5–6.0	1.6–2.0

Sharma et al. (2018)

(hydroponics). The yield of spinach cultivated in an aquaponic system was slightly higher than that of spinach grown in a hydroponic system. Salinity inhibits plants growth, while spinach has a five-ppt tolerance for saline water (Mwazi et al. 2010). Because spinach is a short-duration crop, the absence of aeration and hypoxia did not have a significant impact on yield and yield components when cultivated in a floating system. Nonetheless, the quality was harmed in some way (Lenzi et al. 2011). Plant spacing of 10 × 45 cm with a harvest interval of 14 days boosted crop production, leaf area, biomass, and fresh leaf weight in hydroponic Swiss chard cultivated in gravel film technology (Maboko and Du Plooy 2013). In comparison to soil-grown, hydroponically grown Swiss chard, lettuce, and sweet basil have a higher mineral content, a higher root/shoot ratio, and low nitrate content; however, their nutrient uptake and output are lower (Bulgari et al. 2017). In contrast to soil-grown, hydroponically produced Swiss chard, lettuce, and sweet basil have higher mineral content, a higher root/shoot ratio, and a lower level of nitrates; however, their nutrient uptake and output are lower (Awad et al. 2017).

Drip system This is popular among both household and commercial crops. Water or nutrient solution from the tank is delivered to individual plant roots in an appropriate proportion using a pump (Rouphael and Colla 2005). Plants are frequently grown in a medium that is relatively porous to allow the nutrient solution to drip slowly. It is possible to grow a variety of crops systematically while conserving water.

Aeroponic System

Above the expanding medium, the N.F.T. device is primarily air, making it the most high-tech technique of hydroponic gardening (Maboko et al. 2011). The roots are kept in the air and misted with a nutrient solution. The misting is usually done every several minutes. Because the roots are exposed to the air with the N.F.T. approach, the roots might soon dry out if the misting cycles are broken. The nutrient pump is controlled by a timer in the same manner that conventional hydroponic systems are, but the aeroponic system requires a short cycle timer that runs the pump every few minutes for a few seconds.

Types of Aeroponic System

There are two types of aeroponic systems:

1. Low-pressure units
2. High-pressure units

1. *Low-pressure units*: This system delivers the mist to thc plants at low prcssure with large droplets. The plant's roots are hung above a nutritional solution reservoir or an interior and connected to a reservoir channel. A low-pressure pump pushes the nutrient solution into the reservoir using jets or ultrasonic transducers. When plants reach maturity, dry parts of the root systems affect the units, preventing adequate nutrient uptake. Because of the high cost, these units lack features that purify the nutrient solution as well as remove dirt and microorganisms. These units are usually suited for benchtop cultivation and are also used to demonstrate aeroponics principles.
2. *High-pressure units*: This system delivers the mist to the plants at high pressure. Air and water filtration, nutrient sterilization, low-mass polymers, and a pressured nutrient delivery system are all part of this system. High-value crops are grown with this method.
3. *Commercial system*: The commercial system includes biological systems and high-pressure device hardware. The biological systems matrix includes an improvement in longer plant life and crop maturation (Fig. 19.6).

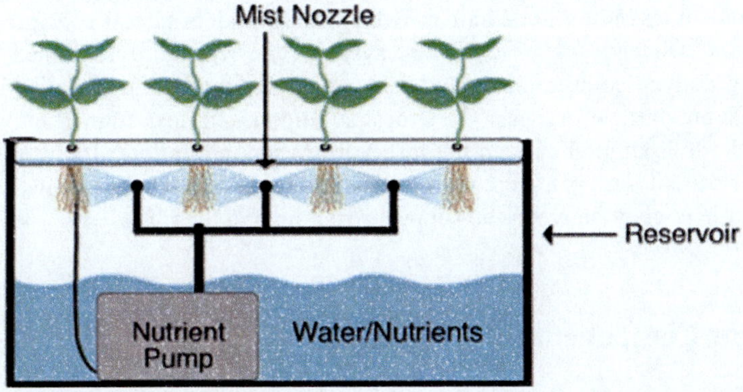

Fig. 19.6 Illustration of aeroponic system

Working

This system reduces labors requirement because the aeroponic system is a continuous operation in a small space. This is mainly used to grow the vegetables in containers filled with flowing plant nutrients but is not suitable for hydroponics or soil cultivation. The main premise of the aeroponic system is to grow plants hung in a closed or semi-closed environment by spraying a nutrient-rich water solution onto the plant's hanging roots and lower stem. The aeroponic setup comprises a proper monitoring and control system for water and nutrient supply to maximize the benefits of aeroponic production. A distribution system with pipes, spray nozzles, a pump, and a timer is required to transfer the nutrient solution from the storage tank. It employs a small internal microjet spray that sprays a fine, high-pressure mist containing nutrient-rich solutions from the nutrient reservoir to the roots in the rooting chamber. The high-pressure aeroponic pump is activated by a controlled cyclic timer. Nutrients are mixed with water in a reservoir basin before being filtered and pumped into a pressure holding tank where they are misted onto the root system on an intermittent basis. Root hairs that have developed aid in the absorption of nutrients from the water. It is also easy to deliver various plant nutrients through the roots because the spray particles are so minute; hence, there is very little wastage of the nutrients solution, and root rot is completely prevented with an abundant supply of oxygen.

Misting occurs every several minutes around the dangling roots. Normally, the system is only turned on for a few seconds every 2–3 minutes because the roots are exposed to the air, and they will quickly dry out if the misting cycles are disrupted. A timer controls the nutrient pump in the same way that it does in other hydroponic systems, with the exception that the aeroponic system requires a short cycle timer that runs the pump for a few seconds every couple of minutes. However, the chamber must be made of lightless materials from all over the place. The roots are functionally sound in the dark and restrict algae growth, which hinders plant growth and contaminates the system. The droplet size of a nutrient mist is an important factor in aeroponics. A large droplet may decrease the availability of oxygen. A droplet that is too small may induce root hair growth, which hinders lateral root growth and reduces the efficiency of an aeroponic system. The water droplets must be large enough to convey enough nutrients to the roots yet small enough not to instantly precipitate out of the root mass. The excess solution is strained, filtered into the base of the unit, and pumped back to the reservoir. Aeroponics allows for simple monitoring of nutrients and pH. Aeroponics allows for unrestricted plant growth because there is little contact between the support structure and the plant.

Different Components of Aeroponics

1. *Nutrients*: $N\text{-}NH_4$ (0.54 g/L), $N\text{-}NO_3$ (0.35 g/L), P (0.40 g/L), K (0.35 g/L), Ca (0.17 g/L), Mg (0.08 g/L), Na (0.04 g/L), Fe (0.09 g/L), Zn (0.03 g/L), and B (0.03 g/L) are the most often used.

2. *Water*: The pH of water can be used as a parameter. Water with a pH of more than 8 is not suited for aeroponics. The EC of water should not be more than 1 mS/cm. Even if the EC and pH values are within acceptable limits, a water chemical analysis is essential. Another potential issue is biological water contamination. Deep well water is usually contaminant-free. Water from shallow wells is likely contaminated with coliform bacteria, particularly *Pectobacterium*, especially in places close to major cities. Water from dubious sources should be microbiologically examined. Certain filters can be used to lessen this risk. If a nutrient tank is installed, water should be filtered before entering it. If none of the other solutions are available, another option is to boil.
3. *The plant materials*: Generally, in vitro plants are advised due to safety concerns. They must, however, be handled with care by a qualified individual. The plants must be of the appropriate age and size before being moved into the greenhouse, and proper hardening off must be performed to allow the plants to acclimatize. The planting materials such as rooted cuttings and sprouted tubers should be disease-free and clean. The roots that emerge from the sand trays should be clean and sand-free.

Advantages of Hydroponic and Aeroponic System

Growing plants in hydroponic and aeroponic systems have numerous advantages (Savvas 2002). The hydroponic system is useful in locations where extreme heat, cold, and desert environmental challenges are common (Polycarpou et al. 2005). The vegetable crops can be grown round the year in hydroponics because climate change has no effect on crop growth in this system (Manzocco et al. 2011). Hydroponically and aeroponically grown vegetables are superior in quality, healthy, and yield higher (Silberbush and Ben-Asher 2001). As in hydroponic and aeroponic systems, the plant roots absorb nutrients effectively because they are applied to the root zone directly. Therefore plants grow much faster and provide better yield. A hydroponic system requires only 1/5th of the overall space and 1/20th of the total water as compared to soil cultivation (Silberbush and Ben-Asher 2001). The aeroponic system uses 98% less water, 60% less fertilizer, and 100% less pesticide and herbicide and maximizes the plant yield by 45–75% than hydroponic or geoponic system (Stoner and Schorr 1983; Spinoff 2006). As hydroponically and aeroponically grown vegetables are cultivated in a controlled environment, they are less prone to insect, pest, disease, and weed attack. In hydroponic and aeroponic systems, plants are grown closer, so plant density is high in these systems. These systems are established inside an enclosed structure where the ambient temperature is controlled, and it allows crops to be grown all year-round despite any season, weather, and climate. Moreover, these systems have been seen to be the best solution for increasing horticulture/agriculture crops in areas with infertile soil and non-arable land (Sonneveld 2000).

Table 19.3 Advantages and disadvantages of hydroponic system

Advantages	Disadvantages
A solution for desert, cold climate region, infertile and nonarable land where soil cultivation is not possible	Management, capital, and labor-intensive
Full control over nutrient content, pH, and growing conditions	A high level of technical skill is required
Water conservation	Daily supervision is necessary
Higher yield and faster growth	Always utilize carefully formulated soluble nutrition
Saves money by recycling nutrients and water	High maintenance cost
More control over the plants rooting environment	All horticulture/agriculture crops cannot be grown in this system
Insect-pest and disease problems can be solved easily, and weed is practically nonexistent	Prone to water-borne diseases

Disadvantages of Hydroponic and Aeroponic System

Despite many advantages, hydroponic and aeroponic systems still have few disadvantages or limitations (Sonneveld 2000). Commercialization of hydroponic and aeroponic systems requires abundant technical knowledge, proficiency, and a hefty amount of money for investment (Sonneveld 2000; Resh 2012). Due to high setup costs, it is observed that these systems are mostly used to grow high-value crops (Van Os et al. 2002). These systems are more prone to water-borne diseases. The spread of diseases and pests occurs quickly in these systems. The hydroponic and aeroponic system demands constant supervision. Despite several disadvantages, researchers have said that hydroponic and aeroponic systems will play a crucial part in the production of vegetables soon (Table 19.3).

Conclusion

As the population increases, growers need to adopt new growing techniques that would help them meet the increasing food demand of the world. To meet the alarming food demand of the world, hydroponic and aeroponic systems can be utilized. These systems use vertical space to grow vegetables. Therefore the per area yield of horticulture/agriculture products increases by 45–50%. By using these systems, growers will grow horticulture/agriculture crops anywhere around the world despite the weather and soil conditions. These systems are quickly gaining popularity and are the fastest-growing sector of modern agriculture.

References

Awad, Y. M., Lee, S. E., Ahmed, M. B. M., Vu, N. T., Farooq, M., Kim, I. S., Kim, H.S., Vithanage, M., Usman, A.R.A., Wabel, M., Meers, E., Kwon, E.E. & Ok, Y. S. (2017). Biochar, a potential hydroponic growth substrate, enhances the nutritional status and growth of leafy vegetables. *Journal of Cleaner Production, 156*, 581–588.

Beibel, J. P. (1960). Hydroponics-The Science of Growing Crops Without Soil||. Florida Department of Agric. *Bull, 180.*

Bulgari, R., Baldi, A., Ferrante, A., & Lenzi, A. (2017). Yield and quality of basil, Swiss chard, and rocket microgreens grown in a hydroponic system. *New Zealand Journal of Crop and Horticultural Science, 45*(2), 119–129.

Butler, J. D., & Oebker, N. F. (1962). Hydroponics as a hobby: growing plants without soil. *Circular; 844.*

Domingues, D. S., Takahashi, H. W., Camara, C. A., & Nixdorf, S. L. (2012). Automated system developed to control pH and concentration of nutrient solution evaluated in hydroponic lettuce production. *Computers and electronics in agriculture, 84*, 53–61

Fallovo, C., Rouphael, Y., Cardarelli, M., Rea, E., Battistelli, A., & Colla, G. (2009). Yield and quality of leafy lettuce in response to nutrient solution composition and growing season. *J. Food Agric. Environ, 7*(2), 456–462.

Frezza, D., León, A., Logegaray, V., Chiesa, A., Desimone, M., & Diaz, L. (2005). Soilless culture technology for high quality lettuce. *Acta Horticulturae, 697*, 43.

Gruda, N. (2009). Do soilless culture systems have an influence on product quality of vegetables?.

Lenzi, A., Baldi, A., & Tesi, R. (2011). Growing spinach in a floating system with different volumes of aerated or non aerated nutrient solution. *Growing Spinach in a Floating System with Different Volumes of Aerated or non Aerated Nutrient Solution*, 21–25.

Maboko, M. M., & Du Plooy, C. P. (2009). Effect of plant spacing on growth and yield of lettuce (Lactuca sativa L.) in a soilless production system. *South African Journal of Plant and Soil, 26*(3), 195–198.

Maboko, M. M., & Du Plooy, C. P. (2013). Effect of plant spacing and harvesting frequency on the yield of Swiss chard cultivars (Beta vulgaris L.) in a closed hydroponic system. *African journal of agricultural research, 8*(10), 936–942.

Maboko, M. M., Du Plooy, C. P., & Bertling, I. (2011). Comparative performance of tomato cultivars cultivated in two hydroponic production systems. *South African Journal of Plant and Soil, 28*(2), 97–102.

Maharana, L. & Koul, D.N. (2011). The emergence of Hydroponics. Yojana (June). 55: 39–40.

Manzocco, L., Foschia, M., Tomasi, N., Maifreni, M., Dalla Costa, L., Marino, M., Cortella,G. & Cesco, S. (2011). Influence of hydroponic and soil cultivation on quality and shelf life of ready-to-eat lamb's lettuce (Valerianella locusta L. Laterr). *Journal of the Science of Food and Agriculture, 91*(8), 1373–1380.

Mwazi, F. N., Amoonga, S. Y. L. V. I. A., & Mubiana, F. S. (2010). Evaluation of the effects of salinity on spinach (Beta vulgaris var. cicla) grown in a hydroponic system along the coast of Namibia. *Agricola, 20*, 14–17.

Nielsen, C. J., Ferrin, D. M., & Stanghellini, M. E. (2006). Efficacy of biosurfactants in the management of Phytophthora capsici on pepper in recirculating hydroponic systems. *Canadian Journal of Plant Pathology, 28*(3), 450–460.

Osvald, J., Petrovic, N., & Demsar, J. (2001). Sugar and organic acid content of tomato fruits (Lycopersicon lycopersicum Mill.) grown on aeroponics at different plant density. *Acta Alimentaria, 30*(1), 53–61.

Polycarpou, P., Neokleous, D., Chimonidou, D., & Papadopoulos, I. (2005). A closed system for soil less culture adapted to the Cyprus conditions. *F. El Gamal, AN Lamaddalen, C. Bogliotti, and R. Guelloubi. Non-conventional water use*, 237–241

Ranawade, P. S., Tidke, S. D., & Kate, A. K. (2017). Comparative cultivation and biochemical analysis of Spinacia oleraceae grown in aquaponics, hydroponics and field conditions. *International Journal of Current Microbiology and Applied Science*, *6*(4), 1007–1013.

Resh, H. M. (2012). *Hydroponic food production: a definitive guidebook for the advanced home gardener and the commercial hydroponic grower*. CRC press.

Reyes, J. L., Montoya, R., Ledesma, C., & Ramírez, R. (2011, May). Development of an aeroponic system for vegetable production. In *II International Symposium on Soilless Culture and Hydroponics 947* (pp. 153–156).

Rouphael, Y., & Colla, G. (2005). Growth, yield, fruit quality and nutrient uptake of hydroponically cultivated zucchini squash as affected by irrigation systems and growing seasons. *Scientia Horticulturae*, *105*(2), 177–195.

Savvas, D. (2002). Nutrient solution recycling. *Hydroponic production of vegetables and ornamentals. Embryo Publications, Athens, Greece*, 299–343.

Sharma, N., Acharya, S., Kumar, K., Singh, N., & Chaurasia, O. P. (2018). Hydroponics as an advanced technique for vegetable production: An overview. *Journal of Soil and Water Conservation*, *17*(4), 364–371.

Shrestha, A., & Dunn, B. (2013). Hydroponics. Oklahoma Cooperative Extension Service, Oklaoma State University, Division of Agricultural Sciences and Natural Resources.

Silberbush, M., & Ben-Asher, J. (2001). Simulation study of nutrient uptake by plants from soilless cultures as affected by salinity buildup and transpiration. *Plant and soil*, *233*(1), 59–69.

Spinoff, N. A. S. A. (2006). Progressive plant growing has business blooming. *Environmental and Agricultural Resources. New York: NASA Spinoff*, 64–77.

Singh, S., & Singh, B. S. (2012, May). Hydroponics–A technique for cultivation of vegetables and medicinal plants. In *Proceedings of 4th Global conference on Horticulture for Food, Nutrition and Livelihood Options. Bhubaneshwar, Odisha, India* (p. 220).

Sonneveld, C. (2000). *Effects of salinity on substrate grown vegetables and ornamentals in greenhouse horticulture*.

Stoner, R., & Schorr, S. (1983). Aeroponics versus bed and hydroponic propagation [The process of propagating and growing plants in air]. *Florists' review (USA)*.

Touliatos, D., Dodd, I. C., & McAinsh, M. (2016). Vertical farming increases lettuce yield per unit area compared to conventional horizontal hydroponics. *Food and energy security*, *5*(3), 184–191.

Trejo-Téllez, L. I., & Gómez-Merino, F. C. (2012). Nutrient solutions for hydroponic systems. *Hydroponics-a standard methodology for plant biological researches*, 1–22. ISBN 978-953-51-0386-8.

Van Os, E. A., Gieling, T. H., & Ruijs, M. N. A. (2002). Equipment for hydroponic installations. *Hydroponic production of vegetables and ornamentals*, 103–41.

Wang, L., Chen, X., Guo, W., Li, Y., Yan, H., & Xue, X. (2017). Yield and Nutritional Quality of Water Spinach (Ipomoea aquatica) as Influenced by Hydroponic Nutrient Solutions with Different pH Adjustments. *International Journal of Agriculture & Biology*, *19*(4).

Chapter 20
Sustainable Production of Underutilized Vegetables

Vrince Vimal and Savita

Introduction

In the last two decades, the development of the Internet has brought a bounty of advantages for both organizations and individuals. The significant advantage includes the productivity with which it manages farmers and customers. Internet of Things (IoT) has been promising to give comparable benefits by its innovation that additionally upgrades the client's view and which likewise can change the workplace. This IoT offers various arrangements in various fields, such as security, traffic, smart cities, smart homes, medical services, agriculture and retail. The critical usages of IoT, explicitly used in agriculture, are greenhouses, precision farming and tamed animals cooperated with the specific research areas, hydrophony, aerophony, and automatic irrigation.10.4. As the population in increasing exponentially it is expected that by 2050, '70%' of the world's population would migrate to urban areas. This will lead to elevated requirement of food supplies.

In agriculture, IoT is utilized to increase production, promote energy conservation, assure self-protection by continuous structural monitoring, achieve self-optimization through different data sources and allow self-configuration in response to environmental changes. Much has been accomplished regarding IoT innovation in rural areas to support effective farming systems (Gupta 2022). IoT has created an outstanding discontent farming climate by examining several inconveniences and challenges in expanding. With the digital revolution, subsistence farmers and technologists are increasingly turning to the Internet of Things (IoT) to solve problems that farmers

V. Vimal (✉)
Computer Science and Engineering, Graphic Era Hill University, Adjunct Professor, Graphic Era Deemed to be University, Dehradun, Uttarakhand, India
e-mail: vvimal@ec.iitr.ac.in

Savita
Graphic Era Hill University, Dehradun, Uttarakhand, India

Savita et al. (eds.), *Production Technology of Underutilized Vegetable Crops*,
https://doi.org/10.1007/978-3-031-15385-3_20

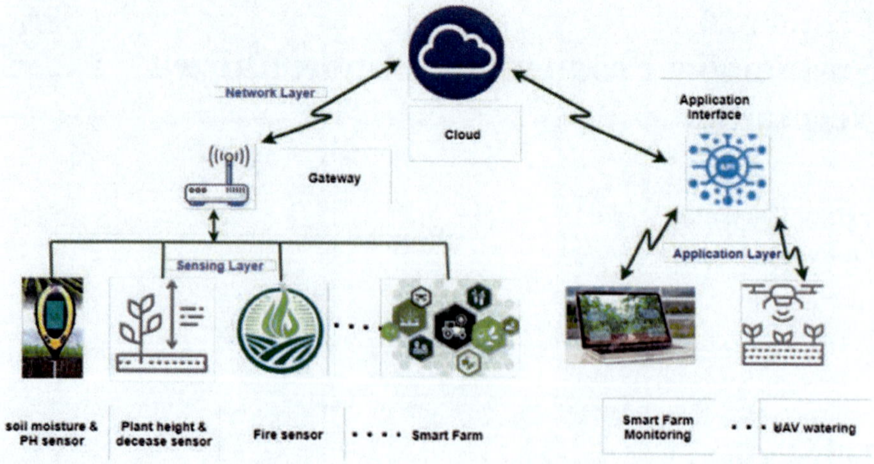

Fig. 20.1 IoT general practices (not exhaustive)

encounter on a daily basis, such as water shortages, fertilizers, financing and awareness (Elijah et al. 2018). Cutting-edge IoT innovations have identified this pressing need and increased productivity while lowering costs. Remote sensor networks enable us to collect data from detecting devices and deliver it to virtual servers known as base station (BS) (El-Sayed and Bayatti 2021). Sensor data provides data on various natural conditions, allowing the entire framework to be screened properly. Observing natural circumstances or yield efficiency is not the only criterion for evaluation, of various parameters such as soil health crop production etc., unwanted growth, damage done by wild animals, burglaries and so forth (Devlet 2021). Also, IoT gives competent scheduling of confined assets, which ensures the best utilization of available resources. Figure 20.1 provides a general (not exhaustive) schematic outline of rural patterns that offer practical and uncomplicated cooperation through a safe and unblemished availability across unique greenhouse, livestock, farmer, field checking, etc. Whereas IoT agricultural companies that use remote devices ensure constant yield and monitoring, along with IoT artificial intelligence, deep learning techniques are also playing a vital role in estimating and planning crop production (Santos et al. 2020).

Suppose the world has to meet the increased requirement of food supplies. In that case, the following measures have to be followed: (i) orthodox farming methods will have to be replaced by intelligent farming tools, (ii) ensuring that smart gadgets reach the last farm across the globe, and (iii) unprivileged/underutilized crop production will have to be improved.

Internet of Things

The Internet of Things (IoT) is a concept that describes the establishment of actual objects—or 'things'—that are implanted with sensors, programming and other advances that enable users to communicate and exchange information between devices and frameworks over the Internet (Gabbai 2015; Garbis and Chapman 2021). These devices include everything from traditional family protests to cutting-edge mechanical equipment. Today, there are more than 20 billion connected IoT devices, and experts predict that this figure will skyrocket by 2025 (Garbis and Chapman 2021).

IoT has made one of the most major developments of the twenty-first century in recent years. Regular products, cooking machines, automobiles, indoor regulators and child screens can be connected to the Internet via implanted gadgets, allowing for continual communication between people, cycles and things (Dudeja et al. 2022).

IoT Pertinent Technologies in Agriculture

Because there are so many solutions used in IoT farming applications that listing them all would be impossible, this chapter concentrates on a few significant technologies that have aided in the modernization of IoT agricultural services. We will briefly discuss the prominent technologies facilitating smart farming.

(i) ***Cloud, FOG and Edge Computing***: In agriculture, coordination between IoT and cloud computing allows ubiquitous access to shared resources. Cloud computing is essential for serving various agricultural needs on-demand and executing operations over the network (Látečková and Trnková 2021). In decades past, agriculture has benefited greatly from cloud computing. Data collection and remote storage, reduced access to ICT infrastructure, online agriculture expert consultations, land information automation and weather prediction are the primary features of cloud computing in agriculture that are being examined (El-Omari 2021).

With low-cost analysis, the cloud, massive data, analysis and portable improvements, tangible things may communicate and acquire information with minimal human intervention. Computerized frameworks can record, screen and adjust every collaboration between related things in this hyper-connected environment. The physical and virtual worlds collide and collaborate.

In Liu et al. (2019), the Internet of Things (IoT), cloud computing, data mining and other technologies are reviewed, and a novel strategy for their application in the field of modern farming is offered, based on the new generation of information technology (IT).

In Raju and Veeramani (2021), authors proposed long-distance communication with no loss of data and no communication interference The system includes an NRF24L01 transceiver module that operates at 2.4 GHz for long-distance communications and agronomic parameter monitoring.

In Feng et al. (2022), an intelligent raising system based on the Internet of Things (IoT) cloud platform is presented for rice seedlings in greenhouses. Using IoT and smart control technologies, the system can deliver suitable temperature and moisture for rice seedlings during three different growth stages. Extreme temperatures or a lack of water impacting the growth of rice seedlings is avoided. The SVR technique is introduced to forecast the operation period of actuators to conserve limited water resources.

(ii) ***Fog Computing:*** It helps IoT devices and sensors, vehicles, drones, applications, robots and users process large amounts of data. The main distinction between cloud computing and fog computing is that with fog computing, certain data is stored in a local data centre, whereas this is not the case in cloud computing. In comparison to cloud computing, the fog computing paradigm requires less power and has lower operating costs (Naha et al. 2018). Fog computing has low latency, real-time interaction, mobility assistance, increased security, efficiency and network bandwidth reduction. Farmers and agricultural stakeholders benefit from these distinct traits.

The amount of data produced by the agricultural IoT is likewise growing, putting more strain on the cloud server. To share the cloud server's offload, the cloud computation is offloaded to a fragment of the sensing layer for execution. This computation or processing of the data in the sensing layer is known as edge computing. Also, as the edge node is in the sensing layer itself, it is beneficial for real-time monitoring as latency is very low compared to the cloud (Zhang et al. 2020). Researchers can find more details about these existing and evolving technologies and their impact on farming practices in Rahbari and Nickray (2019) and references therein.

(iii) ***Data Analytics Techniques:*** Sensors produce large quantities of critical data, known as big data. Crop monitoring can be done in a variety of ways using big data analysis (Osinga et al. 2022; Talaviya et al. 2020). Soil sensor data, remote sensing data, weather stations, historical information and datasets, Geographic Information Systems (GIS) and data gathered from humans are crucial data sources for agricultural process analysis. These data sets obtained from multiple entities are stored in a central server and analysed for various functions. For instance, collecting soil samples from the field and sending these samples for a lab test for evaluating nutrient value is standard practice. The same information can be obtained and sent to the central server (cloud) via sensors and gateways. The farmers can enter his farm's location coordinates and easily retrieve their farm's nutrient contents. Based on this information, Farmer can plan the next crop, decide the amount of fertilizer required, etc. Similarly, weather data helps predict rainfall, which is the primary source of irrigation in countries like India.

Using big data technologies, various agricultural challenges such as crop selection, irrigation methods, fertilizer selection and yield prediction can be addressed (Wolfert et al. 2017). Big data analysis has the potential to be very beneficial to agricultural operations, but it has yet to be employed in agriculture and is still in its early stages of development. There are a variety of causes for this, including a lack of human resources and experience, a lack of reliable infrastructure and a lack of agricultural data standardization and control (Wysel et al. 2021). Agriculture-specific big data platforms on a large scale are becoming increasingly extensively utilized and standardized. However, in order to co-develop big data solutions to tackle agricultural systems and alleviate uncertainties and food security concerns, a transdisciplinary approach is required. For further details on data analytics for agriculture and challenges associated with it, refer to Wolfert et al. (2017).

Software-Defined Networks (SDN) and Network Function Virtualization (NFV)

These are two technological developments that are anticipated to offer IoT deployments with the fully programmable and flexibility they require. The goal of SDN is to separate network control from system transmission capabilities (i.e. the data plane) (Bakhiet et al. 2022). In Friha et al. (2021), the authors proposed an agricultural IoT data management system that gathers, evaluates, depicts and controls IoT data in real time. They offer a blockchain-based integrity monitoring mechanism to prevent inaccurate control and information services. Finally, they suggest that virtual switch software that supports software-defined networking technologies be deployed for better network administration.

Underutilized Vegetables

Underutilized vegetables are crops that are neither developed for enormous scope monetarily nor exchanged broadly. Because the absence of accessibility of planting material, absence of mindfulness on vital significance and absence of mindfulness on healthful and absence of data on creation methods of such yields are reasons, they are considered less-used crops notwithstanding their perceived significance. Examples of underutilized vegetables are broccoli, kale, Chinese cabbage, winged bean, sword bean, gourds, etc. To guarantee food and nourishing security, there is an earnest need to take on hereditary asset investigation and use these underutilized crops' executives. Additionally, the soil and the environment of many countries like India are much ideal for delivering such yields.

Vertical farming is one such practice that helps the crops grow in a vertical stacked manner and requires continuous monitoring of the encompassing parameter. It has gained much relevance in the recent years because the existing agricultural land fails to meet the need of the growing population. This kind of farming is generally practiced in large warehouse and the LED lights are put in rather than the sunlight. Humidity, ambient temperature, light and soil moisture are the main controlling parameters of vertical farming. For optimizing resource use, various automation techniques have been suggested and the automated irrigation system is used for accommodation of actuation. Also, smart monitoring ambient parameter in vertical farming can help and improve the quality and also the productivity of the crop.

This chapter focuses on various applications of IoT and other technologies in different stages of farming. We will discuss possible IoT solutions that can help farmers increase yield with improved quality and less efforts.

Segments Influenced by IoT

A lot of work across the globe has been done in the field of IoT applications in agriculture. The potential area is soil moisture and PH monitoring, field/crop monitoring for various parameters depending on geographical and environmental conditions, automatic and drip irrigation, disease diagnosis and pest monitoring, fertilizer control, smart greenhouse, animal tracking, various applications of unmanned air vehicle (UAV), etc. (Alsamhi et al. 2021; Goyal et al. 2020; Kim et al. 2020; Koksal and Tekinerdogan 2019; Raut et al. 2018; Shyamsunder and Mohan 2021; Vimal et al. 2021; W et al. 2020). Table 20.1 gives some idea on available works. Let us briefly understand how these above-mentioned domains work.

Table 20.1 Applications of technology in various segments

Application	Benefit	Network requirement
Soil moisture monitoring (Aniley et al. 2018)	(i) Ideal crop selection and use of fertilizer as and when required	(i) Minimum latency for quick action and (ii) creating a massive amount of network traffic on a regular basis.
Field surveillance (Boursianis et al. 2020)	Large-scale farms benefiting from automation and remote monitoring system	Longer lag could jeopardize the production system.
Irrigation administration (Raut et al. 2018)	Water or other material utilization optimization in the farm or agriculture area	Low latency should be used to minimize losses.
Scientific infection and infestation surveillance (Lee et al. 2017)	(i) Nutrient and insecticide usage (ii) increases crop quality while lowering costs	(i) Longer lags are not acceptable.
Fertilizer check ((Bhende et al. 2020) and (Jadeja et al. 2021)	Plant growth and crop quality are maintained, and soil nutrition is balanced.	Massive data may be generated due to the mammoth sensors used.
Cattle movement monitoring (Devlet 2021)	Saving damages from cattle's movements.	Longer delay is intolerable.
Smart greenhouses (Rayhana et al. 2020)	Greenhouse gas regulation.	Delay is a crucial parameter.
Diagnostics and remote control (Deng et al. 2018)	Different farm-related equipment can be controlled and diagnosed remotely.	Longer lags may damage device.
Asset tracking (W et al. 2020)	From any location, a farmer can keep track of the position of his vehicles and irrigation systems in place.	Mobility support and less delay are the primarily required parameters

Table 20.2 Nutrient requirement of underutilized vegetable crops

Crop	Nitrogen (kg/ha)	Phosphorus (kg/ha)	Potassium (kg/ha)
Broccoli	100	80	60
Kale	150	75	75
Chinese cabbage	90	50	30
Chayote	100	50	50
Spine and sweet gourd	120	80	80
Winged beans	40	100	40
Sword bean	63	100	75
Rhubarb	100–150	100	100

I. ***Soil Nutrient Level and PH Monitoring:*** The fundamental goal of the IoT-enabled automation is to give an appropriate amount of water system framework and to check the measure of the three significant macronutrients, nitrogen (N), phosphorus (P) and potassium (K), in the earth in this way saving time, finance and efforts of the rancher. The total N, P and K in the earth, for example, are dictated by contrasting the arrangement and shading diagram. This will depict N, P and K as high, medium and low. The customary homestead land strategies require manual mediation. With the mechanized innovation of the water system, human mediation can be limited. At whatever point any parameter goes out of bound, these sensors sense trigger necessary action. Such an excess of working will be refreshed to the client's mobile/iPad/laptop via a gateway (Raut et al. 2018). Tables 20.1 and 20.2 provide the details of pH and nutrient requirements of underutilized vegetable crops.

Crop selection, land preparation, seed selection, crop yield and fertilizer/manure selection are all related to soil quality. The geographical and climatic parameters of the land in question are intricately linked to soil quality. Anticipating soil attributes primarily entails predicting soil nutrients, soil surface humidity and meteorological conditions during a crop's life cycle. Human activities have had a significant impact on soil quality and, as a result, our ability to grow crops (Parikh and James 2012). In general, as mentioned above, there are three primary macronutrients required by soil, i.e. nitrogen, phosphorous and potassium, commonly referred to as NPK (Gregory and Pickering 2007). The produce directly depends upon available nutrients in the soil, which smart sensors can easily detect. Based upon information gathered from these sensors, farmers can select crops for sowing.

To estimate soil parameters, a detailed analysis of soil nutrients, soil moisture and pH is required. In Estrada-López et al. (2018), the authors present the design of a wireless sensor network (WSN) system for smart soil condition estimation. The WSN is made up of low-power autonomous sensor nodes that build spatial distribution maps of soil attributes at two depths below ground using Internet of Things and cloud service connection protocols. The measured data is analysed using an artificial neural network, and the phosphorus (P) quantities in the soil are estimated. A dynamic power management (DPM) technique is used to account for the changing rate of soil phenomena during the day, enabling the system to strike an adaptive balance between energy usage and phosphorus prediction accuracy. The proposed precision agriculture structure allows for a flexible methodology that can be adjusted to different crop varieties and agricultural areas.

Patil et al. (2021) proposed a portable, real-time, cloud-based soil nutrient detection framework. The suggested arrangement captures soil nutrients and factors such as temperature, humidity and pH measurements, which may then be made available to anyone with appropriate cloud channel verification credentials anywhere in the world. Soil analysis is carried out using an RGB colour sensor and a Soil Doctor Plus kit in the suggested framework. They collected soil samples from various farms to demonstrate the proposed cloud-based, real-time soil analysis (India).

The article by Kashyap and Kumar (2021) has more information on sensing methodologies in agriculture for soil parameter monitoring. Soil moisture measurement metrics and laboratory-based testing; in situ, remote and proximal sensing techniques; and in-situ, remote, and proximal sensing methodologies are the four sections of the soil moisture determination. Each technology presented is explored in terms of its applicability, benefits and limits.

II. ***Vertical Farming:*** The global problems of world hunger, lack of arable land and unhealthy diet drive the need to find innovative solutions in farming where production can be increased within a limited area. Vertical farming addresses those issues as it is a farming solution that provides crops year-round and does not require a large tract of arable land. When the quantity of crops produced per season is considered, a single indoor acre of a vertical farm may create yield equivalent to more than 30 acres of farmland, according to Despommier. Chemical insecticides and weedicides can be reduced or eliminated in a vertical farm (Fig. 20.2) (Goyal et al. 2020).

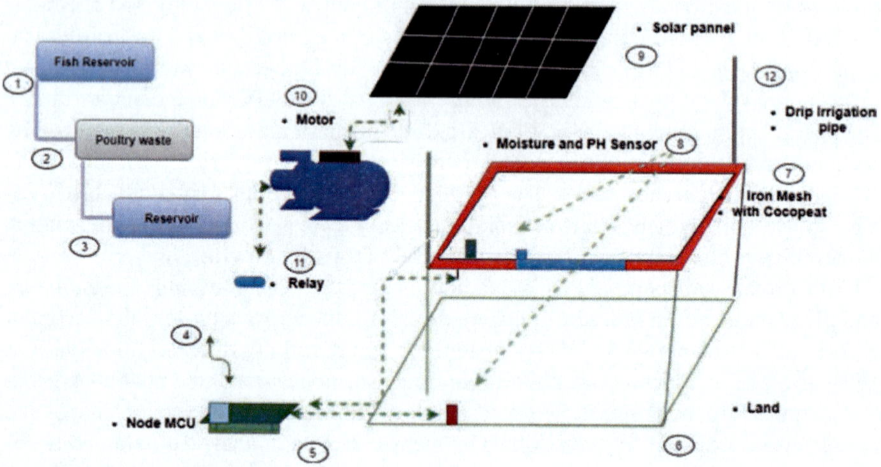

Fig. 20.2 Vertical farming schematic

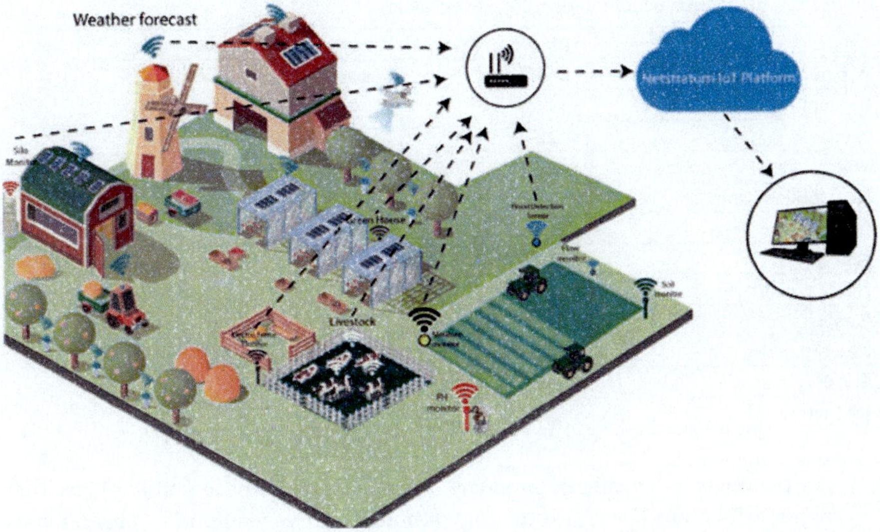

Fig. 20.3 Greenhouse monitoring system (image: postcapes)

Smart Farming It implies implementing another vision of the critical development area where the advancement measures depend on safely incorporating ICT and IoT advances to deal with rural resources. In this vision, the land turns into a substrate where various sensors could secure heterogeneous information. Those sensors are associated with a kind of provincial organization thus connected to the Internet. The continuous streaming information is put away in a complex data set containing all the fundamental knowledge about the land qualities such as imbalance pH, soil nutrient and water status (deficiency or excess). For example, if sprouting broccoli is grown on acidic soil with <5 pH, its growth would be affected as it is susceptible to acidic soil. In that case, a pH-based sensor can be used to regularly monitor soil pH and practical measures like lime, which increases the pH of the soil. Smart projects associated with the options of data processing and sending actuator signals to the base station or to the farmer's smart gadget are gaining popularity (Ayerdi et al. 2020; Colizzi et al. 2020) (Fig. 20.3, Table 20.3).

III. ***Greenhouse:*** Contingent upon the outside climate conditions, the nursery is either cooled or warmed by opening or shutting entryways and adequately enormous windows reasonably positioned at its walls. The standard air ventilation through these gaps protects both warmth and air mass dissemination between the outside climate and inside the house, eliminating the abundance of dampness. This temperature control can be easily achieved by monitoring the temperature using sensors (Liu et al. 2022; Xu et al. 2022).

Nonetheless, yield evapotranspiration contributes to the spread of harmful parasites such as little creepy crawlies, tiny creatures and mushrooms by increasing the dampness concentrated close to the plants exposed to sunshine.

Table 20.3 pH requirement of underutilized vegetables

Crop	pH requirement	Reaction to soil acidity
Asparagus	6–6.8	Less tolerant to acid soil
Broccoli		
Leek		
Water cress		
New Zealand spinach		
Orach		
Kale	5.5–6.8	Moderately tolerant to acid soil
Horse radish		
Parsley		
Chicory	5–6.8	Highly tolerant to acid soil
Sorrel		
Rhubarb		

Agronomists use synthetic products, such as fungicides, to combat these diseases, which are harmful to the environment and vegetation. To prevent contamination, greenhouse windows are usually equipped with networks that function as air routes.

A greenhouse with IoT-enabled frameworks monitors and regulates the environment in real time. It eliminates the need for human intervention in this way. To control the climate in a greenhouse, various sensors that act on the ecological boundaries as specified by the plant requirement are used. Then, when the framework connects with IoT, a cloud worker makes its way to the framework from afar (Shamshiri et al. 2020).

The following are the various sensors that are used to quantify the ecological boundaries to the plant required for controlling the environment in the greenhouse:

(a) *Soil moisture sensor:* The two copper drives operate in the same manner as the sensor testing. They are immersed in a sample soil whose moisture content is being investigated. The amount of wetness in the soil determines its conductivity. It increases as the water content of the dirt increases, forming a conductive path between two sensor tests and stimulating a neighbouring path to allow momentum to pass through (Aniley et al. 2018).

(b) *Light sensor:* In terms of visible light reach, the light sensor is quite sensitive. When the surrounding regular lights are dim, the light sensor connected to the framework displays computerized figures comparable to the light power (Rosell-Polo et al. 2015).

(c) *Humidity sensor:* A humidity sensor is utilized to detect the fumes noticeable all around. The RH (relative humidity) adjustment of the environmental elements would bring about a show of qualities (Alsamhi et al. 2021; Makhlouf et al. 2016; Rayhana et al. 2020; Ullah et al. 2021).

IV. ***Disease Diagnosis:*** As a future of automation in agriculture, precision agriculture addresses difficulties endured by farmers analysing the data gathered via IoT-empowered crop monitoring systems. The ID of crop illness is one of the generally contemplated difficulties. Crop disease cannot be precisely anticipated by just breaking down singular sickness causes. Due to this reason, few researchers have used fuzzy rules for proper prediction. The analysis of data gathered by sensors gathered in a trimming cycle exhibits the connection between temperature, humidity and precipitation with pest breeding. These connections are then worked upon to provide a healthy framework. More details about disease diagnosis can be found in (Lee et al. 2017; Makhlouf et al. 2016) and references therein.

V. ***Fertilizer Control:*** The requirement for the compost is on a high all over; thus, its partner issues include inadequate manure arrangement techniques, lacking nutrients of the compost under the soil, inaccessibility of manure for the yield roots during the mid- and late-season development phases of the plant, and many more (Jadeja et al. 2021). One of the potential answers for these issues is to plan and develop an intelligent machine which can put the manures into various profundities of soil to make it accessible to the crop during all the development phases of the yield. Further, to screen the sum and suitable position of the compost, an IoT-empowered agroCloud module can be utilized to interface the individual through portable registering (Bhende et al. 2020). This module will store the information and will convey to the concerned farmer for the data, for example, soil properties (pH esteem, EC), manure presence (N, P, K) and ecological condition (dampness, temperature) through cell phones and sensors. The full model is advantageous in increasing the yield through multi-profundity, preparing an idea alongside the consolidation of IoT get together for better practices.

VI. ***UAV in Smart Farming:*** The utilization of automated aerial vehicles (UAVs) in detecting as well as communication stages is an advancement innovation with huge potential in precision horticulture. These days, the utilization of UAVs in horticulture is extending to help cultivators with monitoring and decision-making on the farm. UAVs are used in different horticultural practices, like irrigation system, preparation, pesticides, weed the executives and so on (Fig. 20.4).

Images taken by UAVs in agribusiness are analysed at various platforms for various issues and outcomes. In Chebrolu et al. (2018), the authors presented another technique for enlisting pictures of harvests taken by UAVs. In light of their proposed model, they introduced a strategy to adjust three-dimensional cloud on the field, consequently reproducing 3D models of the harvest to monitor growth on a plant-level premise. An original technique to screen the height of a crop of sorghum plants utilizing UAV and 3D model redevelopment was additionally used by authors (Chang et al. 2017). Authors in Deng et al. (2018) mounted a multi-otherworldly camera on a multi-rotor miniature UAV to simultaneously gather multi-unearthly symbolism and Soil Plant Analysis Development (SPAD) upsides of maize. A lot of

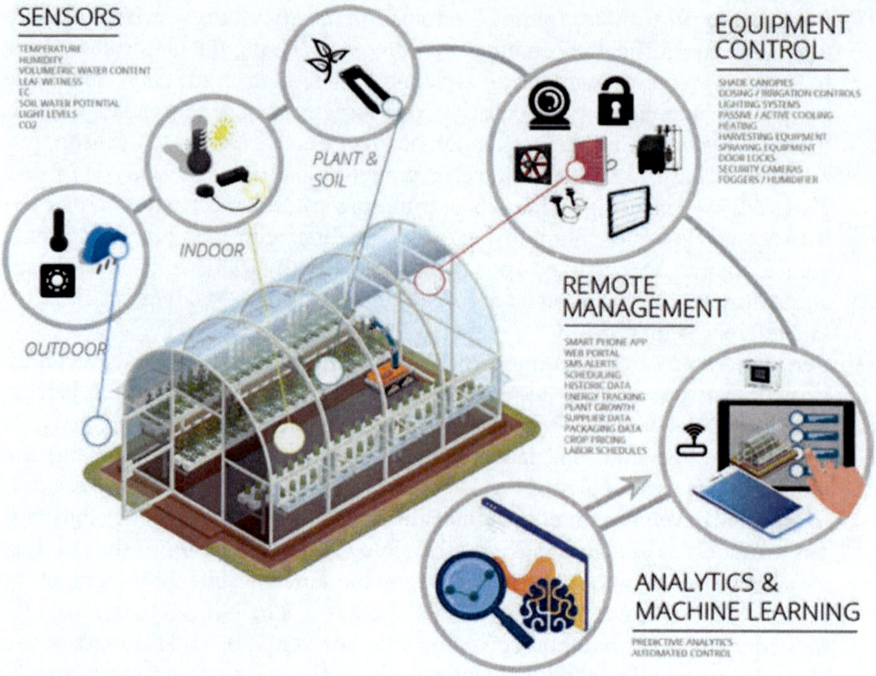

Fig. 20.4 Greenhouse monitoring system. (Image: postcapes)

other works using various cameras and imagery techniques can be found in Aasen et al. (2020), Furukawa et al. (2020), Haque et al. (2018), Lu et al. (2016), Roth et al. (2018) and references therein.

Many articles have investigated multi-UAV systems, which have the potential to redefine agriculture. In Ju and Son (2018), authors fostered a multi-UAV framework for rural fields utilizing the swarm algorithm. They compared the results and a single UAV framework and demonstrated that the multi-UAV framework performed better compared to the single-UAV one. In Zhai et al. (2018), the writers consolidated Particle Swarm Optimization (PSO) and Genetic Algorithm (GA) to resolve the issue of mission arrangement of various UAV frameworks, which is a multi-target advancement issue. They proposed an exactness cultivating framework comprised of a few parts/specialists/robots to achieve composite horticultural missions helpfully. The goal of this work was to use restricted assets of hardware gear in smart farming. In Vu et al. (2018), authors tried the combination of UAVs and automated ground vehicles (UGVs) in different agricultural fields.

In this section, we saw the potential areas in agriculture where IoT applications strengthen their roots and facilitate farming practices. However, this is not a detailed note, as many other applications are not touched above and are way beyond the scope of this chapter. One of such applications is recently presented in Albiero et al. (2021). Authors indicated that though the power of farm tractors had increased a thousandfold, the load on soil and consumption of fuel has also substantially increased, creating a negative impact. They suggested a swarm of tiny electric automated work vehicles with a similar field limit of a large farm truck with an inside burning motor. They conducted a comparison test amid a hefty 270-kW farm vehicle and a swarm of ten multitude farm trucks of 24 kW each. The outcome exhibited a comprehensive benefit for the little robot group.

Smart Vertical Farming

By 2050, the total populace is relied upon to develop to another 2 billion individuals, and taking care of it will be an extensive test. The land is corrupting step by step and becoming costly because of industrialization and urbanization. The worldwide issues of widespread starvation, absence of arable land and an undesirable eating regimen drive the need to discover imaginative arrangements in cultivating where creation can be expanded inside a restricted region. Vertical cultivation resolves those issues as it is a cultivating arrangement that gives crops all year and does not need a huge lot of arable land. As indicated by Despommier, a solitary indoor section of land of an upward homestead might yield identical to in excess of 30 sections of land of farmland when the quantity of harvests delivered per season is considered (platt 2007). The upward homestead permits a decrease or all-out surrender of the utilization of substance pesticides and weedicides.

An aquaponics system can be used to provide additional supplements while reducing the requirement for compost. An aquaponics system takes aqua-farming further by bringing plants and fish together in a biological system (Gosh and Chowdhury 2019). Fish are replenished in lakes, resulting in supplement-rich excrement that is an excellent food source for farm plants on the rise. Hydroponics is utilized in more limited-size vertical cultivating frameworks (Niu and Masabni 2022). Conversely, business vertical ranch frameworks produce a couple of quickly developing vegetable harvests and do exclude a hydroponics part. This utilization of aquaponics frameworks improves on the financial matters and creation issues and amplifies proficiency. Notwithstanding, new normalized aquaponics techniques might assist in making this shut cycle framework more available.

So, most vertical ranches centre on exceptional yield and short-turn harvests, for example, salad greens, with neighbouring eateries frequently purchasing all the produce. High-esteem, fast-developing, little impression, speedy turnover crops, like lettuce, basil and other plates of mixed greens things, can be filled in vertical cultivating. More slow-developing vegetables, such as grains, are not as productive in a business vertical cultivating system.

Also, buyers are worried that routinely delivered food may not be as nutritious, for instance, naturally developed food or food created by more customary techniques, for example, on little, family-claimed ranches or in patio gardens. Moreover, numerous shoppers are worried about the satisfactoriness of such ordinarily delivered food varieties. After collection, food crops are frequently moved to business sectors a huge number of miles away and perhaps weeks old before they are burned through. Newness, flavour and surface frequently endure. Additionally, the utilization of synthetics is thought to influence acceptability unfavourably. Progressively, assortments are getting not as a result of their healthy taste and appearance yet their time span of usability and capacity to withstand taking care of. Furthermore, there is extraordinary worry about the multiplication of such assortments that have been hereditarily adjusted trying to work on their solidness for significant distance shipment.

An organization of sensors appropriated all through an agrarian creation region should have adequate ability to work over significant stretches with nearly nothing and ideally no support. For instance, utilizing remote sensor networks for information assortment has been examined utilizing ZigBee or different conventions. Sensors in remote sensor networks are now and again working utilizing rest and conscious modes to use energy productively and broaden their functional lifetime, arousing during indicated time spans to communicate information. The conscious time should be long enough for the sensor to obtain and advance any information that might be acquired from different sensors in the lattice organization. This outcome is a somewhat enormous number of sensor hubs in the framework being conscious and communicating or retransmitting at generously a similar time. The creators have tracked down this outcome in obstruction and impacts that should be overseen, corrupting the organization's presentation.

I. ***Smart Vertical Farming for Hilly Terraria***: Smart IoT system-based vertical farming gives novel farming procedures, particularly for the farmers of uneven fields. This method utilizes steps and conveys the structure that draws water from the poultry farms containing the excreta of the animals to give dampness to the feed where the hens lay eggs in the system sent on substitute strides to use the greatest from the bountiful accessible daylight (Kosseva 2020a, b; Kristinsson 2007). This mineral-rich water would leak off to the ground underneath, making it rich with supplements and appropriate for natural development. This would be checked by a gateway, for example, NodeMCU, utilizing sunlight-based energy for activity. The framework gathers information from a sensor centre incorporating meteorological information securing framework and a climate information assortment framework. The framework additionally screens components like lighting, moistness, temperature and soil dampness that impact plant development.

There are many advantages of this system (Fig. 20.5). Hill farming suffers from the following obstacles: limited land, water and resources. The implementation of this system would increase the surface area for cultivation and reduce the requirements of chemical fertilizers, which are otherwise costly as the electrical system is proposed to be run by solar energy, so no additional cost of electricity is required. Further cultivation in controlled moisture and PH level helps improve the quality and quantity of crop per square yard.

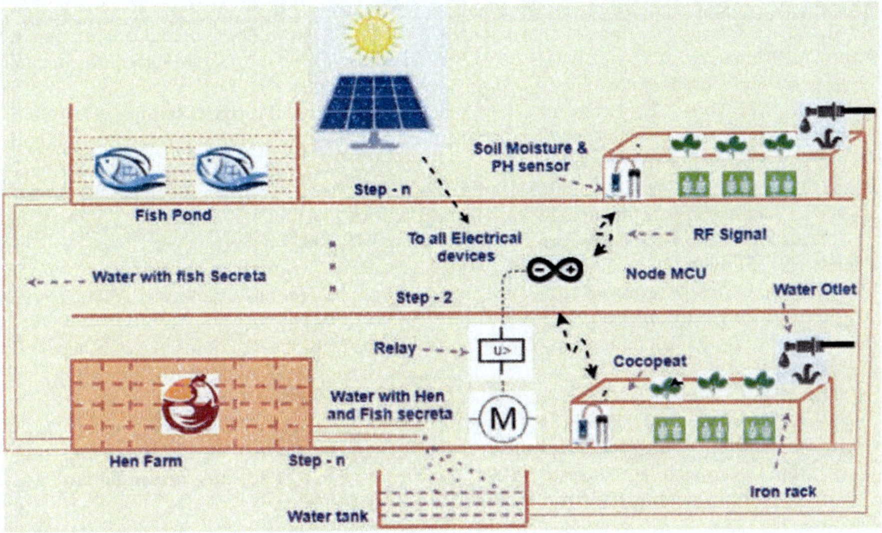

Fig. 20.5 Smart vertical organic farming

Conclusion

The world around us is changing. Interacting via networked devices has become a necessity for survival. On the other hand, as the human population increases every day, the amount of land available for agricultural production is decreasing. It is now necessary to boost productivity on a limited amount of land. As a result, the most effective approach for increasing crop output would be to grow crops on vertical structures in conjunction with the Internet of Things.

On the other hand, underutilized vegetables serve a critical role in addressing food security. They are high in proteins, vitamins and minerals and improve human health by protecting them against cancer, diabetes and anaemia. Therefore, if these crops are sustainably cultivated in a technologically driven environment, their yield can be boosted. These crops can be vital to combating the food insecurity towards which the world is moving.

References

Aasen, H., Kirchgessner, N., Walter, A., & Liebisch, F. (2020). PhenoCams for Field Phenotyping: Using Very High Temporal Resolution Digital Repeated Photography to Investigate Interactions of Growth, Phenology, and Harvest Traits. *Frontiers in Plant Science, 11*, 593. https://doi.org/10.3389/fpls.2020.00593

Albiero, D., Garcia, A., Umezu, C., & Paulo, R. (2021). Swarm robots in mechanized agricultural operations: A review about challenges for research. *Computers and Electronics in Agriculture, 193*(106608).

Alsamhi, S., Afghah, F., Sahal, R., Hawbani, A., Al-qaness, M. A. A., Lee, B., & Guizani, M. (2021). *Green IoT using UAVs in B5G Networks: A Review of Applications and Strategies*.

Aniley, A., Kumar, N., & a, A. (2018). *Review Article: Soil Moisture Sensors in Agriculture and the possible application of nanomaterials in soil moisture sensors*.

Ayerdi, A., Marraccini, E., Leclercq, C., & Scheurer, O. (2020). Precision farming uses typology in arable crop-oriented farms in northern France. *Precision Agriculture*, *21*. https://doi.org/10.1007/s11119-019-09660-y

Bakhiet, M., Alofe, O., Azad, M., Lallie, H., Fatema, K., & Sharif, T. (2022). A comprehensive survey on secure software-defined network for the Internet of Things. *Transactions on Emerging Telecommunications Technologies*, *33*. https://doi.org/10.1002/ett.4391

Bhende, N., Purohit, J., & Junankar, A. A. (2020). *An IoT Based Agri-Cloud Architectural Framework for Monitoring Presence of Fertilizer Under Multilayered Soil Farming* (pp. 144–150). https://doi.org/10.1007/978-3-030-39875-0_15

Boursianis, A. D., Papadopoulou, M. S., Diamantoulakis, P., Liopa-Tsakalidi, A., Barouchas, P., Salahas, G., Karagiannidis, G., Wan, S., & Goudos, S. K. (2020). Internet of Things (IoT) and Agricultural Unmanned Aerial Vehicles (UAVs) in smart farming: A comprehensive review. *Internet of Things*, 100187. https://doi.org/10.1016/j.iot.2020.100187

Chang, A., Jung, J., Maeda, M., & Landivar, J. (2017). Crop height monitoring with digital imagery from Unmanned Aerial System (UAS). *Computers and Electronics in Agriculture*, *141*, 232–237. https://doi.org/10.1016/j.compag.2017.07.008

Chebrolu, N., Läbe, T., & Stachniss, C. (2018). Robust Long-Term Registration of UAV Images of Crop Fields for Precision Agriculture. *IEEE Robotics and Automation Letters*, *3*(4), 3097–3104. https://doi.org/10.1109/LRA.2018.2849603

Colizzi, L., Caivano, D., Ardito, C., Desolda, G., Castrignanò, A., Matera, M., Khosla, R., Moshou, D., Hou, K.-M., Pinet, F., Chanet, J.-P., Hui, G., & Shi, H. (2020). Chapter 1 - Introduction to agricultural IoT. In A. Castrignanò, G. Buttafuoco, R. Khosla, A. M. Mouazen, D. Moshou, & O. Naud (Eds.), *Agricultural Internet of Things and Decision Support for Precision Smart Farming* (pp. 1–33). Academic Press. https://doi.org/10.1016/B978-0-12-818373-1.00001-9

Deng, L., Mao, Z., Li, X., Hu, Z., Duan, F., & Yan, Y. (2018). UAV-based multispectral remote sensing for precision agriculture: A comparison between different cameras. *ISPRS Journal of Photogrammetry and Remote Sensing*, *146*, 124–136. https://doi.org/10.1016/j.isprsjprs.2018.09.008

Devlet, A. (2021). Modern Agriculture and Challanges. *Frontiers in Life Sciences and Related Technologies*. https://doi.org/10.51753/flsrt.856349

Dudeja, R. K., bali, R. S., & Aujla, G. S. (2022). *Internet of Everything: Background and Challenges* (pp. 3–15). https://doi.org/10.1007/978-3-030-89328-6_1

El-Omari, N. (2021). Cloud IoT as a Crucial Enabler: a Survey and Taxonomy. *Modern Applied Science*, *13*, 86. https://doi.org/10.5539/mas.v13n8p86

El-Sayed, H. H., & Bayatti, H. (2021). Improving Network Lifetime in WSN for the application of IoT. *Applied Mathematics \& Information Sciences*, *15*, 453. https://doi.org/10.18576/amis/150407

Elijah, O., Abd Rahman, T., Orikumhi, I., Leow, C. Y., & Hindia, M. (2018). An Overview of Internet of Things (IoT) and Data Analytics in Agriculture: Benefits and Challenges. *IEEE Internet of Things Journal*, *PP*, 1. https://doi.org/10.1109/JIOT.2018.2844296

Estrada-López, J., Castillo Atoche, A., Castillo, J., & Sanchez-Sinencio, E. (2018). Smart Soil Parameters Estimation System Using an Autonomous Wireless Sensor Network With Dynamic Power Management Strategy. *IEEE Sensors Journal*, *PP*, 1. https://doi.org/10.1109/JSEN.2018.2867432

Feng, X., Yan, F., Liu, X., & Jiang, Q. (2022). Development of IoT Cloud Platform Based Intelligent Raising System for Rice Seedlings. *Wireless Personal Communications*, *122*. https://doi.org/10.1007/s11277-021-08967-2

Friha, O., Ferrag, M. A., Shu, L., & Nafaa, M. (2021). *A Robust Security Framework based on Blockchain and SDN for Fog Computing enabled Agricultural Internet of Things*. https://doi.org/10.1109/ITIA50152.2020.9312286

Furukawa, F., Maruyama, K., Saito, Y., & Kaneko, M. (2020). *Corn Height Estimation Using UAV for Yield Prediction and Crop Monitoring* (pp. 51–69). https://doi.org/10.1007/978-3-030-27157-2_5

Gabbai, A. (2015). Kevin Ashton describes "the internet of things": The innovator weighs in on what human life will be like a century from now. *Academia Letters, Article 1003*.

Garbis, J., & Chapman, J. (2021). IoT Devices and "Things." In *Zero Trust Security, An Enterprise Guide* (pp. 193–207). https://doi.org/10.1007/978-1-4842-6702-8_16

Gosh, K., & Chowdhury, S. (2019). *Review of aquaponics system: searching for a technically feasible and economically profitable aquaponics system. 19*, 5–13.

Goyal, S., Kumar, R., & Purohit, A. (2020). Smart Vertical Farming. In *Indian National Patent*.

Gregory, P., & Pickering, J. (2007). Soils and Soil Fertility. In *Science and the Garden: The Scientific Basis of Horticultural Practice* (pp. 83–111). https://doi.org/10.1002/9780470995341.ch4

Gupta, S. (2022). Internet of Things (IOT). In *Internet of Things* (pp. 135–162). https://doi.org/10.1201/9781003181613-11

Haque, M., Uddin, M., & Hassan, S. M. T. (2018). *Application of Unmanned Aerial Vehicle Imagery in Precision Agriculture*. https://doi.org/10.13140/RG.2.2.23372.10882

Jadeja, A., Hirpara, D., Vekaria, L., & Sakarvadia, H. (2021). Fertilizer Storage And Fertilizer Control Order. In *Soil Fertility and Nutrient Management* (pp. 229–234). CRC Press Taylor & Francis Group. https://doi.org/10.1201/9781003200239-11

Ju, C., & Son, H. (2018). Multiple UAV Systems for Agricultural Applications: Control, Implementation, and Evaluation. *Electronics*, *7*, 162. https://doi.org/10.3390/electronics7090162

Kashyap, B., & Kumar, R. (2021). Sensing Methodologies in Agriculture for Soil Moisture and Nutrient Monitoring. *IEEE Access*, *9*, 14095–14121. https://doi.org/10.1109/ACCESS.2021.3052478

Kim, W.-S., Lee, W.-S., & Kim, Y.-J. (2020). A Review of the Applications of the Internet of Things (IoT) for Agricultural Automation. *Journal of Biosystems Engineering*, *45*. https://doi.org/10.1007/s42853-020-00078-3

Koksal, O., & Tekinerdogan, B. (2019). Architecture design approach for IoT-based farm management information systems. *Precision Agriculture*, *20*. https://doi.org/10.1007/s11119-018-09624-8

Kosseva, M. R. (2020a). Chapter 3 - Sources, characteristics and treatment of plant-based food waste. In M. R. Kosseva & C. Webb (Eds.), *Food Industry Wastes (Second Edition)* (Second Edi, pp. 37–66). Academic Press. https://doi.org/10.1016/B978-0-12-817121-9.00003-6

Kosseva, M. R. (2020b). Sources, characteristics, treatment, and analyses of animal-based food wastes. In M. R. Kosseva & C. Webb (Eds.), *Food Industry Wastes (Second Edition)* (Second Edi, pp. 67–85). Academic Press. https://doi.org/10.1016/B978-0-12-817121-9.00004-8

Kristinsson, H. G. (2007). Aquatic food protein hydrolysates. In F. Shahidi (Ed.), *Maximising the Value of Marine By-Products* (pp. 229–248). Woodhead Publishing. https://doi.org/10.1533/9781845692087.2.229

Látečková, A., & Trnková, M. (2021). Cloud Computing in Agricultural Enterprises in Slovakia. *Agris On-Line Papers in Economics and Informatics*, *13*, 87–95. https://doi.org/10.7160/aol.2021.130207

Lee, H., Moon, A., Moon, K., & Lee, Y. (2017). Disease and pest prediction IoT system in orchard: A preliminary study. *2017 Ninth International Conference on Ubiquitous and Future Networks (ICUFN)*, 525–527. https://doi.org/10.1109/ICUFN.2017.7993840

Liu, S., Guo, L., Webb, H., Ya, X., & Chang, X. (2019). Internet of Things Monitoring System of Modern Eco-Agriculture Based on Cloud Computing. *IEEE Access*, *7*, 37050–37058. https://doi.org/10.1109/ACCESS.2019.2903720

Liu, Y., Neal, A. L., Zhang, X., Fan, H., Liu, H., & Li, Z. (2022). Cropping system exerts stronger influence on antibiotic resistance gene assemblages in greenhouse soils than reclaimed wastewater irrigation. *Journal of Hazardous Materials*, *425*, 128046. https://doi.org/10.1016/j.jhazmat.2021.128046

Lu, H., Fu, X., Liu, C., Li, L., He, Y., & Li, N. (2016). Cultivated land information extraction in UAV imagery based on deep convolutional neural network and transfer learning. *Journal of Mountain Science*, *14*, 731–741. https://doi.org/10.1007/s11629-016-3950-2

Makhlouf, S., Laghrouche, M., & El Hamid Adane, A. (2016). Hot Wire Sensor-Based Data Acquisition System for Controlling the Laminar Boundary Layer Near Plant Leaves Within a Greenhouse. *IEEE Sensors Journal*, *16*(8), 2650–2657. https://doi.org/10.1109/JSEN.2016.2518740

Naha, R. K., Garg, S., Georgakopoulos, D., Jayaraman, P. P., Gao, L., Xiang, Y., & Ranjan, R. (2018). Fog Computing: Survey of Trends, Architectures, Requirements, and Research Directions. *IEEE Access*, *6*, 47980–48009. https://doi.org/10.1109/ACCESS.2018.2866491

Niu, G., & Masabni, J. (2022). *Hydroponics* (pp. 153–166). https://doi.org/10.1016/B978-0-323-85152-7.00023-9

Osinga, S. A., Paudel, D., Mouzakitis, S. A., & Athanasiadis, I. N. (2022). Big data in agriculture: Between opportunity and solution. *Agricultural Systems*, *195*, 103298. https://doi.org/10.1016/j.agsy.2021.103298

Parikh, S., & James, B. R. (2012). Soil: The Foundation of Agriculture. *Nature Education Knowledge*, *3*, 2.

Patil, V. K., Jadhav, A., Gavhane, S., & Kapare, V. (2021). IoT Based Real Time Soil Nutrients Detection. *2021 International Conference on Emerging Smart Computing and Informatics (ESCI)*, 737–742. https://doi.org/10.1109/ESCI50559.2021.9396860

platt, peter. (2007). Vertical Farming: An Interview with Dickson Despommier. *Gastronomica*, *7*, 80–87. https://doi.org/10.1525/gfc.2007.7.3.80

Rahbari, D., & Nickray, M. (2019). Computation Offloading and Scheduling in Edge-Fog Cloud Computing. *Journal of Electronic \& Information Systems*, *1*. https://doi.org/10.30564/jeisr.v1i1.1135

Raju, K. L., & Veeramani, V. (2021). *A Self-Powered, Real-Time, NRF24L01 IoT Based-Cloud Enabled Service For Smart Agriculture Decision-Making System*. https://doi.org/10.21203/rs.3.rs-586227/v1

Raut, R., Varma, H., Mulla, C., & Pawar, V. R. (2018). Soil Monitoring, Fertigation, and Irrigation System Using IoT for Agricultural Application. In Y.-C. Hu, S. Tiwari, K. K. Mishra, & M. C. Trivedi (Eds.), *Intelligent Communication and Computational Technologies* (pp. 67–73). Springer Singapore.

Rayhana, R., Xiao, G., & Liu, Z. (2020). Internet of Things Empowered Smart Greenhouse Farming. *IEEE Journal of Radio Frequency Identification*, *4*(3), 195–211. https://doi.org/10.1109/JRFID.2020.2984391

Rosell-Polo, J., Auat Cheein, F., Gregorio Lopez, E., Andújar, D., Puigdomenech, L., Masip, J., & Escolà, A. (2015). Advances in Structured Light Sensors Applications in Precision Agriculture and Livestock Farming. *Advances in Agronomy*, *133*, 71–112. https://doi.org/10.1016/bs.agron.2015.05.002

Roth, L., Aasen, H., Walter, A., & Liebisch, F. (2018). Single image processing in phenotyping with unmanned aerial systems: Taking advantage of multi-view data. *Eucarpia Section Biometrics in Plant Breeding*.

Santos, L., Neves Dos Santos, F., Moura Oliveira, P., & Shinde, P. (2020). *Deep Learning Applications in Agriculture: A Short Review* (pp. 139–151). https://doi.org/10.1007/978-3-030-35990-4_12

Shamshiri, R. R., Bojic, I., van Henten, E., Balasundram, S. K., Dworak, V., Sultan, M., & Weltzien, C. (2020). Model-based evaluation of greenhouse microclimate using IoT-Sensor data fusion for energy efficient crop production. *Journal of Cleaner Production*, *263*, 121303. https://doi.org/10.1016/j.jclepro.2020.121303

Shyamsunder, M., & Mohan, V. (2021). IoT based Soil Quality Monitoring for An Efficient Irrigation. *Alinteri Journal of Agriculture Sciences*, *36*, 13–17. https://doi.org/10.47059/alinteri/V36I2/AJAS21110

Talaviya, T., Shah, D., Patel, N., Yagnik, H., & Shah, M. (2020). Implementation of artificial intelligence in agriculture for optimisation of irrigation and application of pesticides and herbicides. *Artificial Intelligence in Agriculture*, *4*, 58–73. https://doi.org/10.1016/j.aiia.2020.04.002

Ullah, I., Fayaz, M., Khan, M., & Kim, D. (2021). An optimization scheme for IoT based smart greenhouse climate control with efficient energy consumption. *Computing*. https://doi.org/10.1007/s00607-021-00963-5

Vimal, S., Sathish Kumar, N., Kasiselvanathan, M., & Bojan, G. (2021). Smart Irrigation System in Agriculture. *Journal of Physics: Conference Series*, *1917*, 12028. https://doi.org/10.1088/1742-6596/1917/1/012028

Vu, Q., Raković, M., Delic, V., & Ronzhin, A. (2018). *Trends in Development of UAV-UGV Cooperation Approaches in Precision Agriculture: Third International Conference, ICR 2018, Leipzig, Germany, September 18–22, 2018, Proceedings* (pp. 213–221). https://doi.org/10.1007/978-3-319-99582-3_22

W, I., Suryani, V., & Wardana, A. (2020). Narrowband-IoT network for asset tracking system. *IOP Conference Series: Materials Science and Engineering*, *830*, 22087. https://doi.org/10.1088/1757-899X/830/2/022087

Wolfert, S., Ge, L., Verdouw, C., & Bogaardt, M.-J. (2017). Big Data in Smart Farming – A review. *Agricultural Systems*, *153*, 69–80. https://doi.org/10.1016/j.agsy.2017.01.023

Wysel, M., Baker, D., & Billingsley, W. (2021). Data sharing platforms: How value is created from agricultural data. *Agricultural Systems*. https://doi.org/10.1016/j.agsy.2021.103241

Xu, K., Guo, X., He, J., Yu, B., Tan, J., & Guo, Y. (2022). A study on temperature spatial distribution of a greenhouse under solar load with considering crop transpiration and optical effects. *Energy Conversion and Management*, *254*, 115277. https://doi.org/10.1016/j.enconman.2022.115277

Zhai, Z., Ortega, J.-F., Lucas Martínez, N., & Rodríguez-Molina, J. (2018). A Mission Planning Approach for Precision Farming Systems Based on Multi-Objective Optimization. *Sensors*, *18*, 1795. https://doi.org/10.3390/s18061795

Zhang, X., Cao, Z., & Dong, W. (2020). Overview of Edge Computing in the Agricultural Internet of Things: Key Technologies, Applications, Challenges. *IEEE Access*, *8*, 1. https://doi.org/10.1109/ACCESS.2020.3013005

Index

Savita et al. (eds.), *Production Technology of Underutilized Vegetable Crops*,
https://doi.org/10.1007/978-3-031-15385-3

Q

R

S

T